Social Network Analysis of Disaster Response, Recovery, and Adaptation

Social Network Analysis of Disaster Response, Recovery, and Adaptation

Edited by

Eric C. Jones

A.J. Faas

AMSTERDAM • BOSTON • HEIDELBERG • LONDON
NEW YORK • OXFORD • PARIS • SAN DIEGO
SAN FRANCISCO • SINGAPORE • SYDNEY • TOKYO

Butterworth-Heinemann is an imprint of Elsevier

Library of Congress Cataloging-in-Publication Data
A catalog record for this book is available from the Library of Congress

British Library Cataloguing-in-Publication Data
A catalogue record for this book is available from the British Library

ISBN: 978-0-12-805196-2

For information on all Butterworth-Heinemann publications
visit our website at https://www.elsevier.com/

Working together
to grow libraries in
developing countries

www.elsevier.com • www.bookaid.org

Publisher: Candice Janco
Acquisition Editor: Sara Scott
Editorial Project Manager: Hilary Carr
Production Project Manager: Punithavathy Govindaradjane
Cover Designer: Greg Harris

Typeset by TNQ Books and Journals

Contents

PART 1 SOCIAL NETWORK ANALYSIS IN DISASTER RESPONSE, RECOVERY AND ADAPTATION

PART 2 NETWORKS IN DISASTER RESPONSE

CHAPTER 6   **Interorganizational Resilience: Networked Collaborations in Communities After Superstorm Sandy 75**

Jack L. Harris and Marya L. Doerfel

CHAPTER 7   **Shifting Attention: Modeling Follower Relationship Dynamics Among US Emergency Management-Related Organizations During a Colorado Wildfire 93**

Zack W. Almquist, Emma S. Spiro, and Carter T. Butts

CHAPTER 8   The Effect of Hurricane Ike on Personal Network Tie Activation as Response and Recovery Unfolded...........................113

Christopher Steven Marcum, Anna V. Wilkinson, and Laura M. Koehly

PART 3 NETWORKS IN DISASTER RECOVERY

CHAPTER 9   The Family's Burden: Perceived Social Network Resources for Individual Disaster Assistance in Hazard-Prone Florida 127

Michelle Meyer

CHAPTER 10 Interorganizational Network Dynamics in the Wenchuan Earthquake Recovery 143

Jia Lu

CHAPTER 11 Organizational Support Networks and Relational Resilience After the 2010/11 Earthquakes in Canterbury, New Zealand 161

Joanne R. Stevenson and David Conradson

CHAPTER 12 Well-Being and Participation in New Social Networks Following a Day Care Fire in Hermosillo, Mexico...................... 177

Maria L. Rangel, Eric C. Jones, and Arthur D. Murphy

PART 4 NETWORKS IN HAZARD MITIGATION AND ADAPTATION

CHAPTER 13 Networks and Hazard Adaptation Among West African Pastoralists 193

Mark Moritz

CHAPTER 14 Cyclones Alter Risk Sharing Against Illness Through Networks and Groups: Evidence From Fiji...................... 209

Yoshito Takasaki

CHAPTER 15 Stay or Relocate: The Roles of Networks After the Great East Japan Earthquake 223

Young-Jun Lee, Hiroaki Sugiura, and Ingrida Gečienė

CHAPTER 16 Personal Networks and Long-Term Gendered Postdisaster Well-Being in Mexico and Ecuador 239

Graham A. Tobin, Christopher McCarty, Arthur D. Murphy, Linda M. Whiteford, and Eric C. Jones

PART 5 CONCLUSIONS

CHAPTER 17 The Practical and Policy Relevance of Social Network Analysis for Disaster Response, Recovery, and Adaptation 255

Julie K. Maldonado

List of Contributors

Zack W. Almquist
University of Minnesota, Minneapolis, MN, United States; University of Washington, Seattle, WA, United States

Carter T. Butts
University of California, Irvine, CA, United States

David Conradson
University of Canterbury, Christchurch, New Zealand

Fatih Demiroz
Sam Houston State University, Huntsville, TX, United States

Marya L. Doerfel
Rutgers University, New Brunswick, NJ, United States

A.J. Faas
San Jose State University, San Jose, CA, United States

Ingrida Gečienė
Lithuanian Social Research Centre, Institute of Social Innovations, Vilnius, Lithuania

Sherrie K. Godette
North Carolina State University, Raleigh, NC, United States

Jack L. Harris
Rutgers University, New Brunswick, NJ, United States

Eric C. Jones
University of Texas Health Science Center at Houston, El Paso, TX, United States

Naim Kapucu
University of Central Florida, Orlando, FL, United States

Laura M. Koehly
National Institutes of Health, Bethesda, MD, United States

Young-Jun Lee
Hirosaki University, Hirosaki, Aomori, Japan

Jia Lu
California State University-Los Angeles, Los Angeles, CA, United States; University of Southern California, Los Angeles, CA, United States

Julie K. Maldonado
Livelihoods Knowledge Exchange Network, Santa Barbara, CA, United States

Christopher Steven Marcum
National Institutes of Health, Bethesda, MD, United States

Christopher McCarty
University of Florida, Gainesville, FL, United States

Michelle Meyer
Louisiana State University, Baton Rouge, LA, United States

Mark Moritz
The Ohio State University, Columbus, OH, United States

Arthur D. Murphy
University of North Carolina-Greensboro, Greensboro, NC, United States

Branda L. Nowell
North Carolina State University, Raleigh, NC, United States

Maria L. Rangel
University of Texas School of Public Health, Houston, TX, United States

Emma S. Spiro
University of Washington, Seattle, WA, United States

Toddi A. Steelman
North Carolina State University, Raleigh, NC, United States; University of Saskatchewan, Saskatoon, SK, Canada

Joanne R. Stevenson
Resilient Organisations Ltd., Sheffield, New Zealand; University of Canterbury, Christchurch, New Zealand

Hiroaki Sugiura
Aichi University, Nagoya, Japan

Yoshito Takasaki
University of Tokyo, Tokyo, Japan

Graham A. Tobin
University of South Florida, Tampa, FL, United States

Danielle M. Varda
University of Colorado Denver, Denver, CO, United States

Anne-Lise K. Velez
North Carolina State University, Raleigh, NC, United States

Linda M. Whiteford
University of South Florida, Tampa, FL, United States

Anna V. Wilkinson
University of Texas Health Science Center at Houston, Houston, TX, United States

Acknowledgments

Our main debt is to the disaster survivors, practitioners, and first responders whose lives were affected by the disasters and crises that are the subjects of this book. On behalf of all of the contributors to this book, we would like to extend our sincerest thanks to all participants who contributed their time and insights to the studies contained in this collection. We also owe a debt of gratitude to the scholars in this book and elsewhere who have developed means to conceive, collect, and analyze network data while interacting humbly with people in the throes of disaster response, recovery, and adaptation. Their results have been points (and often counterpoints) of departure for our own understandings of networks, hazards, and disasters. Additionally, we would like to thank all of the authors in the book for their contributions as scholars and peer reviewers. Dr. Elizabeth Marino was the only person to serve as a reviewer for a chapter who is not an author, and we are thankful that her comments were helpful in rethinking aspects of Chapters 1, 2, and 17. We are also grateful to several reviewers of the proposal for this book, whose thoughtful comments helped us improve the design and content of the book. We should acknowledge the relevant grants that have supported our work with networks in disasters over the years. These include grants on which we have worked in some capacity from the National Science Foundation (0620213, 0751265, 1123962, 1330070, 1416651, 1560776); National Institutes of Health (MH51278, MH090703); the Mexican National Council of Science and Technology (CONACYT); the University of Colorado Boulder's Natural Hazards Center (Quick Response Grant) and their joint initiative with the Public Entity Risk Institute (Dissertation Fellowship in Hazards, Risk, and Disasters); the University of Florida's Tropical Conservation and Development Program; and the University of North Carolina at Greensboro's Office of Research and College of Arts and Sciences. A big thank you to Cristina Castillo and Sara Masoud, who were extremely helpful in correcting bibliography and citation styles and in tracking down missing or inaccurate references. While we were already anticipating bringing together disaster network research into an edited book, it was the persistent interest of acquisitions editor Sara Scott at Elsevier that made this book happen. We are very pleased to have been able to work with her and with editorial project manager Hilary Carr and colleagues who made the collation and submission of the manuscript materials an efficient and pleasant process.

SOCIAL NETWORK ANALYSIS IN DISASTER RESPONSE, RECOVERY AND ADAPTATION

AN INTRODUCTION TO SOCIAL NETWORK ANALYSIS IN DISASTER CONTEXTS

1

Eric C. Jones[1], A.J. Faas[2]

[1]*University of Texas Health Science Center at Houston, El Paso, TX, United States;*
[2]*San Jose State University, San Jose, CA, United States*

CHAPTER OUTLINE

INTRODUCTION

Hazards become disasters as the opportunities and constraints for maintaining a safe and secure life and livelihood become too strained for many people. Anecdotally and through many case studies, we know that social interactions exacerbate or mitigate those strains. However, we need a concerted intellectual effort to understand the variation in how ties within and between groups, communities, and organizations both respond to and are affected by hazards and disasters. Network research allows us to be incredibly creative in the ways that we capture the subtle, intricate, and dynamic power of relationships and interactions in the areas of disaster response, recovery, and adaptation.

Before, during, and after a disaster, people, agencies, and organizations help and hinder each other, alternatively opening up new possibilities and hamstringing others. To be sure, exposures to hazards become disasters in large part as a result of relational patterns in societies, and our responses to disasters—emergency deployment, mitigation, coping, and adaptation—are likewise facilitated and obstructed by variations in coordination and support between networks of actors. The social network is therefore a

Social Network Analysis of Disaster Response, Recovery, and Adaptation. http://dx.doi.org/10.1016/B978-0-12-805196-2.00001-7

seductive concept in the social science of disasters—a potentially robust tool for investigating complex patterns of social relations and human-environment entanglements.

Social networks are typically invoked in one of two general ways in social science research: (1) metaphorically or heuristically, as descriptive or analytical concepts that summarize researchers' perceived patterns of relationships or interactions in particular contexts, or else as guides for systematic observations of relationships and interactions; and (2) as a formal system of measurement of either personal or whole (i.e., community or institutional) patterns of relationships or interaction employing one or more measures derived from graph theory (e.g., density, centralization, bridging, etc.) and, more recently, examining complexity and network change. This book is principally focused on the second broad approach to network analysis, though contributors draw heavily on concepts and empirical trends identified in studies that engage networks more metaphorically.

This book is the first of its kind; it presents a great deal of the diversity in network approaches to better understanding how people and groups experience extreme events. Since both network analysis and disaster studies are often separated out into the publishing confines of multiple disciplines—and we do not always read outside of these disciplines—this cutting-edge scholarly work would otherwise be scattered in disparate journals and books. Not only does this book bring this work from different disciplines together but it brings together the relatively minimally interacting fields of interpersonal disaster networks and interorganizational disaster networks. By doing so, we can start to explore the theoretical disconnects between the ways networks are deployed by scholars. For example, we note that the study of bridging—or being a relatively unique connection between at least a couple of parts of a network—in the field of interpersonal networks usually means we are examining power, vulnerability, or assets/liabilities, while in the field of interorganizational networks we are typically examining the effectiveness of communication and, to a lesser extent, goal alignment.

We have tried to keep the studies in this book to those that revolve around identifiable disasters and that use systematic qualitative or quantitative analysis of dyadic data—the specific kinds of relationships between specific individuals, roles, or institutions. This said, we would like even more studies to go beyond dyadic analyses and to engage the entire web of relationships involved in disasters. In other words, exchange and dyadic interaction is a minimum in our mind for the study relationships and disasters, but it does not fully leverage the possibilities of social network analysis because the dependence of a dyad on other individuals and dyads and cliques (and the rest of the network) is left unexamined when only paying attention to dyads. While we have chosen studies for this book that we consider to use the methodologies of social network analysis, our interest is less on the importance of conducting social network analysis and more on the kinds of things it can tell us about human behavior in extreme settings.

THE CONTENT AND LAYOUT OF THE BOOK

This first section provides orientations to theory on individual and organizational networks in disaster settings plus a methodological overview of the loosely organized field. Then empirical work comprises the sections on response, recovery, and adaptation/mitigation. Each empirical chapter also addresses methodological and theoretical concerns. Since we are addressing a fairly new and somewhat specialized field, we end the book with a chapter that provides insight into how this new and specialized knowledge can be applied to improve response, recovery, and adaptation.

The authors in this book cover some major events of our times like Hurricanes Katrina, Sandy, and Ike in the United States, the Wenchuan earthquake in China, the Kobe earthquake in Japan, and the Canterbury earthquakes in New Zealand, plus some lesser known events, like volcanic eruptions in Ecuador and Mexico, landslides in Mexico, a day care fire in Mexico, a typhoon in Fiji, wildfires in the United States, and drought in East Africa. These studies include those in contexts of geophysical and hydrometereological hazards (volcanoes and mudslides—Tobin and colleagues; earthquakes—Stevenson, Lee and Sugiura, Lu; hurricanes—Marcum and colleagues, Harris and Doerfel, Meyer; typhoons—Takasaki; droughts—Moritz; wildfires—Nowell and colleagues, Almquist and colleagues), and those caused by human carelessness or recklessness (a building fire—Rangel and colleagues). We purposefully avoided including events of conflict or mass terror since—although we see them theoretically as extreme events, too—intentionality in such events often involves some very different responses and adaptations.

Table 1.1 presents the general focus and methodological approach of each empirical chapter. These chapters cover considerable variation in the kinds of networks that are studied. There is not always a clear division between egocentric networks that focus on the ties in each individual's or organization's own network versus sociocentric networks that focus on an identifiable group or interaction sphere. However, it is worth noting the degree to which this book's chapters cover networks that are more methodologically relying on ties focused on a given individual/organization versus ties that exist within a group of people/institutions that are likely to interact. Among the book's 12 case study chapters, only two chapters break the pattern of egocentric networks being about individuals and sociocentric networks being about organizations. Similarly, few chapters cover network structure (e.g., density, cohesion, bridging, path distance) or employ longitudinal design. It was not our goal to have all chapters cover structure or network dynamics/change, but it is clear that these approaches are not in the majority for disaster network research.

To the extent the data allowed, the authors have tried to address temporal aspects of networks, although we only marked them as longitudinal when comparable data was collected at two or more points in time. Since people may rely on different kinds of people for different things at different points as hazards, the contributors have attempted to account for dynamic shifts in needs, constraints, and help seeking.

Table 1.1 Approaches and Foci of the Empirical Chapters

Chapters	Egocentric/Sociocentric, and Individuals/Organizations	Social Support	Coordination	Structure	Longitudinal
5	Sociocentric—organizations		x	x	
6	Sociocentric—organizations		x	x	x
7	Sociocentric—organizations		x	x	x
8	Egocentric—individuals	x			x
9	Egocentric—individuals	x			
10	Sociocentric—organizations		x	x	x
11	Egocentric—organizations	x		x	
12	Sociocentric—individuals	x		x	
13	Egocentric—individuals	x			x
14	Egocentric—individuals (households)	x			
15	Egocentric—individuals	x			
16	Egocentric—individuals	x		x	

HIGHLIGHTS AND THINGS TO WATCH FOR

We want to highlight a few things that we generally think are missing in disaster network research. This observation is less of a criticism than it is an encouragement to those of us in this field to undertake an exercise in systematically thinking beyond a study to hopefully imagine what might provide the next contribution or breakthrough in the field.

ROLES

When we're studying organizational networks, we should remember that many of those network ties or relationships are actually interpersonal relationships. Some of those relationships are more like friendships or relationships based on personality and common history, while other interpersonal relationships between organizations really are between roles—the relationship can continue if one or more people leave their role and are replaced by someone else. Yet other relationships between institutions are very far removed from individuals and roles. Financial institutions that transfer money have individuals in roles transferring the money, or even deciding to transfer the money, but these interactions have no interpersonal and no interrole qualities.

LEVELS OF NETWORKS

Levels of analysis are construed various ways, sometimes depending on the discipline or the objective of a project. Rather than propose a concrete list of levels of networks, we want to point out that levels of networks can be joined to arrive at new insights in disaster response, recovery, and adaptation. For example, personal networks that influence individual behavior can be joined together into a representative community network that provides greater context for the distribution of the various kinds of personal networks. Nowell and colleagues in Chapter 5 use their case study to tackle head on this issue of intersecting levels of analysis.

Importantly, several authors have lamented the low level of integration between research in the area of social network analysis with research on interorganizational networks (Bergenholtz & Waldstrøøm, 2011; Knobben & Oerlemans, 2006; Roijakkers & Hagedoorn, 2006). Some of this is due to real distinctions between the organizational and individual levels of analysis, but, while interorganizational networks can be conceived of as metaphors for multiagency interdependencies, it is feasible to analyze the social structure between organizations. This book therefore includes studies both of social networks and of interorganizational networks to attempt to synthesize some of the cutting-edge developments in both hitherto disparate bodies of research.

INTERDISCIPLINARITY

The theoretical approaches used by the authors in this book are sometimes constrained by them being single-authored studies, and because most of the coauthored papers come from a single discipline. That said, the following disciplines are represented in this book: public administration (Chapters 3–5 and 10), sociology (Chapters 7–9), communication (Chapter 6), anthropology (Chapters 2, 12, 13, 16, and 17), geography (Chapter 11), and economics (Chapters 14 and 17). A few of these chapters are somewhat interdisciplinary, but that is not the norm. As you read the book, imagine how the authors, other scholars, or you yourself might arrive at more complete understandings in any of these studies by increasing the input from some other disciplines.

POTENTIAL VERSUS REALIZED TIES

In Chapter 2, we provide a very brief discussion of how the measurement of network potential (i.e., existing ties concerning affect or orientation) produces somewhat unclear sets of predictions for a variety of postdisaster outcomes, while the measurement of network activation (i.e., instrumental ties) indicates more clear relationships and mechanisms between network behavior and the post-disaster outcomes of interest. Additionally, the relationship between network potential and network activation is often not clear. As you read through the various chapters, imagine what the network potential is or might be, how it compares to network activation, and what might be some useful ways to bridge the gap between the two. Additionally, when trying to make these observations, thinking in terms of a full network instead of just sets of dyadic ties makes the puzzle harder but probably more interesting.

THE ETHICS OF NAMING VULNERABLE PEOPLE

Concerns of confidentiality are not unique to studies of disasters nor to studies of social networks. However, both kinds of studies present added issues to consider above those that many other social science studies may not encounter. First, people in disaster settings are typically more vulnerable than in other settings. This means that people might be more suggestable when consenting to and answering interviews, and it means that the information they share might be more sensitive than it would be in another setting since the information highlights their level of vulnerability. Second, social network research usually involves the naming of people who have not consented to being named or who have not been approached to participate in the study. A respondent might be asked to give information on faith, wealth, attitude, health issues, politics, and other characteristics of each person that they name in a network study. They also often report on the types of interactions and exchanges they have had with these other people. This means that researchers often possess information (whether accurate or not) about various people that don't know they are now in your study. One protection on this data is the de-identification of people in the networks by giving them unique identifiers. This is not always a quick or easy process, since the same person might be named by several people and, thus, spelling issues, nicknames, variation in reported characteristics, and lack of information may make it hard to determine if they are the same person.

As with all social science research, the process of protecting research subjects from adverse consequences of participating in a study can present some procedural challenges. We feel strongly that every effort should be made to maintain a primary ethical obligation to research subjects and prevent and minimize any potential harm that may result from participation. Beyond minimizing harm, the best research endeavors—especially in disaster contexts—should attempt to apply study findings to maximize the benefits to our subjects and to others who endure the vulnerabilities of disaster and environmental crisis. In the case that network findings are used to directly engage respondents to improve agency or belonging, there is an incredible opportunity for self-evaluation—like looking in a mirror—by the respondent or group of respondents. However, whether the goal is social justice, process improvement, or some other outcome, such an application requires careful management of the interaction with respondents over the results so that people have buy in but also are guided toward solutions that avoid competitive comparisons or jockeying to improve one's situation at the expense of others.

APPLICATIONS

The International Federation of Red Cross and Red Crescent Societies' (2014) World Disasters Report 2014 pays extensive attention to the roles of community networks in disaster contexts around the world. Such large and important humanitarian organizations are looking to shape policy and practice that is sensitive to and supportive of informal community networks. A 2011 United Nations International Strategy for Disaster Reduction report refers to social networks as a type of "community competence" that is a driver of effective response, recovery, and adaptation (UNISDR, 2011). One hope is that disaster risk reduction policy and practice will help build and support such competencies—not simply conclude that community problem-solving capacity is hindered by poor networks or social capital. The care to such detail of support, exchange, coordination, and emergent network dynamics provides one avenue to find a better fit between local cultural dynamics, relational patterns, and broader efforts of disaster risk reduction, prevention, mitigation, response, and recovery. Network heuristics and metaphors may be "good to think" (Descola, 1994), but if we want to see networks effectively engaged in policy and practice, we have to be sure that we do better than "satisficing" (Simon, 1957).

DISASTERS AND THE QUESTION OF NETWORK ONTOLOGIES

We feel compelled to state at the outset that networks, strictly speaking, cannot properly be said to exist outside of formal organizations. They are, instead, analytical abstractions—sets of researcher-specified relationships around which analysts impose theoretically derived boundaries. Social networks, as objective things, do not exist except as formal organizational structures (in contrast to informal organizational structures), plans for action, cognitive models, or forensic traces of patterns of interaction. Networks themselves have no agency, except perhaps in terms of Andrew Pickering's (1995) loose definition of material (i.e., nonhuman) agency as things "doing things in the world." But networks are scarcely even "things" or any category of objective being at all. Networks, rather, are cumulative—often emergent—patterns of dyadic and triadic relationships that are specified by the researcher.

To the extent that there is any agency to networks—again, according to Pickering's relaxed criterion for material agency—there can be no telos, no *intentionality* to social networks. Instead, we examine patterns of connection and interaction between actors—either individuals or organizations—whose actions and intentions are facilitated or constrained by emergent patterns of connectivity. Even with these connections, it is tricky to speak of the kinds of flows that occur across the connections. For example, does information really *flow*? Or does it get upgraded, downgraded, revised, and obfuscated as much (or more so) than it gets passed on as some bit of information? Actors in networks broker and constrain the transmission of information and materials. How do the authors in this book translate these phenomena into and out of social network analysis in disaster response, recovery, and adaptation?

CONCLUSIONS

We hope this book starts to congregate the confirmed generalizations, the conceptual devices, the modes of translation, the definitions, the recognized facts, and the applications of studying relationships and interactions in disaster settings. We especially hope it begins to showcase the nature of variation in human, group and community interactional and exchange behaviors in the face of disaster and as its aftermath unfolds. We do not expect major patterns to be immediately apparent nor do we expect so many patterns to hold in most places or among most populations. However, applying a networks understanding to disasters means that we are able to gather the greatest insight possible into interactions that are ephemeral in and of themselves and that belong to that very liminal space of disasters—where normalcy seems distant even though that normalcy is reeled in daily through those social networks.

SOCIAL NETWORK ANALYSIS FOCUSED ON INDIVIDUALS FACING HAZARDS AND DISASTERS

2

A.J. Faas[1], Eric C. Jones[2]

[1]San Jose State University, San Jose, CA, United States;
[2]University of Texas Health Science Center at Houston, El Paso, TX, United States

CHAPTER OUTLINE

INTRODUCTION

Our objective in this chapter is to examine the benefit to disaster theory of collecting and analyzing social network data focusing on individuals. Simply put, what are the conceptual, theoretical, and empirical trends in social network analysis approaches to studying the behaviors, perceptions, and well-being of individuals in disaster contexts? As disaster researchers, we have found it useful theoretically and methodologically to systematically explore the ties or networks between various individuals and entities to better understand disaster preparedness, response, recovery, adaptation, and mitigation. This chapter focuses on individuals and communities, while in the next chapter Naim Kapucu and Fatih Demiroz focus more on the networks of groups and organizations in disaster settings. This chapter is more conceptual than methodological. The fourth chapter in this section by Danielle Varda gives more detailed coverage of what kinds of network data can be useful in various disaster settings. Additionally, each of the chapters following this introductory section is an empirical study with appropriate attention to theory, methodology, and results.

Network analyses of disasters attempt to deal with the patterns of relationships that enable or inhibit individual, group, or organizational capacity to prepare for, cope with, adapt to, resist, or recover from potential or actual risks, hazards, and disasters. This full range is rarely considered in any one study. Instead, most researchers tend to focus on one or more contextually specific settings such as planning and preparedness, response, mitigation, recovery, or adaptation. The core issues of vulnerability, social capital, or resilience run throughout these investigations of specific settings—these scholars want to know how social networks can reduce vulnerability and how they can improve resilience following exposure to chronic or acute hazards. Importantly, since the network concept is often a core element in social science policy recommendations, it deserves careful scrutiny in its application.

Five major sections comprise this chapter. Research on networks and disasters is largely about social support, so we begin with a section on social support and types of ties, but then follow it with more in-depth sections on what kinds of people belong to these networks and the types of support exchanged (i.e., network composition), and on how these people in a network are connected (i.e., network structure). The final two sections deal with particularly challenging theoretical problems in studying disasters and networks: how do we conceptualize disaster phases or time frames? What is the relationship between network patterns that perhaps structure behavior in ways unperceived vs networks and relationships with which people consciously engage? Each section in this chapter provides an orientation to key network approaches and findings in different areas of disaster research. Each subsection includes discussions of studies that systematically collected some kind of data on dyadic relationships experienced by individuals—that is, data for social network analysis. Usually, these are interpersonal relationships, but sometimes there are organizations providing support to individuals analyzed as network data. Though some studies we cite do not explicitly involve disaster social network analysis, these theoretically more general works have motivated us in the last decade to examine facets of networks in disaster settings. We also consider community networks in disaster settings in this chapter, although few studies have systematically captured and then analyzed interpersonal data across a whole community for disasters.

SOCIAL SUPPORT, SOCIAL CAPITAL, AND TYPES OF TIES

Social support—an interest of many social scientists studying disasters—is primarily concerned with the relations that facilitate or inhibit access to both formal and informal resources in preparation, response, recovery, and long-term adaptation. Social support has been found to enhance individual and group recovery as a complement or alternative to institutional aid (Ibañez, Buck, Khatchikian, & Norris, 2004; Kaniasty, 2012; Perry, Williams, Wallerstein, & Waitzkin, 2008). Although research has found that the poor tend to access less social support than do the wealthy (Norris, Baker, Murphy, & Kaniasty, 2005) and probably perceive less support than do the wealthy (Jones, Gupta, Murphy, & Norris, 2011), poorer individuals had more access to informal support for job searches than did middle class individuals following Hurricane Andrew (e.g., Hurlbert, Beggs, & Haines, 2001). Common facets of social support relevant to disaster research include exchange, reciprocity, help seeking, types of support (e.g., informational, emotional, tangible/material), and types of people giving support (e.g., family, friends, neighbors). Some networks and disaster research have begun to paint a picture of the variation in disaster-related behaviors and outcomes based on network characteristics under the rubric of social capital, or the resources embedded in social relations and potentially accessible through ties in a network (Bourdieu, 1985). In social capital research, researchers distinguish between *bonding social capital*, or strong ties between individuals in a group (e.g., kin, neighborhood), *bridging social capital*, or ties that connect individuals to different groups or link different groups together, and *linking social capital*, or ties to higher levels in social hierarchies.

BONDING SOCIAL CAPITAL: KIN AND OTHER CLOSE TIES

When bonding social capital is operationalized at the individual level, kinship is often a key relational variable. Kinship and reciprocal exchange practices have been found to be intimately bound to one another, and Marshall Sahlins (1972) identified an important link between kinship and the practice of generalized reciprocity, or giving without the calculation of expected returns—although returns of some form or other would ultimately be expected at some point. These kinds of support exchanges are often vital in disaster recovery. Jeanne Hurlbert, Valerie Haines, and John Beggs's (2000) study of recovery from Hurricane Andrew in Louisiana found that kin ties were key sources of informal support. Furthermore, respondents with greater degrees of kin in their personal networks accessed informal support more frequently than those with lesser degrees of kin in their networks in the period following the disaster. In his study of the collapse of the Tasman Bridge in Hobart, Tasmania, Trevor Lee (1980) found that, when newly separated by geographical distance, kin ties tend to prove more robust than nonkin ties in the provision of support. The geographic and spatial distribution of ties is often a key variable in the provision of social support in disaster contexts. A study of Turkana livelihood strategies in the 2005–2006 drought in Kenya (Juma, 2009) found that one key strategy employed by Turkana pastoralists was to split families, sending some members of family units away to stay with relatives.

Importantly, some studies have differentiated the types of support provided by kin and nonkin. As reliable and robust as kinship ties may be, several studies have found that kin ties are not associated with all types of support needed and may even inhibit access to vital resources. In their study of support networks in Haifa, Israel, during the first Gulf War, Yossi Shavit, Claude Fischer, and Yael Koresh (1994) found that, while respondents reported greater degrees of emotional and psychological comfort from friends, kin more reliably provided immediate and direct aid. In their study of social support during and after the Midwestern United States floods of 2008, David Casagrande, Heather McIlvane-Newsad, and Eric Jones (2015) found that people tended to look to family in life-threatening situations, but reached out to neighbors, friends, volunteers, and professionals in preparation and in short-term recovery. Some studies outside disaster contexts have found that kin-centered networks are associated with decreased life satisfaction (Acock and Hurlbert, 1993, p. 323), while others have found that, although kin networks reliably provide intimate and material support, they are inversely related to accessing formal support from institutions (Tausig, 1992, p. 88). This finding was supported by Hurlbert et al.'s (2001) study of support networks in Hurricane Andrew as well. These studies indicate that weak, bridging ties and network diversity can facilitate access to resources as well as information that could increase network members' likelihood of accessing more diverse sources of institutional support.

HOMOPHILY AND DIVERSITY IN NETWORK COMPOSITION

Homophily—the sociological principal that holds that people are more likely to interact with those with whom they share similar characteristics (e.g., race, class, gender, ethnicity, etc.) (McPherson & Smith-Lovin, 1987)—is often contrasted with network diversity, or ties between actors who do not share similar characteristics. Numerous studies have found that the formation of groups and organizations is often underwritten by homophilic patterns in the formation of social ties (cf. McPherson, Smith-Lovin, & Cook, 2001; Petev, 2013). Others have pointed to the ways in which shared ties to groups and organizations (Frank, 2009) or ideologies (Anderson, 1983) often serve as powerful proxies for attributional homophily in facilitating relational patterns. Diversity of relational ties, by contrast, can also mean diversity of resources and information (Granovetter, 1973), which can have important implications in disaster contexts. Thus, we look for similarities and differences among people in a network, and we expect, for example, that similarities between people might reduce misunderstandings and that differences between people might support novel behaviors and interactions.

BONDING THROUGH SIMILARITIES

So, what aspects of homophily and variation in network composition do we expect to be important in disaster settings? Ron Burt's (1992) general link between homophily and trust might help explain why Hurlbert and colleagues (2000, 2001) found that homophilic ties were nearly equal to kin ties as key informal support conduits for survivors of Hurricane Andrew. Lee's (1980) work on the Tasman Bridge disaster—where a bridge collapsed between communities on two sides of a river in Tasmania, Australia—also pointed to the common finding that homophily is often on par with spatial proximity in facilitating the development of social support networks. Homophily, however, is certainly not a relational magic bullet.

As social scientists, we know that people who appear similar according to some obvious trait may differ wildly on a host of other traits. In their study of patterns of interaction on three large wildfires in the American Southwest, Branda Nowell and Toddi Steelman (2015) were interested in factors influencing the frequency and quality of interaction among responders during large wildfires. They focused on homophily as shared affiliations or roles (similar professions or agencies) and prior, direct personal relationships between actors as independent variables. They found that, while homophily was significantly associated with frequency of interaction during wildfire incidents, it was not a significant predictor of quality of interaction. Instead, where homophilic ties lacked prior, direct personal relationships, communication problems were more likely. In a related study, Faas and colleagues (2016) asked key individuals in wildfire response networks in the Pacific Northwest to nominate specific people they would trust in bridging roles in the event of a large wildfire involving outside incident management teams. They then followed up with a survey following 21 large wildfires in the region later in the same year and asked respondents to indicate who actually assumed bridging roles during actual fires. Nearly half of all prefire nominations were consistent with homophily—operationalized as same agency type, e.g., fire department, law enforcement, county government. However, while homophily indicated how people expect to organize, it did not hold up for how they organized during actual incidents, where only 3% of bridging actors named were consistent with the homophily principle.

We know from the limited social network studies of risk perception that an individual's likelihood of perceiving a given risk increases with the number of people in their network who also perceive this risk (Helleringer and Kohler, 2005) and that the strength of ties between actors is associated with similar risk perceptions (Scherer and Cho, 2003). While most research on risk perception has thus far focused on the individual level of analysis, Jason Gordon, Al Luloff, and Jason Stedman (2012) research supports a social cognition perspective in which risk perception is best viewed as a social and cultural phenomenon created through interactions with others in a network (see also Jones et al., 2013). Recent experimental studies of risk perception have found that positive peer feedback via social media was significantly related to behavioral intention to take protective action against a hazard (Verroen, Gutteling, & De Vries, 2013). This body of research suggests that risk perception is a product of interactions within a social network. It is therefore worth revisiting prevailing models of risk perception, which are built upon the individual level of analysis, and developing approaches for investigating risk perception at the community and network level. To this end, some studies have examined what kinds of messages are passed on (e.g., Sutton et al., 2014) and the length of time it takes various kinds of messages to be passed on (e.g., Spiro, Dubois, & Butts, 2012). Studies of these phenomena can inform efforts to craft messages that resonate with focal audiences and that are remembered/used longer, and to strategically disseminate these messages in targeted networks.

BRIDGING ACROSS DIFFERENCES

While community bonding links based on homophily (however construed) can help establish trust and strengthen social support networks, links to diverse "bridging" ties can connect actors to a variety of resources and nonredundant sources of information (Faas et al., 2015, 2016; Newman & Dale, 2005). Moreover, tight-knit networks can potentially inhibit improvisational coping and adaptation strategies by imposing constraining norms on group members. Considering the role of networks in adaptive resource management in the face of climate change (adaptation qua disaster risk reduction), Lenore Newman and Ann Dale (2005) point to the possibility that network diversity increases the capacities for collective decision making and the ability to adapt to sudden change.

Network diversity has also been significantly associated with short-term recovery, as evidenced by studies of social support in Hurricane Andrew (Haines, Hurlbert, & Beggs, 1996). Hurlbert and colleagues also found that individuals with close-knit networks exhibited the lowest levels of access to support from outside the group, especially formal support from institutions, which points us to another range of issues associated with the balance of network homophily and diversity in disasters (Hurlbert et al., 2001). One of these issues is class differences, and the extent to which people of different socioeconomic statuses interact when facing disasters, recovery, or adaptation. A study considered socioeconomic status and found reciprocity between people and alters in their personal networks varied in Mexico vs Ecuador, highlighting the potential role of patron–client relationships in postdisaster recovery, particularly in more class-based, industrial, labor market settings (Jones et al., 2015).

In a study of community adaptation to the 2007 Hebei Spirit oil spill in Korea, So-Min Cheong (2012) identified a prevalence of bonding capital as a factor inhibiting the meaningful exchange of knowledge between locals and experts. She points to the fact that high degrees of bonding capital, or a prevalence of kin and homophilic ties in a network, can constrain innovation, resource diversity, and knowledge dissemination critical to adapting to crisis and change. Insulated local networks, she points out, "can set high barriers against external ideas, information, and resources, and may not foster innovation" (Cheong, 2012, p. 26).

LINKING VIA SOCIAL HIERARCHY

Bridging is fundamentally about connecting disconnected groups or subgroups and is often associated with types of power and influence, since bridging actors often have unique access to information and resources as a result of their unique ties. It is, therefore, theoretically and practically interesting to examine a network to see who has unique ties outside the network. In disaster research, some have called this *linking social capital* (Aldrich & Meyer, 2015), or ties connecting local people with people in power at higher levels of regional and global scale. In our study of roles of linking ties in recovery in postdisaster resettlements in the wake of the devastating 1999 and 2006 eruptions of Mt. Tungurahua, we looked for key individuals who provide and/or receive information, since providing information implies that one has a nonredundant source for such information.

Analysis of personal networks collected within two years after resettlement found in a more urban resettlement with more splintered networks that men were significantly more likely to receive information as a form of social support, while women were significantly more likely to provide informational support, yet there was no such association in the more rural resettlement with networks that were somewhat more intact (Faas et al., 2015). Yet two years later in a related study in the same sites, when respondents were asked whom they would consult for information or opportunities with institutions, the pattern was reversed and men outnumbered women in both the more urban resettlement (7:1) and in the more rural resettlement (2:1) as likely or preferred sources of information (Faas et al., 2014). This may reflect methodological differences; however, it is also possible that women were indeed more important early on as information sources in the resettlement with more fractured networks. It is also possible that women were important links to powerful outsiders but men retained the prestige or reputation for having such outside ties.

COMMUNITY BONDING BECAUSE "WE'RE ALL IN THE SAME BOAT"

Conceptualized at the community level, bonding social capital is often operationalized as the degree of ties between individuals through informal associations and of individuals in formal organizations (Coser, 1975). In his comparison of two similar neighborhoods in the wake of the 1995 Kobe earthquake, Daniel Aldrich (2011) measured community-level social capital according to the number of nonprofit organizations per capita. He found that one high social capital community was remarkably effective in organizing to put out the flames tearing through their neighborhood and providing care to one another in the immediate aftermath. The other community–of low social capital—did not evince the same degree of coordination and many people "stood helplessly as flames consumed their businesses and homes" (Aldrich, 2011, pp. 599–600).

A few other studies that conceptually cover ties while not measuring dyadic relationships are worth mentioning. By modeling risk sharing (i.e., redistribution or transfers through informal social insurance), Yanos Zylberberg (2010) captured the tensions that exist between the postdisaster altruism when all are in it together vs the stressful isolation and rupture of ties and informal transfer—but found that communities with prior experience of trauma are often able to bounce back better in many ways. In other community-oriented research not strictly focused on dyadic ties, Sabrina McCormick (2012) examined the role of crowdsourcing in citizen science for both risk assessment and recovery following the 2010 Deep Horizon oil spill. Cheong (2012) found external ties to be necessary for community recovery, but noted both positive and negative aspects of community ties to external resources in a study of the 2007 Hebei Spirit oil spill. Craig Colten, Jenny Hay, and Alexandra Giancarlo (2012) similarly examined external resources for community resilience for the Deep Horizon oil spill as compared to prior coastal Louisiana oil spills.

SOCIAL NETWORK STRUCTURE: MOVING BEYOND SOCIAL CAPITAL AND *COMMUNITAS*

While it is intuitive to think of homophily and diversity primarily in terms of network content, several disaster studies have pointed to ways in which these concepts have important implications for network structure as well. The analysis of network structure has become extremely popular in the multidisciplinary field of social network analysis (Borgatti, Mehra, Brass, & Labianca, 2009), although relatively few structural analyses of networks in disaster studies exist. Examining network structures in disaster contexts could potentially reveal important relational dynamics. Moreover, network structure is not merely an unconscious phenomenon; people at times knowingly manipulate the structures of their own networks, particularly in hazard and disaster contexts.

DENSITY

We know that network size and density may even be consciously calculated by individuals encountering risk and hazard. Around the world, people consciously build networks to mitigate risk, as when the Pokot of western Kenya gift livestock to increase the size and density of their social support networks for food aid and reciprocation in the event of crisis (Bollig, 2006; Moritz et al., 2011, p. 287). In their research on Hurricane Andrew, Hurlbert and colleagues (2001, p. 212) found that individuals with smaller, denser networks were more likely to receive informal social support, while individuals with larger, less dense social networks were more likely to receive support from institutions (Hurlbert et al., 2001). They found also that individuals embedded in dense networks that reliably provided informal support also exhibited the lowest levels of access to support from outside the group, especially formal support from institutions.

With colleagues, we studied personal and whole social networks in relation to social support exchanges and well-being in two Mexican sites (one resettled by landslide and one evacuated by the erupting volcano Mt. Popocatepetl) and five Ecuadorian sites (two resettled, two evacuated, and one not evacuated by the erupting volcano Mt. Tungurahua) from 2007 to 2011 (Faas et al., 2014, 2015; Jones et al., 2013, 2014, 2015; Tobin et al., 2011). In one analysis, we focused on our subsample of 142 personal networks in the two Ecuadorian resettlements to examine variation in personal network structure. We found that personal network density was positively associated with both the giving and receiving of every category of material, work, informational, and emotional social support in a dense, small urban resettlement of agro-pastoralists displaced by the eruptions of Mt. Tungurahua in Ecuador (Faas et al., 2014, 2015). In contrast, in a smaller, rural agricultural resettlement, network density was only associated with giving and receiving work invitations. This may be due to the fact that rural resettlers in the urban resettlement, lacking productive resources, must rely more heavily on the dense networks of relationships from their prior village for material support. In the small rural resettlement, however, resettlers were less reliant on prior village contacts for support, finding support perhaps through new contacts and branching out and forming less-dense networks. Interestingly, density was not associated with well-being in either of these sites. Moreover, in both sites in our Mexico research, the densest networks were those with the worst levels of well-being (Jones et al., 2014).

Focusing on bonding social capital or dense social networks as critical facets of disaster recovery seems to make good sense. We agree that it is necessary to develop pre- and postdisaster policies and practices that are sensitive to and supportive of informal practices of solidarity and prosocial behavior and seek to prolong these phases to the extent possible, avoiding the risk of "social disarticulation" that Michael Cernea (1996, 1997) has often emphasized in resettlement contexts. However, by placing such emphasis on bonding and *communitas* or a strong sense of community, we run the risk of reifying cultural hierarchies of race, class, gender, and ethnicity that often inhere in "close-knit" networks. We also risk conceptually miscasting those groups that do not manifest solidarity or any dense network patterns of altruism as dysfunctional at best, or undeserving of aid and support, at worst.

Across the social sciences, some of the most common prescriptions in crisis recovery are for maintaining tightly knit kin groups and already-established integrated communities (Cernea, 1996, p. 305) where individuals interact over common interests and even share common histories (Cernea, 1997, p. 1575). The thesis is that these social groups sustain cultural integrity and relatively equitable relations—particularly concerning the exchange of labor and material for agricultural production or goods and services when coping with the scarcity and isolation that often accompany catastrophic events. Indeed, there is compelling evidence that the stress of disaster is mitigated by clinging to familiar practices and relations of shared identity and mutual obligation in order to manage the risks of cultural and economic survival. And yet, throughout the disaster literature, we find descriptions of latent conflict, weakened alliances, exploitation, opportunism, and resistance to social hierarchies in otherwise integrated communities (Faas, 2015a,b, in press; Hoffman, 1999; Oliver-Smith, 1979, 1992).

But when cooperation is seen as a marker of deservingness by the state and nongovernmental organizations, pathologizing a lack of cooperation can and does often serve as a prelude to abandonment. For instance, in his studies of disaster-induced resettlements in the Ecuadorian highlands, (Faas, 2015a, 2012, in press) found that state agencies and nongovernmental organizations both made various types of cooperation and participation—typically appropriations of traditional cooperative labor practices, or *mingas*—conditions for benefit inclusion (housing, irrigation, various forms of aid). One nongovernmental organization using this approach in Ecuador wound up structuring the spaces and times of participation in such a way that it made it virtually impossible for certain households (those with wage labor employment) to participate and demonstrate their deservingness of benefit inclusion (Faas, 2015a, in press). Moreover, local leaders seized upon these policies to amplify their own power to enforce participation and exclusion (Faas, 2015a). The cumulative effect was an often vitriolic politics of deservingness, in which people competed within and between communities to mark themselves and others with deserving and undeserving of aid.

STRUCTURAL DIVERSITY

Because of concerns with an overemphasis on network density discussed earlier, we were interested in moving beyond network size and density to consider other structural variables as well. For example, in various contexts unrelated to disasters, bridging actors may connect dissimilar actors and organizations precisely because they connect those who are not already connected to one another

(Coser, 1975, p. 257; Granovetter, 1982, p. 108). In disaster settings, diversity can be particularly beneficial when brokered by people who know the different groups (e.g., Stevenson and Conradson, Chapter 11).

We have looked at a number of structural facets of personal networks for long-term disaster recovery settings. A typology approach (of dense, extending, subgroups, and sparse networks) highlighted the ways in which personal networks that were dense and cohesive predicted different well-being outcomes—including inverse relationships—in different settings (Tobin et al., 2011, 2012, 2014). With more standard metrics including various measures of centrality and centralization, we similarly found that similarly structured networks did not act the same in all settings (Mexico vs Ecuador, resettled vs not resettled, high impact vs low impact, and each of seven study sites), specifically for outcomes like levels of risk perception (Jones et al., 2013; Tobin et al., 2011), gendered analysis of mental health (Jones et al., 2014).

Let us also consider the case of cohesive subgroups. Not all networks necessarily have subgroups. Extremely dense networks appear as a single group, and really sparse networks might not appear to form any group at all, structurally speaking. But when we do see subgroups, their cohesiveness is also partially a matter of whether they are held to other subgroups by one or two ties or by multiple ties. Subgroup cohesion can be measured in personal as well as whole networks. Our expectation was that the existence of subgroups in a personal network would actually be positively associated with well-being, since it would not be so dense as to be suffocating nor so sparse as to be an unreliable source of social support, and not entirely dependent upon key brokers that link network subgroups together. And yet, we found no association between subgroup cohesion and social support exchanges in the resettlements.

Looking at well-being indicators, we found that personal networks with subgroups had the poorest mental health scores overall for resettled women and the highest scores for functioning symptoms of depression in resettled men (Jones et al., 2014). For a whole network—a group of parents and caretakers of children in the ABC Day Care Center fire in 2009—Eric Jones and Arthur Murphy (2015) have started to examine how the network has integrated and fractured over time (see also Rangel et al., Chapter 12), and it appears that the changes in the subgroups of the networks were largely based on two axes: whether the groups had parents of deceased vs parents of injured children, and whether the parents were taking more political and justice-seeking approaches vs less so.

SURFACE STRUCTURE AND DEEP STRUCTURE

Early and mid-twentieth century structuralist thinkers (e.g., de Saussure, Chomsky, Levi-Strauss) each produced different versions of theoretical distinctions between what might be called "deep" and "surface" structure. French anthropologist Claude Levi-Strauss (1969) posited that deep, unconscious semiotic patterns generated or structured conscious patterns of culture and meaning. In her analysis of culture change in disaster contexts, Hoffman (1999, p. 308) contrasted deep structure, or "the invisible rules, how reality is organized and people, space, time, and other material are categorized," with surface structure, or "all the top-lying cultural minutia, particular customs and ceremonies, habits and practices." For Hoffman, deep structure provided a grammar that structured the more mutable patterns of practices in which people consciously engaged. Though much of this variety of structuralism fell out of favor in social science because it was largely ahistorical, often deterministic, and relied upon interpretations of cultures as bounded entities, there is nonetheless some enduring value to considering how unconscious patterns in human life and interactions might structure, at least in part, our conscious activities. For instance, we might think about how patterns in people's social networks might place certain constraints on their behaviors or otherwise afford certain possibilities in ways that they do not consciously make note of in their everyday decisions.

The analysis of total personal networks—or representative samples thereof—arguably arrives as something quite different than those personal networks constructed from context-specific prompts (see Varda, Chapter 4). Our network approach reported in the preceding section was an attempt to get at the deep structure of networks in human behavior—the unconscious, unseen patterns in network ties. Results thus far are inconsistent and highly context dependent (Faas et al., 2014, 2015; Jones et al., 2013, 2014, 2015). A presentation of all of our findings across the full cross-cultural sample of Mexican and Ecuadorian sites would be far less consistent than the smaller Ecuadorian samples we presented. And, while we speak here of just one set of studies, our cross-cultural sample of 460 personal networks in seven disaster-affected sites of varying impact is unprecedented in both design and scope and therefore worthy of consideration. Of course, we focus primarily on the results of network statistics in reporting our findings discussed herein. Our work, unlike most others employing formal network approaches, involved long-term ethnographic research as well. However, while our ethnographic data explain a fair amount of the network variation, they are far from explaining all that we have found.

We can point to surface structure as patterns of relations and actions taken but that may be structured to some extent by the deep structures of personal networks we yield through our context-free name generators. The most common approaches to network analysis in disaster contexts identify more of a surface structure, that is, how people consciously engage in network interactions. This latter approach has demonstrated greater consistency across settings than our attempts at deep structure. Work by Jeanne Hurlbert, John Beggs, and Valerie Haines (2001) on Hurricanes Andrew and Katrina has produced remarkably consistent findings. David Casagrande and colleagues' (2015) work on Midwestern floods likewise identified a surface structure, though one that was processual and rooted in social space and geography.

This inconsistency in the relationship of deep structure to outcomes likely tells us something about the relationship between networks and behaviors—the deeper we probe into the structure of social networks, the more we see how the formations around any individual or group extend to other individuals and groups; the content and patterning of these structures are formed through historical processes of interactions and the structuring influence of institutions. The emergent social networks may reflect constraints and affordances on human behavior, but deep structure cannot truly be envisioned as a grammatical structure for human agency. Thus, although people with similarly structured networks will experience similar behavioral constraints and capacities for at least some settings—such as being able to access particular types of information or resources—the implications of these structures are highly dependent on history, context, and the conceptual and behavioral repertoires we often refer to as culture.

INDIVIDUALS IN NETWORKS ACROSS PHASES OF DISASTER

Some of the earliest efforts in the social science of disasters involved the development of models of disaster phases or processes that could be used to plot disaster processes along spatiotemporal dimensions developed in the interest of systematic comparisons of assorted case studies. In one of the earliest such schemes, Anthony Wallace (1956, p. 1) borrowed a rudimentary temporal scheme—predisaster (stasis), warning, threat, impact, inventory, rescue, remedy, recovery—from psychologists John Powell, Jeannette Rayner, and Jacob Finesinger (1953, cited in Wallace, 1956). Wallace added a heuristic spatial scheme—identifying impact areas, fringe impact areas, filter (i.e., adjacent) areas, organized community aid, and organized regional aid. Overarching these was a cultural-psychological model called the "disaster syndrome." In stage one, people often reacted as stunned and immobilized. Stage two roughly corresponded with initial relief efforts and was characterized as a heightened state of suggestibility, altruistic behavior, and anxiety regarding the well-being of others. Stage three involved extreme euphoria and intense motivation to participate in relief efforts. Stage four was marked by a return to normality, apathy, and criticism of the impacts and responses to the disaster (Wallace, 1956, pp. 149–150). A coherent and sophisticated framework, it nonetheless lacked engagement with human-environment relations, was rooted in a particular American context, and overemphasized acute crises of homeostatic conditions.

While a progression of linear stages or phases in disaster research are somewhat maligned since not everyone experiences the same thing at the same time and because some stages of mitigation, preparation, response, recovery, and adaptation are even skipped, reversed, or seemingly perpetual, it is undeniable that the unfolding of an extreme event places people in a variety of settings with shifting dynamics over time. In her conceptual model of a cultural response to disaster, Susanna Hoffman (1999) described a "highly pervasive" pattern of three dynamic phases through which disaster survivors generally proceed. There is an initial phase of individual experience that proceeds to collective support. A secondary phase of collective support proceeds to factional competition. Finally, a third phase of closure follows in which factional competition gives way to attention to household affairs. Anthony Oliver-Smith (1979, 1992) similarly noted that previously existing stratifications like class and ethnicity can temporarily disappear in a short-lived wave of altruism. Once national and international aid appears (or disappears), or once people must sustain this liminal state of recovery too long, old divisions can reemerge, and conflicts begin again over social norms, expectations, and access to resources (Faas, 2015a; Kroll-Smith, in press). Scholars in many fields have now noted aspects of these dynamics, and, while none of these researchers were explicitly investigating networks per se, they nonetheless tell network stories of isolation, support exchanged, conflict, and factionalization.

Those engaging in more explicit network analyses have found evidence to support some descriptive and heuristic models of relational processes in disaster settings. For instance, there is ample evidence to suggest that volunteerism from core relationships is frequently followed by a decrease in availability of support and an increase in conflict and fewer or weakened relationships (Bolin, Jackson, & Crist, 1998; Norris, Baker, Murphy, & Kaniasty, 2005; Palinkas, Downs, Petterson, & Russell, 1993; Ritchie, 2012). While studying the role of social capital in response to and recovery from Hurricane Katrina, however, Weil, Lee, and Shihadeh (2012) found some contrary evidence in that those with the most social ties (i.e., the most socially embedded) had the greatest social support provision responsibilities and subsequently experienced more stress than others. However, they also found that these more socially embedded individuals recovered from stress more rapidly than others, indicating a process whereby social embeddedness can result in greater exposure to stress early on but a more likely recovery over time.

Several contributors to Ronald Perry and Enrico Quarantelli's (2005) volume, *What is a Disaster*, called on researchers to alter their focus from an emphasis on chronological time and geographic space to an interpretation of social time and social space. In their study of citizen response and recovery in the extensive flooding of the Mississippi River in western Illinois in 2008, Casagrande and colleagues (2015) attempted to operationalize the concept of social time and space. They found that instances of support from kin far outnumbered support from all other sources (neighbors, friends, professionals, elected officials, volunteers) in the *vital* phase (securing physical safety) of disaster. Their findings suggest a processual model of support networks with important spatiotemporal dimensions: though kin are most common sources of support in the vital phase and are important features in other phases as well, professional (i.e., institutional) support was most common in nonvital preparatory, nonvital postimpact, and long-term recovery phases, while the contributions of other support providers changed through each phase, as well. In this model, the phases of disaster are relative to individual experience and the ties whose support is most common or vital to each.

One reason that it has been difficult to develop reliable spatiotemporal models of disaster phases is that longitudinal analysis of individual networks in disaster situations is rather uncommon, and less common than longitudinal analysis of organizational networks in disaster situations.fn11[1] Jones and Murphy (2015) added to the general dynamic of individualism-altruism-conflict-routine by looking at a community of parents and caregivers who lost children or whose children were injured in the 2009 ABC Day Care Center fire in Hermosillo, Mexico. This new network of individuals created two major alliances initially by year one (one of parents of deceased, one of parents of injured), then subdivided into four by year two—parents of deceased divided into a more political and a less political group, and parents of injured children did the same. Subsequent ethnographic work by the team suggests two tensions in opposite directions—further splintering into one or two more groups, and attempts by some leaders to pull the groups back together. Larger political processes powerfully imposed on these dynamics, and these processes included recent elections in the state, renewed investigations of culpability for the fire, and the annual anniversary of the event that includes academics, nongovernmental organizations, and supportive local residents who did not have children at the day care center.

CONCLUSIONS

The study of informal networks provides insight into not only social support and social constraints but also indicates some persistent issues in disaster research. Primary among these issues are (1) the nature of changing settings as disasters unfold and are addressed, often referred to as stages, (2) differences between the potential resources that people can tap into and the perceptions and behaviors that actually occur, and (3) the challenge of capturing the kinds of ties that create relevant networks for disaster response, recovery, and adaptation, especially when involving multiplex relationships.

We know that the network concept is powerful and useful for disaster research, but we also note the wide variation of methodologies employed, from those that are heuristic to those that are descriptive to those that involve formal network analysis focusing on ties and nodes. The formal network analyses are often based on measurements taken at one point in time, whereas the heuristic and ethnographically descriptive approaches are often based on some kinds of observations over longer periods of time. Additionally, findings from formalized network analyses tend to be less consistent than are other theoretical models and policy prescriptions in the fields of disasters and disaster-induced resettlement.

---

[1]In terms of more systematic network research on change over time, we recognize that it is logistically challenging and expensive to collect multiple waves of data on networks in disaster settings, and we have also found it is quite difficult to justify the same study again to funding agencies that are used to funding new and novel research topics and settings.

The fact that more formalized social network studies point to inconsistent outcomes associated with similar networks in various contexts should give us pause and care in our interpretation of network structures and compositions.

For example, networks variously described as close-knit and tight may, in some circumstances, prove to be invaluable resources for individuals and households in need of short-term, immediate support in a crisis. However, such networks can also be experienced as constraining, with endemic material scarcity and redundant sources of information that may limit opportunities for response, recovery, and adaptation (Marino & Lazrus, 2015). Dense groups may be rich in support while poor in well-being. In some cases, we find women in bridging roles, but the power and prestige of bridging is retained by men. We may expect networks with subgroups to be resilient, and yet we have found them to be insufficiently associated with important indicators of well-being in some cases.

Findings from formal network analyses tend to be less consistent than theoretical models or policy prescriptions tend to be in the fields of disasters and disaster-induced resettlement. However, social network analysis is often based on formal measurements taken at one point in time, whereas the descriptive and heuristic approaches are often based on a series of ethnographic observations over time. Nonetheless, the fact that formal studies point to inconsistent outcomes associated with similar network compositions in various contexts should give us pause and suggest a more critical approach to our interpretation of network structure and composition.

This review is an attempt to foster the development of dialogue and theory across disparate bodies of disaster literature that evaluate the influence of relationships on disaster response, recovery, and adaptation. These approaches include metaphors, qualitative descriptions, what networks are composed of, how networks are structured, and how to address complexity and change. We find in our disaster research that many scholars, including ourselves, often want to tell network stories—or stories about the importance of relationships—often when not intentionally looking to address issues of formalized social network analysis. Thus, we seek a more critical and nuanced vocabulary to more carefully delineate major facets of the field of networks and disasters—particularly to more fully explore the influences of context.

INTERORGANIZATIONAL NETWORKS IN DISASTER MANAGEMENT

3

Naim Kapucu[1], Fatih Demiroz[2]

[1]University of Central Florida, Orlando, FL, United States;
[2]Sam Houston State University, Huntsville, TX, United States

CHAPTER OUTLINE

INTRODUCTION

Disasters are complex and challenging situations in which people often expect governments to solve problems swiftly and effectively. Managing catastrophic disasters is fundamentally different than addressing routine emergencies. Although such extreme events happen relatively rarely in any one place, they cause significant disruption in the society. Successful management of catastrophic incidents typically requires a broad set of skills and resources that could only be provided by multiple stakeholders in a horizontal and collaborative web of relationships. In other words, interorganizational networks are the norm in dealing with disasters. Disaster management is not the only business of government that benefits from interorganizational networks. There is a wide variety of services from health care to social services that are delivered through cross-sector networks and partnerships. Transition of service delivery from hierarchical bureaucracies (direct government) to cross-sector partnerships dramatically influences the position of networks in disaster management. Understanding how networks are designed and developed and function and evolve before, during, and after disasters helps us get better disaster management outcomes when networks are used.

Because disasters create conditions in which altruistic behavior and need for collaboration becomes salient, networks may occur naturally in handling stressful situations (whether they are effective or not is a different subject). Disaster management requires efforts from all levels of government[1]; however, government's capacity and resources are usually inadequate for handling major disasters, and collaboration with private and nonprofit organizations becomes vital. This multilevel, multisector, and collaborative nature of disaster management makes it an ideal context to research networks (McGuire & Silvia, 2010). The purpose of this chapter is to provide a theoretical overview of interorganizational networks in disaster response and recovery. We outline the nature of disaster management as being a layered function that involves multiple networks and cross-sector agencies.

First, we present a brief historical perspective on interorganizational networks including the disaster management field. We then continue with explanations of various types of interorganizational networks and how they are used in the context of collaborative emergency management. Finally, we develop a basic framework that explores the various factors that facilitate and hinder effective collaboration between agencies in disasters.

INTERORGANIZATIONAL NETWORKS

Networks are becoming increasingly important in policy implementation. They are not new in the American governmental system; however, the use of *network* as a term is relatively new. Intergovernmental relationships and the relationship between different branches of government can be viewed as a form of network. Networks today, on the other hand, mostly refer to intersector relations in the US governmental system. As traditional bureaucracies shrink and recede in policy implementation and service delivery, nonstate actors complement, in some cases even replace, them by operating in network structures. Some of the indicators of this change include privatization, contracting out services, growth of nonprofit sector, and creation of partnerships for solving complex problems.

Several types of networks exist in the interorganizational networks literature based on how they function, what kind of goals they try to accomplish, and how they organize (Kapucu, Hu, & Khosa, 2014). The formation of networks can be voluntary (i.e., organizations join partnerships without coercion or targeted incentives) or mandated by higher levels of government. For example, the Federal Emergency Management Agency (FEMA) in the United States requires local emergency management agencies to adopt the National Response Framework and National Disaster Recovery Framework in their plans to receive federal funding for certain programs. These frameworks require emergency management agencies to build cross-sector partnerships and collaboration.

INTERORGANIZATIONAL NETWORKS IN DISASTER MANAGEMENT

Recently, the United States emergency management system was developed upon and now operates on the "whole community" approach. This approach is built on the notion that disasters are complex problems that cannot be solved or managed by any single agency or an actor. It requires involvement of governmental as well as nongovernmental organizations and citizens (Kapucu, 2009). The concept of community disaster resilience is built on the same idea that disasters cannot be handled by a group of public servants with the resources typically at their disposal. Instead, a community should work together to reduce its vulnerabilities

---

[1]These efforts might be mandatory or voluntary.

(social and physical), share resources, and utilize informal networks for minimizing impact of disasters and bouncing forward after an incident (Edwards, 2012). These vulnerabilities emerge and grow as communities evolve over time. Immigration, growth in vulnerable populations (elderly, disabled, and minors), and the changing nature of hazards (e.g., climate change) can increase risk levels and risk types in a community.

Despite *disaster resilience* being a widely used term, its definition needs some clarification. Fran Norris and her colleagues (2008) define resilience as the process of linking community resources (including individual and organizational) to a set of adaptive capacities. Community resources and adaptive capacities include social capital (Aldrich, 2012), organizational networks, human capital, economic development dynamics, and community capacity (Norris et al., 2008). Interorganizational networks in a community for emergency management purposes are mostly developed and maintained by emergency management organizations (e.g., emergency management agencies, sheriff's office, health departments, etc.). They are expected to be operational in all phases of emergency management (i.e., preparedness, mitigation, response, and recovery),[2] although each phase needs different organizations to work together for variety of goals.

Each emergency management phase possesses specific characteristics, challenges, and requirements, which means interorganizational networks are shaped by the needs of each phase. For example, networks for responding to disasters differ fundamentally from ones for the mitigation phase. Response networks rely much more on instant and simultaneous interactions, decision support systems, and constant flow of information through network ties for a relatively short period of time. Mitigation networks, on the other hand, require organizations to collaborate for long periods of time (as long as decades) for implementing certain pieces of laws and regulations such as flood insurance, building codes, and moving the people away from high-risk areas. For example, Florida counties develop and use local mitigation strategies to minimize impacts of natural and man-made hazards. Documents outlining these strategies highlight partnerships in ways that are different than disaster response. Local mitigation strategies are prepared in coordination with local economic development plans, comprehensive emergency management plans, and continuity of operations plans. Their operations include public education and community outreach, reinforcing critical facilities through federal grants, and implementation of strict building codes (Seminole County, n.d.). Also, private sector partners are included in the system where needed (especially for local economic development planning) (Miami-Dade County, 2015).

It is noteworthy that all interorganizational networks—be they response, recovery, or mitigation— require time and effort to build trust, interoperability, and harmony among network members. That means even a successful response network that functions only a few hours often requires a long history of partnership among its participants. Ad hoc participants in networks and helpful bystanders always have some role in managing disasters (e.g., providing information, creating new connections, supporting charity services), but lack of such coordination and accountability can easily lead to chaos and may hinder success of the formal network actors.

NETWORKS OF NETWORKS: MULTILAYERED DISASTER NETWORKS

Disaster management systems in the developed countries rely on myriad agencies for effective disaster management. There exists an overall national network comprised of subnetworks that

---

[2]Preparedness and mitigation can be thought of as facets of adaptation. Most of the disaster network research is focused on response phase of disaster management.

include national organizations such as FEMA, local and state/regional level emergency management agencies, primary response agencies such as fire departments, police departments, and search-and-rescue units, and nonprofits such as the Red Cross, civic organizations, and churches (Waugh, 2003).

Each level of government may play a crucial role in managing disasters and emergencies. At the national level, in the United States, for example, the role of the National Response Framework and National Incident Management System guides different levels of government and nongovernmental players to work together in managing disasters (McGuire & Silvia, 2010). National funding initiatives such as the Urban Area Security Initiative (UASI) have facilitated collaboration in managing disasters across jurisdictions or topical foci as well. The UASI is an initiative by the United States Department of Homeland Security to provide local communities funding assistance to improve and develop response capabilities and preparedness planning by developing regional councils and working group networks. Alexandra Jordan (2010) finds that UASI indeed brings government agencies at urban and peripheral settings (suburban and rural) together to form collaborations for managing disasters. Funding is the primary motive for collaboration in these cases, and it shapes how the relationships are developed and structured. For example, metropolitan areas demand bigger shares from the funding streams because they need to protect far more people and property compared to smaller jurisdictions, whereas smaller jurisdictions rely on resources of metropolitan cities when they are overwhelmed by disasters. The relationships between urban and rural jurisdictions are shaped by the needs of individual jurisdictions (particularly urban jurisdictions). Jordan (2010, p. 12) summarizes these relationships as follows: "Collaboration in this instance did not originate from common regional needs, but from power imbalances between urban, rural or suburban communities."

At the state or provincial levels, many partnerships and networks have also developed to effectively guide the management of emergencies and disasters. One such partnership model in the United States is the Emergency Management Association Compact (EMAC). This compact is a national interstate mutual aid agreement that helps to facilitate the sharing of resources, equipment, information, and personnel across states during disasters. Naim Kapucu, Augustin, and Garayev (2009) studied the role of the EMAC to understand interstate relationships and partnerships in Hurricanes Katrina and Rita in 2005. Their study concluded that horizontal partnerships and disaster networks such as the EMAC between states in the United States are essential for providing administrative oversight and the resources and technology for quick problem solving in disaster response.

At the local and county levels in the United States, emergency managers have a strong network of their own, which is comprised of first responders, emergency operation centers (EOCs), local nonprofits, etc. These networks allow them to access resources of critical private companies, nonprofits, and community organizations that otherwise would be neglected in a traditional command-and-control emergency management structure (Waugh, 2003). These relationships can be structured several different ways. Naim Kapucu and Vener Garayev (2014) categorize local disaster response networks into three groups: vertical or hierarchical, horizontal or decentralized, and a combination of the two. These categories are the reflection of three response systems in practice in the United States. The first is based on the Emergency Support Function (ESF) system, which is considered horizontal. The second is based on the American organizational strategy for emergency response operations, the Incident Command System (ICS), which is considered vertical. The third is a hybrid combination of these two systems (Kapucu & Garayev, 2014; Kapucu & Hu, 2014). For example, Orange County, Florida, uses the emergency support function system in its comprehensive emergency management plan. Primary and secondary organizations are identified to

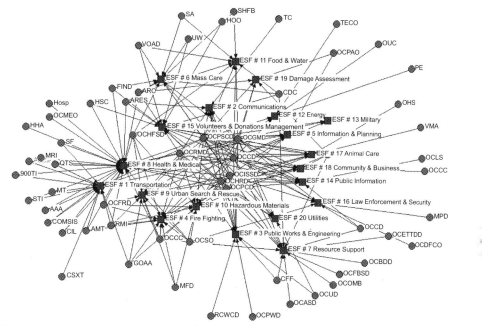

FIGURE 3.1

Orange County, Florida, emergency support function affiliation network.

carry out specific support functions in response to disasters. Fig. 3.1 visualizes how primary and secondary organizations are clustered around certain support functions. Circles in the image represent organizations and squares represent specific support functions.

EFFECTIVE INTERORGANIZATIONAL DISASTER NETWORKS

One of the clear distinctions between traditional bureaucratic government and networks is in the ways we assess their effectiveness. Network effectiveness can be defined as the "attainment of positive network-level outcomes that could not normally be achieved by individual organizational participants acting independently" (Provan & Kenis, 2008, p. 230). Networks have complex, vague, and fluid structures as opposed to the clear and relatively rigid structure of bureaucracies, which makes it difficult to assess their effectiveness. Measuring network effectiveness and performance using tools for traditional performance management tools (i.e., tools for single agency performance management) would not be possible. Alex Turrini, Daniela Cristofoli, Francesca Frosini, and Greta Nasi (2010) conducted a meta-analysis of the network effectiveness literature and explored how researchers define network effectiveness[3] and what are its determinants. After reviewing 93 relevant articles that were published after Keith Provan and Brinton

---

[3]Apparently each network has its own goals, environment, and dynamics; so there is no one-size-fits-all rubric for assessing effectiveness.

Milward's (1995) seminal article on interorganizational networks, Turrini and colleagues found that scholars assess network effectiveness at three levels. These are client-level effectiveness (whether clients are satisfied with the final results), community-level effectiveness (whether the community is benefiting from the operations of the network), and network-level (network capacity of achieving stated goals, network sustainability and viability, and network innovation and change) effectiveness. They also identify three umbrella concepts as determinants of network effectiveness. These are network structural characteristics, network functioning characteristics, and network contextual characteristics.

Network structural characteristics include external control of network (network dependence on its constituencies), integration mechanisms (having common communication system and creating a synergy for service delivery), size and composition of network, formalization and accountability (formalized rules and functions and being accountable to external stakeholders such as funders), and network inner stability (stability of the personnel working in the network). Functional characteristics of public networks refer to management craft within the network. Turrini and colleagues (2010) include traditional managerial work, generic networking, buffering instability and nurturing stability, and steering network processes in the functional characteristics. Network contextual characteristics refer to environment that the network operates in. Stability is the key factor for network effectiveness regarding the contextual factors. Stability in the environment helps networks to improve its effectiveness up to a certain point, but it may hinder it after that certain point (Turrini et al., 2010).

Disaster response networks are a special type of interorganizational network that possess some distinctive dynamics stemming from the conditions that they operate within. For example, disaster response networks are more emergent than planned. This means that although they are not completely independent from previously established relationships, disaster responders may not be able to follow predisaster institutional arrangements (Nowell & Steelman, 2014). Chaotic and uncertain natures of disasters make them particularly challenging for hierarchical organizational structures. Top–down mechanisms and standard operating procedures function well up to a certain point, but complexity of disaster situations requires respondents to be flexible and adaptable to environment. Also, even a particular network structure (e.g., centralized) that is effective in delivery of certain services (Raab, Mannak, & Cambré, 2015) may not necessarily generate positive results under every disaster incident. Thus, researchers need to keep in mind the dynamics of disasters when assessing the effectiveness of disaster management networks.

Although the literature on network effectiveness has been growing in the last few decades, research concentrating specifically on disaster management networks is relatively limited. The research pertaining to effectiveness of disaster networks mostly concentrates on factors shaping the structure of response networks. For example, Branda Nowell and colleagues make important contributions in assessing performance of disaster response networks (also see Steelman et al., 2016). Branda Nowell and Toddi Steelman (2014) examined the factors shaping communication networks in wildfire response. They concentrated on how institutional and relational embeddedness impact formation and structure of wildfire response communication networks. They looked at predisaster social ties, common stakeholder affiliation, and shared functional role as relational embeddedness. They found that relational embeddedness, which refers to the familiarity of actors in a network, predicts formation of wildfire communication network. Their findings also partially support that institutional embeddedness contributes to network density and more frequent communication among network actors.

Sang Ok Choi and Ralf Brower (2006) examined network effectiveness using Provan and Milward's (1995) theoretical model for network effectiveness, which suggests having a leading central organization in a network may lead to better network effectiveness. They looked at how people's perception of leading network actors corresponds to actual network structure. They studied emergency support functions in the Delta County, Florida, emergency response plans. These plans identify lead agencies for each support function. In their research, Choi and Brower (2006) compared the lead agencies listed in support functions, the most influential actor in the given support function as perceived by the actors in the network, and actual network position (i.e., actor with the highest centrality). They found four types of cognitive accuracy in their study network. Type I networks are the most coherent networks in which the leading actor is same across emergency response plan, network participants' perception, and actual network. Types II and III are partially coherent networks in which the leading network actor is the same only in the plans and perceived structures or perceived structures and actual networks. Type IV networks are the incoherent networks in which the leading agency is inconsistent across three levels. Their findings are useful for emergency managers to explore to what extent "participants have a clear operational picture of the actual networks in use" (Choi & Brower, 2006, p. 671), and their method is also a valuable tool for assessing network effectiveness.

In a recent study, A.J. Faas and colleagues (2016) compare similarity and dissimilarity theories identification of bridging actors in wildfire response networks. According to the similarity thesis, people will likely choose similar actors (homophily) as bridges between them and other organizations (e.g., between local fire service and federal land agency). Dissimilarity thesis, on the other hand, claims that bridging actors in the network will emerge based on operational priorities rather than familiarity. Their study partially confirms both these perspectives. The study findings show that although similarity and familiarity predicts how people expect to organize in dealing with emergencies, the actual network structure forms in a different way. Faas and colleagues (2016, p. 20) note "dissimilarity thesis might hold some weight over the similarity thesis in actual practice."

COMMUNICATION

A critical component of disaster and crisis management is communication within networks especially within uncertain environments. Researchers have examined barriers to collaborations and produced ideas for best practices to increase effectiveness and performance; however, gaps in the literature still exist and will continue to exist due to the ever-changing context of response and recovery (Jordan, 2010; Kapucu, 2006a; Kapucu et al., 2009; McGuire & Silvia, 2010; Moynihan, 2008; Varda, Forgette, Banks, & Contractor, 2009).

Focusing on interorganizational communication and its effect on decision-making, Kapucu (2006a) mentions that uncertain environments, like those of disaster response, hinge on the connection of a network and interoperability. Boundary spanners provide a unique aspect of information sharing within a network as they provide a bridge between sectors, which become a necessity during the response-and-recovery efforts. For example, the September 11, 2001 (9/11) airplane hijackings in the United States were an eye-opening situation showcasing the need for response organizations to plan communication strategies between specified internal and external organizations (Kapucu, 2006a). The findings from Kapucu's research suggest the importance of interoperable (shared language and systems) communications in disaster response. During the 9/11 responses, the New York police and fire departments were unable to communicate effectively since the fire department continued to use analog radios that were not compatible with the police radio system. Moreover, experience working with other agencies in disaster preparedness activities—such as training and exercises and planning—facilitates collaboration within disasters (Jordan, 2010).

Naim Kapucu and Qian Hu (2014) found results similar to those of Nowell and Steelman (2014) regarding the relationship between formal and informal relationships. They examined the relationship between friendship networks, disaster preparedness networks (as predictors), and disaster response networks (as dependent variables). Their findings show that both disaster preparedness networks and friendship ties are predictors of formation of disaster response networks. However, collaborations during disaster preparedness are stronger predictors of collaborations during disaster response. Furthermore, according to their findings, organizations with more central positions in preparedness networks have a higher degree of information communication technology (ICT) utilization whereas there was no relationship between such technology utilization and position in friendship networks. Analysis of response networks results in slightly different findings. Influential and central organizations in the network do not necessarily have higher levels of communication technology use when compared to preparedness network. Also, organizations that have an intermediating role do not have a higher information communication technology use compared to other organizations. Figs. 3.2A–C visualize friendship (i.e., informal) ties and disaster response collaboration networks of Orange County, Florida.

The findings of Hu and Kapucu's research (2014) tap some important points. The level of information communication technology use in leading and influential organizations in emergency management networks should match their organizational mission and goals. They also add that, despite the rapid growth of information communication technology, other means of communication such as email, phone calls, and face-to-face interactions remain functional.

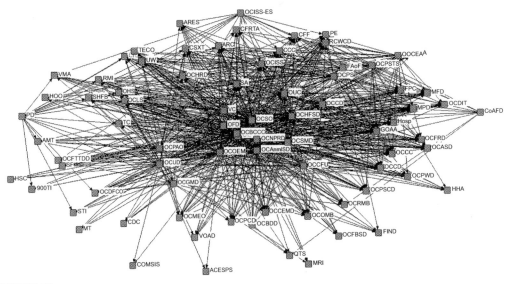

FIGURE 3.2A

Orange County, Florida friendship network.

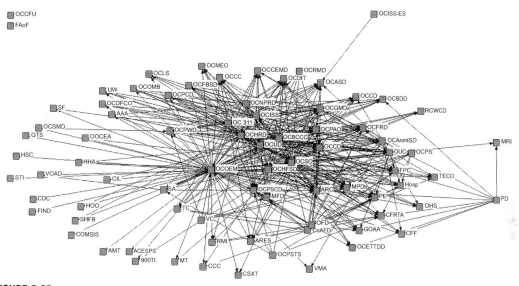

FIGURE 3.2B

Orange County, Florida disaster preparedness network.

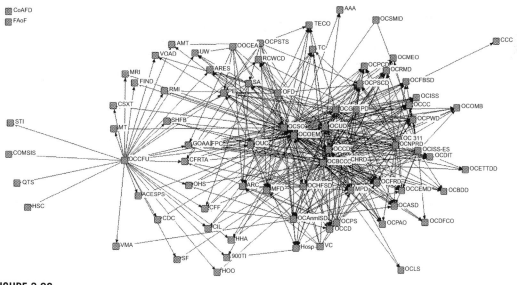

FIGURE 3.2C

Orange County Florida disaster response network.

TRUST AND SOCIAL CAPITAL

Disaster management networks consist of formal ties as well as informal relationships. Formal ties are often established through intergovernmental relations (vertically) by signing memorandums of understanding (MOUs) between government agencies at the same level of government or from different jurisdictions (horizontally), and between public and private sector actors. Formal networks are functional for sharing resources, information, and even human capital between partners of the network. Michael McGuire and Chris Silvia (2010) studied county-level emergency management agencies across the United States to study the factors that impact and influence intergovernmental collaboration. Their findings show that the degree of problem severity, past experiences with disasters, the capacity and capabilities of public agencies and managers (measured through training in multiple functional areas and state certifications and managerial and technical capabilities of managers), and the agency structure (well-defined program and mission) all contribute positively to intergovernmental collaboration. Factors such as trust, respect, and regular interaction are important in facilitating effective response in disasters (Kapucu et al., 2009; Waugh, 2003). The capacity to collaborate is also relevant since financial, technological, and human resources facilitate disaster response (Kapucu et al., 2009).

Informal ties, on the other hand, have greater variation in their scope than do formal ties. In some cases, the informal ties blend with formal ones. Informal ties' type, role, size, participants, and capacity are likely to change more than are those of formal ties depending on hazards, community capacity, and social capital available at micro, meso, and macro levels (Halpern, 2005). Individual social capital of community and organizational leaders complement formal ties by helping organizations build trust quickly and work together smoothly. Grouping public managers together with administration, researchers have analyzed relationships and connections for aspects, which encourage strong ties. Some identified are boundary spanners and trust (Kapucu, 2006a).

In a recent study, Michael Siciliano and Clayton Wukich (2015) examined the influence of homophily—organizations having more ties with organizations from the same jurisdiction and sector—and tie sharing (transitivity) on disaster response networks by using data from Hurricane Katrina. Their findings show that nonprofits are more likely to work with other nonprofits than with for-profit or government agencies—especially when they did not have access to government resources in a timely and effective manner. Similar tendencies occurred with public agencies. Local agencies were more likely to work with other local agencies. Given that the federal and state response to Katrina generally was not effective, dense collaboration between local organizations and organizations from the similar sectors is understandable. Siciliano and Wukich (2015) also found that organizations are likely to establish ties with others that they have as common friends. That means organizations remain in networks in which they have a feeling of trust under uncertain conditions like Hurricane Katrina.

An example of the role of social capital in collaboration is seen in the ad hoc connection between public and nonprofit organizations during response and recovery efforts (Nolte & Boenigk, 2013). Operating in a reactive and changing environment already places stress on an organization to provide for the community. Bringing in support services adds another layer in coordination efforts and can negatively impact the situation. For example, Isabella Nolte and Silke Boenigk (2013) conducted a study on the public and nonprofit relationship to research communication, task coordination, degree of mutuality, common relations, and preconceived positive collaborative experiences and their impact on performance during a disaster. Their results showcased the positive impact of task coordination, degree of mutuality, and preconceived positive collaborative experiences and did not highlight a statistically significant effect of communication and common relations, which is the opposite to the researchers' preliminary ideas.

LEARNING AND ADAPTATION IN DISASTER NETWORKS

Crisis learning is used as a means to judge the effectiveness of response by some scholars. Crisis learning in disasters is defined as the recognition and implementation of practices and behaviors by the disaster network to make crisis response more effective. Some barriers to effective crisis learning may be, for example, the high consequentiality of crises, the need for intergovernmental learning, lack of prior relevant experience, or limited information (Moynihan, 2008). Don Moynihan (2008) uses the Exotic Newcastle disease that hit California in 2003 as a case to illustrate possibility of network learning. Exotic Newcastle disease was highly contagious among poultry and other birds. The disease can kill an entire flock of unvaccinated birds, and even vaccinated birds are vulnerable (Moynihan, 2008). To manage the spread of this disease and minimize its effects, over 7000 workers from several state and federal agencies worked in a network setting for about a year.[4]

Network teams visited residential areas[5] and commercial areas and were able to find 932 sites that were infected with Exotic Newcastle disease. It is noteworthy that Exotic Newcastle disease differs from several other disasters in an important way. Many hazards such as tornadoes and earthquakes (even hurricanes) happen very quickly in a very short period of time and do not leave a considerable time for response. Even in situations in which responders have time to respond (e.g., hazardous material spills and mine accidents), they have to operate under chaotic and uncertain conditions. Exotic Newcastle disease spanned nearly a one-year period (October 2002–September 2003), which gave enough time for people and organizations that deal with the disease time to think about and make sense of the situation, and adapt to changing conditions. These conditions allowed network participants to develop relationships and trust among each other over time. Acknowledging interdependencies and building trust among network participants helped reduce environmental and institutional uncertainties and enhance learning in the network. Moynihan discusses six tools for learning in a crisis network, which are relevant both for short-term (e.g., tornadoes) and long-term (e.g., Exotic Newcastle disease) emergencies. These tools are virtual experience (using preplanning, simulations, role-playing, and training), learning from other organizations in the network, learning from information systems (using reliable and interoperable communication systems for situational awareness), learning forums (open discussion platforms for sharing knowledge and experiences), creating memory from standard operating procedures (developing routines for institutionalizing learning), and learning from the past.

This section discusses the determinants of network effectiveness particularly for disaster response. These determinants are similar to findings of Turrini and colleagues (2010) and Raab and colleagues (2015), as noted earlier. However, what we include here pertains to network-level

---

[4]Not all 7000 people worked together all the time. Moynihan (2008) notes that maximum number of people working together in this network was 2500 at its peak.

[5]These residential areas were mostly premises where a small number of poultry were kept in the backyard.

effectiveness only and does not include client and community levels. The following section discusses the factors that hinder network effectiveness.

FACTORS HINDERING COLLABORATION IN FORMING DISASTER NETWORKS OF ORGANIZATIONS

There are some factors that clearly hinder effective collaboration within disasters. Formal missions, cultural conflicts, role ambiguity, disagreements between organizations and emergency management strategies, and lack of plans that guide communications across sector boundaries all can hinder collaborative response (Jordan, 2010). One major challenge in bringing diverse organizations to the same table is to tackle the issue of developing a common group identity and a shared understanding of decision-making since diverse missions, organizational cultures, and operational functions can pose major challenges. For example, national security and law enforcement organizations are known for their secretive behaviors and centralized decision-making processes while, on the other hand, nonprofits more often operate as open systems that invite the help of volunteers and utilize decentralized decision-making models (Waugh, 2003). In some cases, however, even organizations with similar goals and operations fail to collaborate. After the 9/11 plane hijackings, for example, national intelligence agencies in the United States were criticized for not sharing information with each other before and during the attacks. Consequently, fusion centers at local, state, and national levels were established for effective intelligence gathering and distribution (DHS, 2015).

The cost of collaboration might discourage organizations that are expected to build partnerships. Smaller organizations might lack staff, expertise, and other resources for building and maintaining collaborative ties. Especially in rural areas, where resources are very scarce and emergency management is often handled by nonemergency management professionals (i.e., law enforcement, health care professionals), building partnerships are very likely to be at the bottom of policy agendas. These organizations simply cannot afford to dedicate resources for formal partnerships and are likely to rely on their informal ties with their counterparts in larger urban jurisdictions in case of a major emergency. That being said, the role of power balances between different jurisdictions has had a mixed impact on collaboration. Organizational capacity, jurisdiction size, resource availability, and population dynamics each impacts the development of collaborative ties.

Unequal power between jurisdictions often causes some jurisdictions to be reluctant to include more nontraditional agencies in their decision-making or to alter their power structures. On the other hand, sometimes "perceptions of unequal power between the core city and peripheral jurisdictions (i.e., adjacent suburban and rural communities)" facilitate collaboration (Jordan, 2010, p. 12). Organizational culture, and familiarity (or lack of it), can also impede collaboration and networking among organizations. For example, after the 9/11 attacks in the United States, the Department of Homeland Security was established and 22 different federal agencies were swept under this new cabinet-level institution. People from security backgrounds (i.e., people with guns) and those with nonsecurity backgrounds such as emergency management or finance (i.e., people without guns) were forced to operate together. Cultural barriers between organizations created tension and undermined collaboration and networking. Similar situations happen at the local level as well. Cultural differences, organizational routines, and even the jargon that organizations use make it challenging for organizations to break their silos and collaborate.

The conceptual map in Fig. 3.3 summarizes the factors that facilitate and hinder effective collaboration between actors during disaster management as identified through the readings and literature.

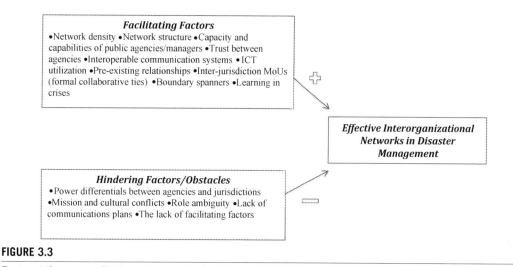

FIGURE 3.3

Factors influencing effective interorganizational networks in disaster management.

APPLICATIONS OF SOCIAL NETWORK ANALYSIS

Social network analysis (SNA) can focus on the relationships between actors (nodes) in addition to the attributes of actors. Five things make the network approach potentially useful as a complement to other approaches employed to capture variation in how communities engage disaster (Kapucu et al., 2014).[6] First, the foci of networks are the relations of the actors involved in a process, in addition to the characteristics of the entities involved. Second, networks are a tool to connect people, groups, agencies, and organizations across their traditional boundaries of geography, jurisdiction, organizational affiliation, sector, expertise, and ideology (i.e., bridging, or boundary spanning, in the parlance of network analysis). Third, networks are conceptualized as volatile systems with structures and members adjusting themselves to ever-changing circumstances. Fourth, effectiveness, rather than predefined processes, should prevail in networks. Networks, therefore, are ideally characterized by some minimum flexibility, improvisation, and innovation. Moreover, authority is addressed not by a hierarchical status but by the actors' acknowledged and recognized knowledge, skills, and experience. Last, reciprocity, mutuality, and egalitarianism can be key elements in networked structures. These facets allow for equitable and responsible input from actors. Depending on the interplay of factors that can create a network (e.g., actors, knowledge, resources, tasks, and organizations), several networks can emerge in any complex environment.

The impact of Hurricane Katrina in 2005 along with the lack of an effective response led to a wave of scholarly research on social networks in disaster events. Studies have focused on patterns of behaviors of evacuees in recovery decisions, the needs and support structures of residents, and emergency response networks in Katrina (Comfort & Haase, 2006; Hurlbert et al., 2005; Kapucu, 2006b; Robinson

---

[6]Kapucu, Hu, and Khosa, (2014), provide detailed application of network methodology in public administration research with several examples on disaster networks.

Table 3.1 Interorganizational Network Meta-Matrix

	Agents	*Knowledge*	*Resources*	*Tasks*	*Functions*
Agents relation	**Communication network** *Who knows whom*	**Knowledge network** *Who knows what*	**Capabilities network** *Who has what resource*	**Assignment network** *Who does what task*	**Affiliation network** *Who is responsible for what function*
Knowledge relation		**Information network** *What informs what is shared*	**Skills network** *What knowledge is needed to use what resource*	**Needs network** *What lknowledge is needed to do what task*	**Competency network** *What knowledge is needed for which functions*
Resources relation			**Substitution network** *What resources can be substituted or shared for which*	**Resource dependence network** *What resources are needed to do what task*	**Community capital network** *What resources are needed for which functions*
Tasks relation				**Precedence network** *Which task must be done before which*	**Sector responsibility network** *What tasks are assigned where*
Functions relation					**Interorganizational networks** *Who collaborates for what functions*

Adapted from Kapucu, N. (2006). Interagency communication networks during emergencies boundary spanners in multiagency coordination. The American Review of Public Administration, *36(2), 207–225.*

et al., 2006; Varda et al., 2009). Table 3.1 represents possible networks based on several tasks and factors. The matrix in Table 3.1 can be a source for content analysis for generating networks out of available archival data such as policy documents and after action reports. This method was used in collecting data for several interorganizational network studies (Comfort & Kapucu, 2006; Kapucu, 2006a; Kapucu et al, 2009). This matrix can be used in collecting several subnetworks focused on either organizational and interorganizational networks or function/task-based networks such as communication and knowledge-sharing networks. The network data collection tool can also be used in collecting data at different phase of disasters such as response and recovery.

CONCLUSIONS

This chapter provides a brief theoretical overview of the role of interorganizational networks in disaster research. Although disaster management has inherently been a collaborative effort, a hierarchical structure has been prevalent particularly during the Cold War because of the nature of that era. Harnessing interorganizational networks is now a vital element of effective disaster management, and even of other government businesses. Kapucu and his colleagues (Kapucu et al., 2010; Kapucu & Demiroz, 2011; Kapucu & Hu, 2014) have approached disaster management networks within a collaborative public management framework, and have focused on how collaborative partnerships are built for accomplishing all phases of emergency management (i.e., preparedness, mitigation, response, and recovery). Such a whole-community approach goes beyond formal interorganizational partnerships for disaster response. It includes individual and social factors that would contribute to reducing physical and social vulnerabilities of communities and building disaster-resilient communities. As often cited in the literature, these factors include social capital, economic diversity, and adaptive capacities (i.e., resourcefulness and rapidity, robustness, and redundancy of resources) (Bruneau et al., 2003; Norris et al., 2008) that a community might possess.

Within disaster and crisis responses, it is common to find a plethora of tips and tricks for the response-and-recovery efforts. However, every effort is quite contingent upon the context of the situation and the ability of the disaster management stakeholders to adapt and apply the best-fitted framework, model, or practice. The ways people manage disasters have similarities and differences. The ways disasters occur—particularly those resulting from natural hazards—are similar across the world. When they hit, they are not bound with jurisdictions and can easily overwhelm a single governmental unit. Because of this situation, even in the most hierarchical governmental systems, disaster response happens as emergent, collaborative, and horizontal (Khosa, 2013). On the other hand, significant differences exist in the way people manage disasters. Social and physical vulnerabilities in communities, culture of preparedness, level of social capital (cf. Aldrich, 2012), and resource munificence make significant impacts on the way that disasters are managed. Although disasters are managed collaboratively in the most hierarchical governments such as Pakistan (Khosa, 2013), collaborative networks are shaped by these dynamics. To summarize, each community's experience—be it in New Orleans, Pakistan, or China—with disasters will be similar in their roots but will differ dramatically when it comes to fruits.

Change is inevitable and so are disasters and crises; therefore, the highly complex and multilayered field of disaster-and-crisis response is a highly complex and multilayered field, which cannot be avoided. Learning becomes an essential component to increase a community's capability to respond and, hopefully, lower the negative impact on affected areas.

Although building networks and maintaining them are highly desirable goals, they are not often easy to accomplish. There are many factors that might encourage organizations to engage in partnerships or abstain from them. The chapter briefly touches many indicators of effectiveness such as network structure, organizational capacities, trust, information communication technology utilization and interoperability, preexisting relationships, formal agreements (e.g., memorandums of understanding), boundary-spanning actors, and learning in networks. Also, we seek to highlight factors that hinder network effectiveness. These factors are power differentials between agencies and jurisdictions, mission and cultural conflicts, role ambiguity, lack of communication plans, and lack of facilitating factors. The chapter highlights factors of both sides and provides an overview of what was discussed on network effectiveness.

STRATEGIES FOR RESEARCHING SOCIAL NETWORKS IN DISASTER RESPONSE, RECOVERY, AND MITIGATION*

4

Danielle M. Varda

University of Colorado Denver, Denver, CO, United States

CHAPTER OUTLINE

*Excerpts: With permission from Springer Science Business Media: Varda, D.M., Forgette, R., Banks, D., & Contractor, N. (2009). Social network methodology in the study of disasters: issues and insights prompted by post-Katrina research. *Population Research and Policy Review*, 28(1), 11–29, and any original (first) copyright notice displayed with material.

INTRODUCTION: WHY SOCIAL NETWORK ANALYSIS METHODOLOGIES?

When a disaster affects a community, dissipated (and ad hoc) social relationships often result, leading to new questions about how to identify and study social networks. Disasters have the ability to disorganize an entire social infrastructure, turning what is known about the way people relate, the way organizations behave, and the system of social and resource support into new questions that, if understood, could greatly impact people's abilities to deal with the consequences of disasters. These networks arise on different scales, from the individual to the community level, so researchers exploring network dynamics need to draw on a range of methodologies. To explore the impact of a disaster on social networks at all levels requires innovative research designs and methodological tools that account for the unique social structures and their accompanying dynamics. What follows is a representation of a methodological framework for designing disaster network studies. That framework is followed by attention to units of analysis, levels of analysis, types of network links, approaches to analysis, network dynamics, and the main challenges faced in research on disaster networks. Specific types of questions that can be answered using a network analysis approach are identified throughout this chapter.

A NETWORK PARADIGM FOR DISASTER CONTEXTS

Various research designs for applying social network analysis in disaster settings exist. In one approach, Danielle Varda, Rich Forgette, David Banks, and Noshir Contractor (2009) introduced the in/out/seeker/provider (IOSP) framework that attempts to identify and measure various aspects of networks for multiple levels of investigation and analysis. It is an approach that can also be applied for understanding impacts both *on* the networks and *to* the networks in disaster settings. Examples of network actors that might fall into each category—any person or group can be in more than one category—whether individuals, groups, or communities, are as follows (see also Fig. 4.1):

- In/seekers: Network members who are in the disaster area, seeking something from others in the network (e.g., flood survivors).
- Out/seekers: Network members who are out of the disaster area, seeking something from the network or providing something (e.g., outside relief organizations).
- In/providers: Network members who are in the disaster area, providing something to other network members (e.g., emergency workers).
- Out/providers: Network members who are out of the disaster area, providing something to other network members (e.g., hospitals).

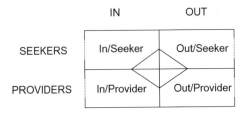

FIGURE 4.1

Social network actors in a postdisaster setting, based on the in/out/seekers/providers (IOSP) framework.

The framework is used to identify network boundaries, the unit and level of analysis, options to operationalize the ties between network members, the framing of research questions, instrument development, and data gathering. Each of these dimensions are first steps to next measure and analyze networks. While this framework serves as a tool to simplify identification of specific dimensions of a network—particularly in a disaster setting—more detailed nuances of networks, such as their dynamic nature, multiple roles and locations that actors take, or specified outcomes of the network such as the degree and nature of service needs or provisions, are not specified, as those are typically specific to a certain disaster setting/community. The following sections discuss the various dimensions of this framework in more detail including the unit and level of analysis, types of ties of interest to disaster researchers, and methods to operationalize these category distinctions using techniques of social network analysis, particularly in a dynamic nature.

UNITS OF ANALYSIS: INDIVIDUALS, GROUPS, AND COMMUNITY

Enrico Quarantelli and Russell Dynes (1977) noted the importance of attention to both individual and group characteristics in a disaster context. Valerie Haines, Jeanne Hurlbert, and John Beggs (1996, p. 261) expanded this, pointing to the importance of the community context within which social networks are embedded. Their study found that all variations of units of analysis (*individual*, or personal characteristics of providers; *group*, or characteristics of a personal network; and *community*, or the characteristics of the community in which they live) affected support provision in both the preparation and short-term recovery phases of a hurricane.

The inclusion of individuals as units of analysis involves identifying single actors, independent of the group to which they belong, and asking questions about how individuals activate their social networks or play a role within a social network. This unit of analysis can be used to answer such questions as: Who seeks social support in an emergency? Who provides the support? How do patterns of support in a crisis compare to those in everyday life for people involved in disasters (both seekers and providers) (Shavit, Fischer, & Koresh, 1994)? Much of the social network disaster research to date that looks at the individual seeks to identify their typical social networks, the mobilization and activation of their social ties, and the impact of the various social support configurations.

At the group level, the most common type of study includes organizations as the unit of analysis (e.g., church groups, the Red Cross, or federal relief agencies). The questions addressed at this level include: What is the probability that relationships forged in a crisis persist and become formalized? Do stable relationships serve to create norm gaps or facilitate communication and interactions (Comfort & Haase, 2006; Farmer, Tierney, & Kung-McIntyre, 2003; Robinson, Berrett, & Stone, 2006)? What suites of relationships between what types of organizations help to reduce vulnerabilities during emergencies (Kapucu, 2006a; Perrow, 1999; Robinson et al., 2006)?

The emergence of interorganizational networks in a disaster setting is sometimes particularly difficult to study because many organizations that respond to a disaster transcend multiple organizational types and do not necessarily have classic structural dimensions of formal organizations, e.g., the decentralized, volunteer-driven structure of the Red Cross, or the emergent transitory network of religious volunteers (Quarantelli & Dynes, 1977). The use of network analysis for organizational studies in a disaster context—several in this book—has alleviated some of these difficulties in studying organizational interaction.

Not only is it important to characterize individuals and groups in their community context but the community context itself can become a third unit of analysis. For example, Robert Stallings and Henry Quarantelli (1985) studied informal emergent citizen groups; that is, private citizens working together in pursuit of collective goals relevant to actual or potential disasters without a formal organizational identity. This community response was documented and identified in three types: damage assessment, operations, and coordinating groups. Each collectively plays a role in the overall emergent community of local citizens that react post disaster.

LEVELS OF ANALYSIS: PERSONAL, WHOLE, AND MULTILEVEL NETWORKS

In addition to categorizing social network actors, the framework distinguishes between egocentric and sociocentric levels of analysis. The egocentric approach is focused on an individual (people or organizations) and the kinds of ties they have to others. The sociocentric approach is focused on a bounded group of actors. Describing networks at each of these levels provides different ways to answer questions posed by disaster researchers. Additionally, joining multiple levels of networks in disaster settings helps to understand myriad, complex, and complicated suites of social support and constraint that people, groups, and communities face.

EGOCENTRIC OR PERSONAL NETWORKS

Egocentric network data, sometimes known as personal network data or egonets or ego networks, consists of information on the local social environment surrounding the ego and is often used to predict the consequences of a specific network structure (e.g., if an egocentric network is densely connected, then communication may be more efficient). The "individual" that is the focus of an egocentric or personal network need not be a single person but could also represent single organizations, agencies, and other nodes from a given unit of analysis with specific structural characteristics. Each of these types of nodes (individuals, organizations, and agencies) can be analyzed by their particular network connections with each other (people as well as organizations and agencies with which they link).

Egocentric data are gathered by asking the respondent (also known as the ego) to list a certain number of other people/organizations (also known as the alters) to whom they related in specified ways; this is referred to as the "name generator approach" to data collection. Then a number of questions are asked about the network members (*alters* in the language of network analysis); this is called the name interpreter or name generator. Personal network data elicitation methods sometimes prompt the naming of network alters bounded by context-specific prompts (e.g., "seven people who provide you support," "five people with whom you work," "all the people to whom you owe money," etc.). Then these lists and their characteristics are examined for patterns across the data set or even summarized per interviewee for other analyses at the case level. This approach, while useful in testing specific hypotheses about the relationship of specific subsets of personal networks to variables of interest (e.g., well-being, support access), may not tell us about relationships between members of one's personal network (e.g., Does your sister interact a little or a lot with your friend? Does your partner know your boss?). Now, collecting structural ties on a network of 50 alters requires prompts for 1225 alter–alter pairs, or $(50 \times 50 - 50)/2$, while a network of 100 would require 4950 (assuming that partner interacting with boss is same as boss interacting with partner).

Given the challenging task of collecting data (especially the alter–alter ties necessary to measure structure) on total personal networks, Christopher McCarty (2002b) developed a name generator approach that would yield a representative sample of an individual's total personal network structure. This method may well additionally provide likely representation of network composition when averaging all respondents. By simply asking respondents to free list 60 network members (e.g., "tell me the names of 60 people you know and know you by sight or by name with whom you've interacted in the past two years or could have if you wanted to") and provide tie strength and relationship type on all 1770 unique pairs, McCarty and colleagues found they could replicate observed variation by randomly selecting a subset of 30 alters from the free-listed 60, which resulted in a decrease in respondent burden from evaluating 1770 ties to only 435 (McCarty, Killworth, & Rennell, 2007). It appears that a personal network list of 40–60 subsampled randomly to 20–30 names, and then eliciting ego's estimation of ties between them, provides accurate structural representation of a personal network.

Egocentric measures may also be constructed from complete network data. That is, if a researcher has gathered data about a whole set of network members and their network ties, analyses can be applied that measure characteristics by only evaluating an individual's direct connections and the connections between those neighbors (path distance = 1) of the individual.

Egocentric analysis allows one to sample individuals from the population in order to draw conclusions about patterns of social support, constraint, and behavior. Examples of questions that disaster research asks of egocentric data include:

- Q1: What kinds of routine networks allocate resources in nonroutine situations (Hurlbert, Haines, & Beggs, 2000)?
- Q2: What is the size of informal helper networks mobilized during a serious personal crisis (Chatters, Taylor, & Neighbors, 1989)?
- Q3: How do people use their social networks during a mortal threat (Casagrande, McIlvaine-Newsad, & Jones, 2015; Shavit et al., 1994)?
- Q4: How might you mobilize an effective coordination system using identified brokers in a network of embedded ties (Saban, 2014)?

SOCIOCENTRIC OR WHOLE NETWORKS

Whole network analysis focuses on the characterization of ties between a set of nodes (actors) in a network. At the individual level, the focus is on the ego (i.e., a single node of interest for a given outcome or variable) as it is positioned within the larger social network, such as the number of incoming and outgoing ties an individual has, whether they serve as bridges or gatekeepers, if they are part of a subgroup, or how easy it is for that individual to access the rest of the network. Research questions posed by disaster researchers at this level include:

- Q4: Can central actors in the network serve as a means for increased information dissemination (Kapucu, 2004)?
- Q5: Can redundancy in network ties be reduced to produce a more efficient means of communicating timely information (Comfort & Haase, 2006)?

The second level of sociocentric analysis is dyadic, which examines the relationship between any two network members in the larger network. For example, a relationship between two disaster survivors who help each other after an event. The third level of sociometric analysis is triadic; this studies any three nodes and the relationship between them. At this level, one examines transitivity, or the level of information flow within a network based on the connections between any three members, for example, the relationships between three disaster survivors or, alternatively, three disaster response organizations. Beyond this are subgroup analysis and the global or overall analysis. Examples of questions that disaster researchers might ask from a sociocentric perspective include:

- Q6: Are emergent networks lasting and how do they evolve over time? Do these networks formalize and continue to function postrecovery?
- Q7: How important is reciprocity in a coordination and communication emergency network? Do information and resources simply need to be disseminated, or are responses or reciprocation necessary?
- Q8: Does the level of centralization within a decision-making network relate to the ability of network members to make timely decisions?
- Q9: Do network members that are structurally equivalent have similar communication patterns pre and post disaster?

MULTILEVEL NETWORKS

At the highest level of analysis, a combination of the various units and levels of analyses are used to examine important theoretical questions through hypothesis testing. The multitheoretical, multilevel (MTML) model developed by Peter Monge and Noshir Contractor (2003) allows the researcher to apply analytical frameworks (such as theories of contagion, social support, proximity, homophily, resource and exchange, collective action, and evolutionary) to analyses at the egocentric and sociocentric levels. It can employ both descriptive and statistical analyses, for example, p^* models and quadratic assignment procedure (QAP), which are conducted with corrections for dependent observations, to explore multiple substantive theories across the various analytical levels previously described. An example question that might examine networks at these various levels is:

- Q10: How do characteristics of providers, their personal networks, and the community contexts in which they live facilitate or impede their ability to provide support (Haines et al., 1996)?

Although Haines and colleagues (1996) did not use predictive network statistics, for a question like Q10, parameter estimates could be used to indicate whether graph realizations that contain hypothesized properties have significantly higher probabilities of being observed. By observing characteristics such as the number of links observed, reciprocated links, transitive triads, and the overall network centralization, one can test the extent to which the observed graph exhibits structural features such as reciprocity and density. For example, we could assess the degree to which a disaster network has members who typically help one another (reciprocity) or measure the degree to which members of a network have formed possible ties (density).

The MTML framework posits that multiple theories operating at multiple levels explain the emergence of networks. Since networks emerge from multiple theoretical motivations (for example, one theory might posit that disaster victims from a similar geographic area will likely seek each other out post disaster (a proximity theory), another might suggest that disaster victims will seek out others who have similar needs (a homophily theory), multiple measures of how and why ties develop can be measured, and these types of competing theories can be tested to see which ones make the most sense in a particular disaster setting. While these types of relationships can be assessed descriptively, statistical techniques (e.g., p^* or exponential random graph models, discussed in Section "Analyzing

Disaster Networks") enable us to detect the extent to which distinct multiple structural configuration—and by extension, distinct theoretical mechanisms—are driving the emergence of a particular network (Contractor, Wasserman, & Faust, 2006).

Additionally, there is considerable variability across theories that appear to be relevant in explaining the emergence of networks. These could include, but are not restricted to, *exploring* new ideas, *exploiting* existing resources, *mobilizing* toward collective action, *bridging* across boundaries, *bonding* for social support, or *swarming* wherein a latent network is rapidly activated. When in disaster settings do each of these social drivers occur? For example, in an emergency response, networks might be most geared toward mobilizing and swarming. Clearly there is a need for a contextual meta-theory of the social drivers for creating and sustaining communities in disaster settings that explains when and why some subset of the theories outlined in the MTML model is more important than others.

CONCEPTUALIZING TIES AND COLLECTING DATA

As in all social network analyses, the identification of the various types of relationships is the root of understanding the connections between nodes. While relationships between individuals do in fact exist, a network with its specific boundaries and membership and ties is operationalized by the researcher. At each level of analysis, the researcher decides whether to categorize connections as communication, advice, kinship, friendship, professional (work-related), membership-based (e.g., clubs, the gym, school), religious, proximal (e.g., neighbor), or various types of resource sharing, to name a few. Naim Kapucu (2005) cites five general categories of the kinds of relationships that network scholars often capture (adapted from Carley, 2002).

- People/agent relations that can answer questions regarding who knows whom, who knows what, who has what resource, who does what, and who works where.
- Knowledge relations that answer what informs what, what knowledge is needed to use what resource, what is needed to do that task, and what knowledge is where.
- Resource relations that answer what resources can be substituted for which, what resources are needed to do that task, and what resources are where.
- Task relations that answer which task must be done before which and what tasks are done where.
- Organizational relations that answer which organization works with which.

Scholars identify network links by observation, collecting survey data, or determining ties based on formal chains of command. Scholars increasingly infer types of ties based on electronic traces, such as links between Websites and by mining the text on these Websites, as well as the increasing use of social media sites such as Facebook, Instagram, and Twitter, to observe and collect network data pre, during, and post disasters.

Examining news media reports or blogs provides the basis for observational studies of network links that are also unobtrusive and quick, although not always comprehensive—this would be suitable for studying how countries interact in responding to a disaster, such as multinational relief efforts following the 2004 Indian Ocean earthquake and tsunami. In these cases, the researcher must rely upon domain knowledge and expert judgment to assess the nature and importance of the ties. Other observational data about network links include watching interactions between categories of people or noting co-occurrence of types of people at disaster response settings.

Survey data related to disasters is collected in various ways, ranging from random sampling to convenience sampling to opportunistic sampling. For disaster response and recovery data, snowball sampling is relatively common; here one asks a respondent to name their contacts, and then goes to those contacts and repeats the question. For example, a survey can collect information on where a disaster victim got support from in the days following the event. However, the statistical properties of all but the random sampling procedure are complex, and this can limit the generalizability of whatever inferences the researcher makes. As a respondent lists people and organizations that they look to for support, those alters can be added to the sample and sought out to also answer the survey, getting a more complete picture of the network.

Chain of command data is often derived from organizational charts and reporting structures. These studies often also involve text mining, since a researcher wants to use the text in the email or other communique to automatically classify the kind of tie that the communication represents (e.g., Diesner & Carley, 2005). For example, during any major operation such as disaster response, organizations report status through standardized written communications. The situation report is one predominant form of these communications and is often posted on the Web during disasters to provide information about an agency's response. Researchers use these data to unobtrusively infer networks in near real time using text mining tools such as Data-to-Knowledge and Text-to-Knowledge (D2K and T2K).

Mixed-methods can be a useful and comprehensive approach to collecting social network ties and understanding the contexts in which the ties are formed. These data collection techniques might include surveys, focus groups, meeting with key informants (for example, to review hierarchies or chain of command), data mining, and coding of social media (Islam & Walkerden, 2014). This is especially appropriate if the context in which a set of network relationships (disaster victims seeking services) is not well understood across different context (for example, in culturally different settings). For example, to better understand the patterns of disaster victims seeking services in a rural versus urban community, focus groups or interviews can provide additional context to explain variations in the reporting of network connections.

ANALYZING DISASTER NETWORKS

Just as in most fields using quantitative analysis, social network analysis is amenable to both descriptive and predictive statistics. The descriptive statistics are based on both composition of the network and the structure of the network, and can focus on individual nodes in the network or on the entire network. Predictive statistics in social network analysis has required the development of new techniques and models to avoid the problem of lack of independence of observations that inferential statistics typically requires—in other words, networks are relational and thus nodes and their actions are dependent on one another.

Disaster network analysis differs from nondisaster settings in terms of the likely relevance of various analytical techniques for the outcomes of interest. Support, resource sharing, information transfer, constraint, and various network structures related to these characteristics likely prove more relevant than number of steps across a network or whether the people are similar (i.e., homophily).

DESCRIPTIVE ANALYSIS OF NETWORKS

Descriptive analyses of networks explore the structures, characteristics, and tendencies across the networks by presenting descriptive data in specific and summary of various attributes associated with the network data. These descriptive statistics can include frequencies and cross-tabulations of organizational and network characteristics. While networks can be described in any number of descriptive ways, a couple of examples include:

Representation/Diversity

This type of characteristic is often described as homophily, or the extent to which actors in a network are similar, and even whether the ties between actors are created based on similar features of each actor (McPherson, Smith-Lovin, & Brashears, 2001). For example, one can describe the number of ties between similar members, or what dissimilarities exist between members in a network (i.e., heterogeneity). This has been a topic of disaster research, often to answer questions about the types of social support that people tend to look for before, during, and after disasters. In particular, in culturally diverse communities, it has been used to understand whether disaster victims are more likely to look toward others like them or are equally likely to look/trust outside their own communities for help (Messias, Barrington, & Lacy, 2012).

Reciprocity

Reciprocity will calculate the proportion of dyads in each network that are reciprocated between any two individuals (i.e., reciprocal nominations). This allows us to understand, for example, the degree to which members of a disaster network might seek help from each other. In a study on communications in an emergency department, researchers found that although there were particular individuals being asked for help by other members, assistance was also given and received amongst many of the medical staff (Creswick, Westbrook, & Braithwaite, 2009).

Centralization

Centralization indicates the degree to which a few members hold the most number of connections in the network. A high centralization score represents a highly centralized network (only a few members hold the most central positions). A low centralization score represents a less centralized network (more members hold the most central position). Other forms of centrality (individual level measure) and centralization (network level measure) are also relevant for disaster studies, such as:

- Betweenness, which serves to capture gatekeeping, bridging, and bottleneck functions in disaster settings;
- Eigenvector, which denotes the nodes in disaster settings that are connected to the nodes with most ties.

For example, the role of organizations in disaster networks can be understood through the use of centrality measures, to understand which ones play brokerage and dissemination roles in the network (Moore, Eng, & Daniel, 2003).

A caution about missing data: centrality measures are susceptible to degradation with levels of random missing data beyond 20%, although each measure is differentially susceptible to missing data (Valente et al., 2008). The solution is to modify the design to minimize missing data and/or to carefully choose metrics that are capturing aspects of disaster relationships but minimally subject to degradation to missing data.

Multiplexity

Multiplexity represents the number of different types of relationships among members of a network, describing the extent to which a dyadic pair engages in several types of relationships and the extent to which types of relationships co-occur. Kapucu and Qian Hu (2014) found that the degree of friendships in a network is predictive of the level of involvement of organizations in a disaster network.

STATISTICAL COMPARISON OF NETWORKS

In social network analysis, the statistical or predictive approach compares two or more networks and the degree to which they are different, plus which variables are responsible for those differences. Often this involves comparison of a social network to a randomly generated network or other types of simulations (Guo & Kapucu, 2015). But more useful for disaster research, and most social science research, is: (1) the comparison against a theoretically determined expected value; (2) the comparison of multiplex networks, such as comparing a network of friendship ties to the same set of nodes or people connected by help-seeking ties to see if they are correlated and whether structural features of the network account for the composition differences; and (3) the comparison of the same network at multiple points in time, as discussed in the following section on dynamics.

Regular inferential statistics give a much smaller standard error estimate than that which is provided by an alternate, bootstrap technique because classical statistics assume independence of observations. Network ties are not independent and therefore require a specific set of techniques, which include exponential random graph models, QAP, and multilevel modeling (MLM).

Exponential Random Graph Models

Exponentially parameterized random graph models—often called exponential random graph or p-star (p*) models—provide estimates of organizational and network effects across networks (Frank & Strauss, 1986; Pattison & Wasserman, 1999; Robins et al., 1999; Wasserman & Pattison, 1996). The exponential random graph framework provides a general way of representing probabilistic models that specify the potential sources of dependence and heterogeneity that give rise to network structure. The *statnet* package within the computing environment of *R* calculates results from exponential random graph models (Handcock et al., 2008; R Development Core Team, 2011).

Quadratic Assignment Procedure and Autocorrelations

In social network terms, the procedure examines whether one dyadic variable or network (i.e., an actor-by-actor matrix) is correlated with either another dyadic variable for the same set of nodes or to a monadic or node attribute variable (a vector representing an interval-scaled attribute of each actor). For example, if the dyadic variable is who helps whom, and the monadic variable is the degree to which each node was impacted by the disaster, the procedure tests whether helping relationships are patterned by how hard they were hit (e.g., people try to help those who are most in need). A simple way to perform autocorrelations is to use the Geary autocorrelation in UCINET (Borgatti et al., 1999), which indicates smaller p values as statistically significant autocorrelations (Varda, 2011). The software Barriers 2.2 also calculates autocorrelation of matrices using the Monmonier algorithm (Manni, Guérard, & Heyer, 2004).

Multilevel Modeling

To examine the relationships between the proposed structural signatures and effectiveness across networks, a MLM approach can be used in order to account for the nested features of these data (dyads nested within organizations) (Gooty & Yammarino, 2011), specifically to examine how variables measured at one level affect relationships at another. MLM accounts for the dependent nature of nested data used to estimate, for example, variation on one-on-one organizational relationship (dyad level) outcomes with organizational (organizational level) predictors using a restricted maximum likelihood approach (Raudenbush & Bryk, 2002). Using analysis techniques, pertinent statistics on the network structures and characteristics at the individual organization, organizational dyads, and whole network levels can be reviewed and compared across collaboratives.

THE DYNAMIC NATURE OF DISASTER NETWORKS

The dynamic nature of interorganizational and individual networks over the entire duration of the disaster period will complicate network measurement and analysis, but this is indeed the nature of social networks in disaster contexts, and, often, it is crucial to conduct research in a way that captures these transitional elements. There is no strong theory for the general study of dynamic models of this kind. Currently, the best available tools are to take traditional summaries of the properties of static networks, such as average in degree or density, and plot these against time. Examining the shifting positions of particular—often key—actors or nodes also provides insight into long-term disaster networks.

One conceptualization of a dynamic network framework includes predisaster networks, ad hoc social networks, emergent networks, and stationary ties. Networks are first characterized as *predisaster networks*, that is, those that existed prior to the occurrence of the disaster. Changes in such networks are typically slow and not especially purposeful. David Banks and Kathleen Carley (1996) describe several models for such change, and the process is largely one of "in-filling" the social networks (e.g., friends of friends become friends).

Second, during a disaster, many *ad hoc social networks* form among the seekers and providers both in and out of the disaster area. These types of networks often come together quickly and are likely to dissipate as network members move through the recovery phase. For example, temporary shelters typically connect previously unconnected people. These seekers and providers often form strong relationships based on mutual understanding and trust. These types of ad hoc networks may be influential in the decisions that network members are making regarding their movement in and out of this and other networks.

Third, *emergent networks* are those characterized by new ties (social relations) and new functions (goals or tasks) (Stallings & Quarantelli, 1985). For people who have lost their previous network and are unable to reconnect it, the emergent network is the pathway back to normalcy. For organizations, the emergent network represents new opportunities for collaborations and partnerships.

Finally, *stationary ties* are those relatively unaffected by the disaster. These ties persist and are not degraded (although possibly enhanced) by movement, social infrastructure devastation, or inconsistent communication. These types of ties are often kinship ties (e.g., a father/daughter tie remains as such regardless of the condition of the social context). These stationary ties play an important role in disaster research, with the majority of findings from individual social support studies (not necessarily conceived as social network analysis research) indicating that disaster victims activate their kinship ties when they are in need or in crisis (Chatters, Taylor, & Neighbors, 1989; Haines et al., 1996; Hurlbert et al., 2000; Marsden, 1987; McPherson, Smith-Lovin, & Brashears, 2006; Quarantelli & Dynes, 1977; Shavit et al., 1994).

These sociotemporal characterizations of networks exist for various settings. An illustrative example of how social networks might change over the course of a disaster occurring can be conceptualized as follows.

With individuals in hurricane or flood settings, for example, one expects that during the preparation phase a typical actor provides help to those in their social network (say, in boarding up a house or caravanning out of the city). During the life-threat phase and the refugee phase, a social network may be much sparser, pared down to all but kinship ties (e.g., Casagrande et al., 2015). During the recovery, a fragile emergent network is built, but the diversity within that network is not great and the formation of ties is driven by immediate needs. Finally, some people may successfully reconstruct most of their original network or build a new and robust and potentially more structurally and/or compositionally diverse network in a fresh location. Each of these phases would be visible in a time series plot of density or network activity (Fig. 4.2).

For organizations in these hurricane or flood settings, there is a greater range of behavior than for individuals. A relief organization might draw upon preexisting ties to accumulate resources in readiness, and then have an explosion of new ties to individuals who are helped. At the same time, their ties to other organizations might be competitive or cooperative, which affects the number and kinds of contacts that are made. Additionally, one might find emergent clique or subgroup structure for faith-based groups, federal groups, and local groups. Longitudinal plots of structural properties of such networks could reveal much of this kind of story.

Past research on dynamic networks in nondisaster settings has focused on in-filling in static networks (e.g., Bearman, Moody, & Stovel, 2004), in terms of either descriptive statistics or mathematical models. This process is not as dominant in disaster networks, although it plays a role in the formation of temporary networks among refugees in shelters and coordination among relief agencies. In contrast, disaster research on network dynamics focuses on the kinds of changes that are forced upon the preexisting network by circumstances, and the kinds of adaptive responses that the thinned-out networks make as they attempt to secure and distribute resources or achieve other objectives.

Further progress in the dynamics of disaster networks might proceed in several different directions. One approach is to build visualization tools that create "movies" that show how (small) networks lose

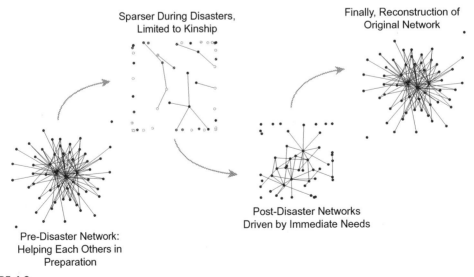

FIGURE 4.2

Evolution of a network pre, during, and post disasters.

and gain ties in response to particular kinds of disruption (e.g., Moody, McFarland, & Bender-DeMoll, 2005). A second approach is to create mathematical models for how social ties evolve through a social network during an event (Phan & Airoldi, 2014). A third approach is to identify statistical tools for testing whether a particular model of network change is corroborated by one's data (e.g., Sanil, Banks, & Carley, 1995). Although none of these is especially easy to realize, the insights and/or the strength of the conclusions can be worth it.

CHALLENGES

Although social network analysis can be used to collect and analyze the data necessary to examine patterns of relationships and interactions in disaster situations, there are significant challenges required to overcome when considering this approach:

BOUNDING THE NETWORK

Within a community, the potential members of a network are diverse and undefined. It is not typically clear how to "bound" the network, that is, to determine whom to include in a survey for collecting data or even whom to consider an essential or nonessential member. Some communities will attempt to start with lists of members of some other type of membership, but it is almost always going to be the case that the researchers have to decide who will be included and who will not—some communities or organizations are just too large or too diverse to help us hone in on answers to our disaster research questions. Do we include active members? Do we include isolates (i.e., disconnected individuals)? Do we want to include several departments inside an organization, or can one respondent adequately respond to a survey on behalf of the entire organization?

The reason this is so challenging is that including additional entities from the community is possible but comes with increased project complexity. Those entities will need to be brought into the process and be prepared to make a commitment to data collection and buy-in for the overall project. Also, each additional entity adds to the complexity of the survey administration. For personal networks, how multiplex do the relationships need to be, and what is a reasonable size for the personal network to capture variation in support and constraint aspects of disaster experience? The respondent burden is increased when the list of respondents grows. Random subsampling of a personal network can decrease that burden.

In some whole-network (non-egocentric) studies, using a "name generator" approach that allows respondents to list their partners can lack reliability (no one respondent has the same list of partners to report relationships about), can tend to result in fewer partners mentioned (often underestimating the amount of actual network ties that exist in the network), and can increase data management costs (by having to join together lists of some overlapping names that have different spellings). In place of name generators, researchers provide respondents with a prepared list of organization or network members and might allow them to add or subtract names.

DEFINING THE RESEARCH QUESTIONS

In addition to thinking about how to bound the list of network members, the next hardest step is to create or customize the survey for data collection. In some cases, the user may have the advantage to simply modify an existing validated, reliable online survey to support the networking of organizations—such as for disaster resilience (e.g., Eisenman et al., 2014). However, even this takes substantial thought to be sure that the survey questions will answer the research questions. Alternatively, developing a survey from scratch will take substantial effort but allows the user total customization for what they want to collect. The challenge in this approach primarily involves refraining from asking too many questions.

Regardless of the survey used, it is encouraged to carefully define the research questions and specifically the types of relationships to survey participants in order to avoid undue burden that can annoy participants and jeopardize data quality. While several forms of connection between entities are possible (described in detail throughout this book), and can be considered as grounds for a relationship, it may not be necessary to just find out who knows whom but rather to focus on more meaningful relationships. This would encourage more variance in the relationships—making analysis and interpretation easier—but potentially increasing the number of questions asked. Potential research questions could include:

- Who is operative in what subject area domains (i.e., emergency management, homelessness, early childhood development, senior services, hunger, etc.)?
- What is the impact on the network of shifting funding to certain emergency management, recovery, or adaptation/mitigation?
- What is the potential influence on the network as key individuals or organizations leave?
- Where do gaps exist in the network that lead to service gaps to the community for emergency management, recovery, or adaptation?
- Where are there strong relationships that can be leveraged to reduce vulnerability or improve community resilience?

One way to decrease burden, especially in personal network studies, is to ask one question that provides for variation of ties rather than asking for multiple ties. For example, ask about the individuals in a list of people with whom someone has exchanged information about the hurricane or response to it: On a scale of 1–5, how strong is the relationship between A and B? Or, less about strength, and more about quality, are A and B connected via friendship, work, religious activities, community service, fraternal organizations, kinship, other, or not at all? The researcher could allow, regarding the latter of these two questions, for people to answer just one or perhaps multiple ties.

One question to consider is regarding how the network's structure influences the experience of the people served. There is a significant methodological difference between studying a network to understand relationships between entities and understanding a customer experience working through the network. This chapter has conceptualized most disaster research as interpersonal networks or as interorganizational networks, and has not touched on the personal or service relationships between organizations and citizens.

SURVEY ADMINISTRATION

Taking a network survey can be demanding. Respondents are asked a series of questions about themselves. They are then asked to select which other entities on the bounded list they have relationships with according to the given definition. They are then asked a set of questions about the nature of their relationship with each entity that they have selected. This can take substantial time, and as mentioned, as the bounded list grows, the burden can become substantive. Additionally, there is variation within the data collection process that cannot be controlled and that impact the value of the data. For instance, whether an organization has a single person or a team of people responding to the survey influences how comprehensive the entity's perspective will be on its external partnerships. When organizational leaders delegate survey completion to a single staff person who has less understanding of those partnerships, the survey response value is limited.

Response rates for network surveys reflect what other types of surveys can expect, with the same challenges for getting people to answer surveys—even if they are generally willing to do so. Despite the failure of some participants to respond with maximum input or at all, the analysis is still able to quantify relationships about which the researcher currently has no data. One way to think about relationships is to trust that any one person answering the survey is reporting actual relationships (making it sufficient if the partner organization does not respond). However, this would not allow measurement of reciprocity or to assess perceptions of partners throughout the network.

While there is no perfect way to administer a network survey, it helps to consider limiting the overall number of questions asked, as well as limiting the number of pages (or screens) a respondent will experience. Danielle Varda (2008) has tested various approaches and developed a validated network survey (the Program to Analyze, Record, and Track Networks to Enhance Relationships, or PARTNER Tool) that minimizes time spent answering questions. Appropriate skip logic in surveys and using targeted subsamples with sufficient statistical power are also very useful for minimizing burden. Finally, increasing response rates can be done through face-to-face interviews, repeated attempts to interview, providing incentives (including drawings for prizes for respondents), or engaging the community in the work from the start of the project to encourage interest and buy-in.

CONCLUSIONS

Disasters present social scientists the opportunity to study human behavior at times in which social adaptation and instinct are often starkly revealed, whether in complex or complicated settings. Just as importantly, social science research has potential value in mitigating disaster loss, improving disaster responses, as well as evaluating for-profit, nonprofit, and government performance.

In this chapter, a methodological framework is presented for describing social networks in disaster settings, accompanied by a review of the appropriate units of analysis, levels of analysis, types of ties, analysis, and network dynamics. The various examples of ongoing social network analysis research generally show that informal personal and group relationships play an important role in disaster relief efforts independent of government aid and survivors' personal conditions (income, education, level of damage). However, major methodological challenges remain, including network bounding and defining the appropriate relationships to form the networks, and dealing with complicated and burdensome survey administration and data management.

The goal of social network analysis in disaster research is a more complete understanding of who is at risk, who recovers, how survivors recover from disasters, and how to adapt in order to mitigate vulnerability and impacts. Vulnerable or at-risk populations are typically defined in terms of personal or physical attributes. Personal attributes typically include an individual's socioeconomic status, employment, disabilities, and age. Physical vulnerabilities may include the housing status and quality, or availability of personal transportation. Network studies capture new variables and measures for identifying vulnerable or at-risk populations. Social or network vulnerability assesses the extent to which socially isolated disaster survivors are less likely to adapt and recover after a crisis.

Interorganizational network studies may provide insights on the rate, nature, and efficiencies of disaster response, recovery, and adaptation efforts.

Future methodological efforts in social network analysis will need to investigate spatial correlates or network ties and behaviors. Understanding the spatial or geographic correlates of socially isolated disaster survivors may allow governmental and nongovernmental emergency management teams to better target relief efforts. A second area of methodological innovation is to seek to capture public-private and local-external synergies for disaster relief efforts. Further study of personal, organizational, and community networks (both public and private) in a postdisaster setting may allow planners to improve the speed, coordination, and breadth of coverage of disaster response, recovery, and adaptation.

NETWORKS IN DISASTER RESPONSE

PERSPECTIVE MATTERS: THE CHALLENGES OF PERFORMANCE MEASUREMENT IN WILDFIRE RESPONSE NETWORKS

Branda L. Nowell[1], Toddi A. Steelman[1,2], Anne-Lise K. Velez[1], Sherrie K. Godette[1]

[1]*North Carolina State University, Raleigh, NC, United States;* [2]*University of Saskatchewan, Saskatoon, SK, Canada*

CHAPTER OUTLINE

INTRODUCTION

Disaster incident response has long been recognized as complex and beyond the scope of resources of any single organization or agency to address single-handedly. This has spurred continued interest in how to accomplish a coordinated, coherent response among a myriad of organizations and agencies that become active during a response phase of a disaster (e.g., Comfort, 2007; Faas and colleagues, 2016; Kapucu, 2006a, 2006b; Nowell & Steelman, 2013, 2015; Steelman, Nowell, Bayoumi, & McCaffrey, 2014). We have found this collection of actors to be productively viewed as an *incident response network* (Nowell & Steelman, 2013; Steelman et al., 2014). A social network perspective provides additional insight for understanding the structure and functioning of this collection of actors.

Social Network Analysis of Disaster Response, Recovery, and Adaptation. http://dx.doi.org/10.1016/B978-0-12-805196-2.00005-4

When incident response is understood as a network response, it necessitates that we extend our focus beyond the performance of any single organization or agency during the incident and toward an understanding of the incident response stakeholders, their relationships, and the setting as a whole. Performance within networks can be conceptualized at four levels of analysis: performance of individuals, performance of organizations/agencies, performance of subgroups, and performance of the whole network. Research on response networks has shown that consideration of these levels is critical because performance at one level does not necessarily equate to performance at higher levels (Mandell & Keast, 2008). In fact, complex interdependencies within networks have been shown to result in high performance at one level while reducing performance at other levels (Mandell & Keast, 2008).

While performance measurement and evaluation is a well-established field, the science of evaluating performance within the constraints of social networks or organizations remains in its infancy (Kenis & Provan, 2009; Mandell & Keast, 2008; O'Toole, 2015). Further, many of the traditionally upheld pillars for what constitutes quality performance measures meet significant challenges when applied to evaluating response networks. Empirical research on the performance of networks in disaster contexts is especially sparse (Magsino, 2009). Understanding what we mean by performance in disaster response networks is essential to realizing the full value of what networks can accomplish.

In this chapter, we focus on performance of whole networks. Specifically, we offer insights and findings from efforts to operationalize performance of incident response networks associated with complex wildfire events occurring in the wildland–urban interface.[1] We first review the existing literature on network performance to identify key considerations and principles associated with evaluating performance in networks. Second, we describe how these principles were applied in a large multisite field study of wildfire incident response networks. Lastly, we empirically examine the relevance of pluralism in conceptualizing and measuring network performance. We conclude with a discussion of the implications of these findings for advancing the study of performance in networks.

ASSESSING PERFORMANCE IN NETWORKS: PRINCIPLES FROM THE LITERATURE

For our purposes, we view network performance as a multidimensional construct that concerns a network's achievement of process and outcome goals relative to the constraints of the context in which the network operates. We take this definition from the diverse public administration literature outside of disaster studies on how network performance can be measured (Kenis & Provan, 2009; Mandell & Keast, 2008; Turrini, Cristofoli, Frosini, & Nasi, 2010). In this manner, we build on Turrini and colleagues' (2010, p. 529) definition of network effectiveness:

> the effects, outcome, impacts, and benefits that are produced by the network as a whole and that can accrue to more than just the single member organizations in terms of increasing efficiency, client satisfaction, increased legitimacy, resource acquisition, and reduced costs.

---

[1]The wildland urban interface is the place where people and forests meet and often create vulnerabilities to hazardous situations in the face of a wildfire.

Understanding network function is essential to understanding network effectiveness and hence performance (Provan & Milward, 1995; Provan & Sebastian, 1998; Turrini et al., 2010). Functionality of the network is important because you need to know what you want the network to accomplish so that it can be measured against appropriate criteria for its given purpose. In thinking about the network from a functionalist outcome perspective, we need to know the *interdependent* actions that only the network can perform (Mandell & Keast, 2008). This is distinct from organizational functionality, which can be described as a function for which effective execution does not require coordination with other network actors outside of the command-and-control structure of one's own organizational structure (Nowell & Steelman, 2013). In effect, this means identifying those interdependent functions and then identifying how those functions can best be measured. It means being very clear about what constitutes performance, including the key tasks we want the network to accomplish as a whole.

Myrna Mandell and Robyn Keast (2008) suggest that in addition to thinking about the specific outcome functions of the given network, we need to consider the processes or relationships that enable these key interdependent functions to take place. This might include measures related to trust, the strength of relationships among members, reciprocity, and levels of participation among others. Finally, outcome and process functions can be affected by the context in which the network operates, so network context characteristics are also important features to consider (Kenis & Provan, 2009; Turrini et al., 2010). If there is agreement in looking at the outcome and process functionality of performance within an appropriately understood context, then the question arises about how to go about measuring these interdependent indicators of performance.

Keith Provan and Brinton Milward (1995) made the case more than 20 years ago that a multimeasure, multiperspective approach to measurement is essential for the most complete understanding of network effectiveness. Multiple viewpoints are important to account for in operationalizing network effectiveness because different constituencies will have different preferences in terms of what goals or outcomes matter to them (Kenis & Provan, 2009; Mandell & Keast, 2008). Given the multilevel nature of networks and the need for coordination and cooperation, it is important to understand perspectives at different levels of operations, which again means conceptualizing and measuring effectiveness in multiple ways and across multiple perspectives (Mandell & Keast, 2008).

Accordingly, a summary of recommendations from the existing literature on conceptualizing and operationalizing whole-network performance would include the following principles:

1. Network-level measures should reflect functional outcomes that require, or can be affected by, the interdependent actions of multiple actors and therefore cannot be achieved by a single entity working alone (Provan & Milward, 1995; Provan & Sebastian, 1998; Turrini et al., 2010).
2. Network performance should be conceptualized holistically rather than using single-indicator assessments. It should focus on the collaborative process as well as the collective outcomes (Koliba, 2014; Mandell & Keast, 2008).
3. Network performance measures should attend to both actor-level performance in contributing to the broader goals of the network as well as network-level performance in accomplishing its goals.
4. Network outcomes should be calibrated to be appropriate to the exogenous conditions within which the network is operating (Kenis & Provan, 2009; Raab, Mannak, & Cambré, 2015).
5. Assessments of network performance should consider the perceptions of all network members rather than single informants (Mandell & Keast, 2008).

MEASURING NETWORK PERFORMANCE DURING WILDFIRE DISASTERS

There is little research directed at understanding how to measure network performance in disasters. Calls have been made to explore whole disaster networks and their interactions (Comfort & Haase, 2006; Kapucu, 2005a,b, 2006a,b; Magsino, 2009), but there is little research on the topic, with the exception of Kapucu and Garayev (2014). While there is empirical interest in performance as it relates to the temporal connection between preparedness before a disaster and how that influences efficacy during a disaster (Kapucu, Arslan, & Collins, 2010; Nowell & Steelman, 2015), empirical measures for functional network performance during a disaster are lacking.

For this study, we sought to assess performance of incident response networks in the context of large-scale wildfire events in the wildland–urban interface. Wildfire events are growing in intensity and scale throughout the United States (Westerling, Hidalgo, Cayan, & Swetnam, 2006) and globally (Bowman et al., 2011; Flannigan et al., 2013; Moritz et al., 2014). Climate change, historical land management practices, and the significant expansion of human developments into previously uninhabited wildland areas have all contributed to an increasing severity and impact of wildfires on communities (Williams, 2013).

Large-scale wildfire events, particularly those that exist at the boundary of a wildland–urban interface, are generally coordinated through an incident management team working within the structure of the incident command system. The incident command system is an organization comprised of designated roles and principles that enable both human and physical resources from multiple agencies to become integrated into a unified organizational structure centrally coordinated by an incident commander and his/her incident management team (Irwin, 1989). Importantly, while these incident management teams have command over fire operations, they generally must work in lateral coordination with other local and state agencies for many other disaster response functions, such as evacuation, road closure, sheltering, and public information.

Managing a large-scale wildfire, like responding to most disasters, typically involves multiple jurisdictions with overlapping responsibilities and mandates (Fleming, McCartha, & Steelman, 2015; Moynihan, 2009; Steelman, Nowell, Bayoumi, & McCaffrey, 2012). When a large-scale disaster strikes, local responder agencies and organizations need to be integrated with extra-local federal disaster management teams under challenging conditions. A network perspective can be useful in conceptualizing this diverse array of actors as a collective entity or whole network (Provan, Fish, & Sydow, 2007). We label this collection of actors as the *incident response network*, which is defined as

> the collection of individuals, organizations, and agencies that have sustained involvement during the event who aim to serve the community in minimizing and coping with damages brought on by the disaster.
>
> **(Nowell & Steelman, 2013, p. 235)**

We focus specifically on the response conditions of the disaster, as opposed to the preparedness or recovery phases. The incident response network is activated to respond to the disaster and assist the community to minimize the disaster's impacts.

Our previous discussion implies that frameworks for measuring network performance in disasters need to be explicit about the specific populations, levels of analysis, phase of disaster, and the disaster context in which performance is to be measured (Nowell & Steelman, 2013). To measure network performance, we first have to ask ourselves what network we are measuring and for what purpose. Accordingly, in this study, we utilized the principles previously outlined to develop a set of whole-network performance indicators for use in evaluating performance during large-scale wildfire events. The performance measurement process is described next as it relates to each principle.

Principle 1: Network Level Measures Should Reflect Functional Outcomes That Cannot Be Achieved by a Single Organization, Agency, or Group Working Alone

At the onset of this project, it was not clear what types of outcomes were organizational or agency-related versus whole network. In other words, while many functional areas and objectives exist in incident response during wildfires, it was not clear which of these would require network or subgroup-level coordination. Therefore, the first stage in developing performance measures was gaining deeper insight into incident response from a network perspective. This involved field observations on seven wildfires in the American West as well as semistructured interviews with 24 key informants who had extensive experience in land management and wildfire response. These individuals represented diverse perspectives in incident management during wildfires and occupied multiple roles. Key informants were identified using a snowball sampling strategy to identify information-rich cases (Quinn, 2015) based on the investigators' more than 10 years of experience in the field. Participants had collective experience of 646 years on a total of 824 large-scale wildfires. These interviews included questions about what constituted an effective incident response.

Principle 2: Network Performance Should Be Conceptualized Holistically Rather Than By Using Single Indicator Assessments

To address this principle, transcripts from the key informant interviews were qualitatively analyzed to identify key themes in characterizing a high-performing incident response network on a wildfire. This analysis resulted in the development of 30 items, which were then reviewed by experts in the field for both clarity and comprehensiveness in representing incident response network performance. These 30 items were then distributed to incident responders on 21 complex wildfire incidents in Oregon, Washington, Idaho, and Montana in the summer of 2013. Exploratory factor analysis on the 30 items revealed five subscales: (1) coordination and fire response, comprising nine items (alpha=0.90); (2) evacuation management, comprising six items (alpha=0.91); (3) sheltering and mass care comprising five items (alpha=0.91); (4) public information management, comprising four items (alpha=0.84); and (5) road closure coordination, comprising three items (alpha=0.88). Survey respondents were asked their level of agreement on a five-point scale for each item (see Table 5.1).

Principle 3: Network Performance Measures Should Attend to Both Actor-Level Performance in Contributing to the Broader Goals of the Network as Well as Network-Level Performance in Accomplishing Its Goals

In addition to the five network performance subscales, which capture overall performance of the incident response network, we also collected data evaluating the performance of individual actors as well as organizations and agencies within the network. These questions were guided and informed by our fieldwork and key informant interviews. For example, at the individual level, we utilized a social network roster questionnaire to ask how responders would rate the effectiveness of other responders in coordinating with the other individuals who are members of the incident response network. At the organization and agency level, we asked responders to rate the effectiveness of the incident management team as a whole in being responsive to *local responders*.[2] For this chapter, we focus exclusively on the five subscales that represented the whole-network performance measures.

---

[2]Local responders are those who reside in the place where the fire is occurring. They have formal responsibilities in law enforcement (sheriff, state police), emergency management, and fire management (volunteer fire departments and fire service).

Table 5.1 Network Performance Criteria Measures
Evacuation Network Performance: 6 Items (alpha = 0.91)
1. Cooperating agencies were able to use existing evacuation plans to quickly establish a coordinated evacuation strategy. 2. Residents received timely notification of evacuation status using clear, preestablished language to distinguish between an evacuation warning and an evacuation notice. 3. Evacuations were executed in a timely and orderly fashion. 4. Cooperating agencies had a prepared plan for how reentry into evacuated areas would be coordinated. 5. Trigger points for when evacuated areas would be opened for reentry were clearly communicated to the public. 6. Reentry was carried out in an organized and orderly fashion.
Coordination and Response Network Performance: 9 Items (alpha = 0.90)
1. A coordinated set of fire management objectives were agreed upon among all affected jurisdictions. 2. All concerned jurisdictions prioritized maintaining good communication across agencies. 3. Credit for success and effort was shared among agencies during public meetings and media events. 4. There was a general willingness across agencies to offer assistance to other agencies or jurisdictions. 5. "Borrowed resources" were released in a timely fashion to minimize burden on the lending agency. 6. Community values at risk from wildfire were readily identified. 7. Efforts to protect community values were appropriate given available resources and risks to firefighter safety. 8. The overall strategy taken in managing this fire was appropriate. 9. Local resources were incorporated into the incident management operations.
Sheltering and Mass Care: 5 Items (alpha = 0.91)
1. Sheltering options were clearly communicated to evacuees. 2. Adequate sheltering options were prepared to house evacuees. 3. Donations for evacuees were well coordinated. 4. Auxiliary care needs of evacuees (e.g., food, water, clothing, transportation, spiritual, or mental health assistance) were adequately provided for. 5. Adequate sheltering options were made available to evacuate pets and livestock.
Public Information: 4 Items (alpha = 0.84)
1. Local resources were leveraged to ensure timely dissemination of public information. 2. Public information was coordinated among cooperating agencies to ensure continuity of the message. 3. Social media was used effectively to provide timely public updates concerning the status of the fire. 4. A system for communication with the media was put in place to ensure timely dissemination of public information.
Road Closure: 3 Items (alpha = 0.88)
1. All cooperating and fire management agencies maintained a timely awareness of the status of road closures. 2. Trigger points for making decisions about road closures were proactively communicated to the local community. 3. A consistent message was provided to the public about the status of road closures.

Principle 4: Network Outcomes Should Be Calibrated to Be Appropriate to the Exogenous Conditions Within Which the Network is Operating

Patrick Kenis and Keith Provan (2009) argued for consideration of variation in exogenous factors when considering appropriate indicators of performance. These included the: (1) governance structure of the network, (2) whether the network was voluntary or mandated, and (3) the stage of development of the network. While important considerations in networks may vary on these dimensions, these exogenous factors had little bearing on our study as all networks were voluntary and were all at the same developmental stage. Similarly, the structure of the incident command system means all incident response networks possess the same basic governance structure. Network performance was measured to capture differences in complexity between incidents as well as to capture perceptions of all actors involved in the responder network. Our focus here was in examining differences between the 21 incidents we investigated as well as variation in evaluations by individuals and subgroups within the same incident.

Principle 5: Assessments of Network Performance Should Consider the Perceptions of All Network Members Rather Than Single Informants

In this study, we identified all agencies and organizations that had a sustained role in responding to each of our 21 wildfire incidents. Identification was done first through interviews with incident management team liaison officers for each of the 21 incidents. This initial list of names of engaged organizations was then validated and augmented by follow-up interviews with other central actors involved in the incident—generally the sheriff and/or county emergency manager. At each stage in identifying the network, we also asked informants to identify the individual who played a leadership role in coordinating their agencies' operations with the rest of the network. This process continued until no new network members were identified. These individuals represent the incident response network for each of the 21 incidents and, accordingly, became the sample for our performance evaluation questionnaire.

INVESTIGATING PLURALISM WITHIN INCIDENT RESPONSE NETWORK PERFORMANCE ASSESSMENTS

The prevailing conventional wisdom related to performance measurement suggests that we should look for differences among incidents. In other words, if we are to take networks seriously (e.g., O'Toole, 1997), we should be able to distinguish high- and low-performing incident response networks. Further, traditional performance measurement advises to prioritize precise, objective measures that are based on quantities that can be influenced or controlled by those being evaluated (Neely, Richards, Mills, Platts, & Bourne, 1997). Such a goal relies on a realist perspective of performance, assuming that there is an objective measure of performance to be uncovered that will be understood and rated by all actors relatively equivalently.

The assumptions underlying the realist perspective have been heavily criticized (e.g., Mandell & Keast, 2008), particularly in the context of complex problem domains in which the ultimate outcomes are impacted by a complex system of factors, many of which are outside of the control of the actors. Further, emphasis on easily quantifiable performance outcomes (i.e., reducing the number of acres burned) can lead to goal conflict among relevant actors. For example, an emphasis in national wildfire policy on measuring effectiveness of incident response based on the numbers of acres burned has been attributed to the escalation in wildfire risk we face today (Williams, 2013). This is because fire-adapted landscapes are ecologically designed to burn in order to renew themselves, and lack of fire on the land leads to an excess of woody fuels that increase fire intensity when the next fire occurs (Mutch, 1970). All of this indicates that, when working in complex problem domains, there are few clear-cut objective measures of effectiveness. Rather, any observed outcome would be judged on its effectiveness based on its consequences relative to alternative outcomes. This requires normative judgments to be made about what is valued. Given that different people may legitimately come to different conclusions in these value judgments due to different goal orientations, scholars have argued that understanding performance in networks requires accounting for pluralism within the network (Mandell & Keast, 2008; Turrini et al., 2010). Accordingly, if we heed the call to look at the diversity of individual perspectives, we need to look at the differences among individuals and stakeholder groups within a network and not simply at differences in performance between incidents.

The preceding discussion suggests that variation among actors within responder networks is just as important to understand as variation between networks in terms of evaluation of performance. However, consideration of pluralism in thinking about and measuring network performance has received scant empirical attention. In this study, we utilized a multilevel approach—individuals, subgroups, whole network—to investigate within- versus between-network variation on network performance measures. Within-network variation refers to differences in network performance evaluations across individuals or between subgroups within an incident response network. Between-network variation refers to variation in the averaged performance ratings for each of the 21 incidents that we studied. We then explored whether within-network variation in ratings of network performance appears to cluster at a subgroup level. Institutional affiliations (e.g., whether you represent the incident management team, or United States Forest Service, or county emergency response) tend to be strong identifiers in incident response to wildfire (Steelman et al., 2014). Accordingly, we operationalized subgroup differences in this study using institutional affiliations.

To build both the theory and empirical base for better understanding pluralism and its implications for assessing network performance, we focused on two research questions:

1. How much variation in performance assessments occurs within versus between networks?
2. To what extent is within-network variation associated with institutional affiliations?

METHODS

The data collected for this analysis were part of a larger study conducted by Nowell and Steelman (NSF # CMMI-1161755; JFS L12AC20571). The study took place during large wildfire events occurring in counties at risk for wildfire within the wildland–urban interface in Idaho, Montana, Oregon, and Washington, in summer 2013. The wildfire events considered in this study met two study requirements: (1) they were of sufficient complexity to require a national or regional incident management team; and (2) the wildfires occurred in the wildland–urban interface where they threatened human populations and infrastructure.

Twenty-one incidents within our four-state sample region met the criteria for inclusion in the survey resulting in performance data on 21 responder networks. We sent surveys containing the network performance measures described previously to all members of the incident response network for each of the 21 incidents. In total, we invited 885 individuals representing 369 organizations/agencies to participate in the survey. As summarized in Table 5.2, survey data were collected from 551 respondents (58% response rate) representing the United States Forest Service, incident management teams, county emergency responders, and other host land agencies (i.e., United States Bureau of Land Management, state forestry, etc.).

Table 5.2 Survey Response Rates by Subgroup

Group	Response Rate
Incident management teams	103 (93%)
County organizations	207 (45%)
Forest service districts	79 (62%)
Forest service supervisor's office	94 (69%)
Other host agencies	28 (51%)
Total	*511*
Total response rate	*58%*

Hierarchical linear modeling, also known as multilevel modeling, is a useful approach to testing relationships between variables at different levels of analysis. In this study, we employed hierarchical linear modeling because of its capacity to help disentangle the amount of variation that occurred in performance indicators at different levels of analysis, as well its ability to test hypotheses regarding the source of that variation (Hox, 2010).

RESULTS
BETWEEN-INCIDENTS NETWORK PERFORMANCE VARIATION

To answer our first research question, we ran a null model for each of the subscales of our network performance measures separately. A null model in hierarchical linear modeling allowed us to test whether there is significant variation both within and between response networks—while accounting for variation in the other levels. For each performance subscale, an interclass correlation was calculated to provide a breakdown of the percentage of variation that occurs at each level of analysis. Descriptive analysis on the overall variability on the network performance subscales (see Table 5.3) indicated a negative skew with scale averages ranging from 3.87 to 4.44 on a five-point scale. However, suitable variability was observed on all outcomes with standard deviations ranging around 1.0. Across responder networks, respondents tended to rate coordination and response network performance ($M = 4.44$) higher than all other network performance subscales.

When modeling the source of that variability, results overwhelmingly indicated that the greatest portion of variance was between respondents rather than between networks. In fact, only one performance dimension—evacuation—approached significant network level variation in the null models. This meant that the difference in the average rating of performance between our 21 wildfire incidents was not significant relative to the differences that existed among raters within incidents. In other words, the incidents were statistically indistinguishable from each other in terms of being higher or lower performing networks on these metrics once we accounted for within-network variation.

There were, however, some differences observed between network performance subscales (see Table 5.4). For evacuation performance ratings, 17% of the variance was between incidents while 83% was between individuals. For public information performance, only 5% of the variance was between

Table 5.3 Network Performance Subscale Descriptive Statistics

Network Performance Dimension	Mean	SD	Min	Max
Evacuation	3.99	0.91	1	5
Public information	4.34	0.77	1	5
Coordination and response	4.44	0.73	1	5
Road closure	4.13	0.95	1	5
Sheltering and mass care	4.00	0.91	1	5

Table 5.4 Hierarchical Linear Modeling Model—Performance Scores by Incident

Network Performance Dimension	Covariance Estimate of Intercept (Std. Error)	ICC	p-value
Evacuation	0.147 (0.076)^	0.174	.053
Public information	0.001(0.001)	0.048	.142
Coordination and response	0.001 (0.001)	0.034	.201
Road closure	0.026 (0.024)	0.028	.287
Sheltering and mass care	0.053(0.065)	0.063	.416

^$p = <.10$.

incidents while 95% was between individuals. The remaining outcome measures did not approach significance, meaning that the mean differences between incidents, after controlling for within-incident variation, was limited. This gives credence to the idea that perspectives within networks matter when conceptualizing performance within networks. The question then becomes, "what drives the within-network variation?"

NETWORK PERFORMANCE VARIATION WITHIN INCIDENTS

In this second phase of analysis, we investigated whether stakeholder affiliation (subgroups) helped to explain differences between individuals in their ratings of network performance (see Table 5.5). In other words, we sought to understand whether individuals representing the same stakeholder group evaluated the performance of their respective response network more similarly to one another. To address this question, stakeholder affiliation was added as a level-one fixed-effect predictor in the null models described in phase one. Here, we find that modeling stakeholder affiliation at level one generated a less than 3% change in variance explained. Interestingly, we found that for evacuation and sheltering, public information, road closures, and mass care network performance scores, incident management teams were the only group that had significant differences ($p < .05$) in ratings compared to other stakeholder groups. Overall, members of the incident management team tended to rate the effectiveness of the network more positively than did the other network actors. For mass care performance ratings, county response agencies were significantly more negative in their evaluations relative to other groups.

Table 5.5 Hierarchical Linear Modeling Model: Network Performance Scores by Respondent Sample Group

Institutional Subgroup	Network Performance Subscales				
	Evacuation	Public Information	Coordination and Response	Road Closures	Shelter and Mass Care
County responders	0.35 (0.30)	0.01 (0.04)	0.01 (0.04)	0.37 (0.22)	0.73 (0.37)*
USFS – District Office	0.29 (0.33)	0.05 (0.05)	0.03 (0.04)	0.42 (0.24)	0.43 (0.41)
USFS – National Forest Headquarters	0.16 (0.34)	0.04 (0.05)	0.02 (0.04)	0.17 (0.24)	−0.15 (0.42)
Incident management team	0.63 (0.32)*	0.110(0.05)**	0.07 (0.04)	0.64 (0.23)*	0.92 (0.42)*
Change in residual variance explained	*>3%*				

Referent class = non-USFS host agency.
Public information and coordination reported as absolute values due to reflected data transformation for normalization.
*$p = .05$; **$p = <.1$.*

DISCUSSION

There is general consensus in the literature that disaster response is best understood as a network phenomenon and that the outcomes we care most about during disasters rely on information flow, coordinated action, and collective impact within that network (Comfort, 2007; Magsino, 2009; Mandell & Keast, 2008). Therefore, to assess performance during incident response requires that one be able to assess the performance of the network. Since performance criteria constitute value judgments about what matters (Simon, 1976), it is important to be clear about what we measure when we want to evaluate performance in networks. Challenges arise when measuring performance in networks because networks violate many of the assumptions related to how and what should be measured. For instance, conventionally we think measurements should be exact, objective, and based on characteristics under the control of the participants (Neely et al., 1997; Sandström & Carlsson, 2008). In contrast, networks take place in messy problem domains that are rarely fully under the control of the network (Provan & Kenis, 2008). Further, differences in perspective and position in the network matter, thereby creating difficulty in measuring a single, objective reality (Mandell & Keast, 2008).

Network scholars argue that efforts to understand performance in networks should both anticipate and attend to diverse perspectives among network members (e.g., Mandell & Keast, 2008; Turrini et al., 2010). Unfortunately, this advice stands in sharp contrast to conventional thinking about performance measurement, which prioritizes interrater reliability and would therefore dismiss variation across informants nested within the same incident as measurement error. Our findings are provocative on this front, offering empirical evidence that illuminates just how significant this issue may be when studying network performance in disasters.

As Mandell and Keast (2008) predicted, we found that responders within the same incident varied significantly in their perceptions of how well the network performed. In fact, less than 20% of the variation in ratings on any measure could be explained by differences between incidents. Interestingly, certain performance dimensions (e.g., public information and evacuation) appeared to distinguish between incidents better than others. To find strong differences between incidents using a multiinformant approach, two things must happen. First, members within networks must share a similar perspective in evaluating that performance dimension. Second, these shared assessments by some networks must be, on average, substantially more or less positive relative to other networks. In our study, these two conditions appeared to be better met in relation to the outcomes of public information and evacuation. However, the differences between incidents on these outcomes—after controlling for within-group variation—only approached a traditional level of significance ($p < .1$).

Perhaps this finding is not surprising if one stops to consider that incident response networks are situated in highly diverse settings. Disasters are complex events that necessitate a wide array of functions to be carried out by a myriad of actors with complementary resources and skills. Within these diverse settings, subgroup identities can become more pronounced (Faas et al., 2016). For example, stakeholder distinctions such as local government versus state or federal government can become more important (Fleming et al., 2015). Therefore, it is reasonable to suspect that individuals within the same subgroup may share more similar perspectives when evaluating an incident relative to others in the network. Surprisingly, however, this was largely not the case.

The perception of how key activities occurred in an incident appeared to be weakly tied to institutional affiliation (subgroup) or role during the incident. When we investigated this further, however, the only group that really made a difference was the incident management team. These team members, as a group, tended to rate the areas of public information, sheltering and mass care, evacuation, coordination/response, and road closures more positively compared to ratings provided by other groups. In the end, we did not find strong evidence of variation at the incident or the subgroup level.

PERFORMANCE MEASUREMENT IMPLICATIONS

Drawing from cognitive theory, we can think of three processes that can affect why and how different individuals might evaluate performance differently. The first suggests that *sensing*, which includes what kind of information one encounters, is important. Actors within an incident response network will have different experiences and access to different sets of information. Therefore, where one is positioned in the network may greatly affect how one perceives the network to have performed.

The second process potentially responsible for variation in perceived performance is *sense-making*, or how one interprets what one sees or experiences. The theory of sense-making asserts that two people exposed to exactly the same experience will likely attend to and interpret that experience differently based on differences in their individual past experience (Weick, Sutcliffe, & Obstfeld, 2005). This would contribute to variation in how someone views effectiveness relative to others.

The third process possibly implicated in different perceptions of performance is *assessing/evaluating*, which includes the kind of normative judgments one makes about the interpretation of the data. Judgments can be influenced by a number of factors including the degree of personal importance one assigns to a given outcome, differences in normative beliefs in what high performance means in a given setting, as well as personal or professional biases stemming from enlightened self-interest in the network being evaluated more or less positively.

This individual difference perspective may assist in understanding why incident management teams rated various elements of performance slightly higher than the other stakeholder groups. We can think of several explanations based on the insights in the previous paragraph. First, in terms of sensing, incident management teams are structurally located in a central position in the network, which may give them access to different information than other positions. This could lead to a different appraisal of the performance criteria. Second, in terms of sense-making, the incident management team will typically have more experience on large-scale wildfires relative to other local participants in the network. Consequently, they engage in a different type of sense-making than less-experienced participants. Third, in terms of assessing/evaluating, they have more experience, which allows them to evaluate performance on a wildfire. This may translate into the incident management team using a different scale upon which to calibrate their experience on a wildfire. For example, for a local responder, an event may be the worst incident that they have ever experienced. For an incident management team that deals with these events multiple times a year, this incident may not even rank in the top 20 of the worst events that they have seen. Alternatively, incident management teams may have an inherent—and potentially tacit—bias to view the incident as having gone more positively because incident management teams are the most centrally involved actors in coordinating activities within an incident response network. Therefore, they may be likely to perceive a stronger association between an evaluation of the network overall and an evaluation of them as an incident management team. Finally, more than any other subgroup, the incident management team engages in daily discussion about the incident as a group.

Research has shown that sense-making and decision-making in a group is unique and can trend toward more intense or extreme ideas (Moscovici & Zavalloni, 1969). In other words, an individual's assessment of an incident may become even more positive or more negative depending on the sentiment of the group as the incident is being discussed. As such, it could be that the relatively more positive sentiments of the incident management team members are rooted in group sense-making dynamics. These explanations and others warrant future research. Given the degree of pluralism that was evident in the 21 incidents studied in this chapter, it is clear that we need to better understand the factors that help to explain differences in evaluations between responders.

However, even if we were able to fully explain actor-level variation in evaluation ratings of network-level performance, we are still left with a quandary. If we are to consider disaster response as a network phenomenon, improving disaster response will require an ability to differentiate and learn from higher- and lower-performing networks. Pluralism within disaster response networks *is not measurement error*—it is an endemic reality to the phenomenon of networks in complex problem domains (Koliba, 2014; Mandell & Keast, 2008; Provan & Kenis, 2008). Our study suggests that significant pluralism exists among actors who comprise a disaster response network and this has implications for how we think about conceptualizing and measuring network performance in disasters. Therefore, we need tools for differentiating between networks in such a way that still does not discount or ignore this reality.

LIMITATIONS AND FUTURE RESEARCH

Findings from the current study are provocative, suggesting that evaluating the performance of whole networks may not be possible when pluralism exists. Data to support this premise are drawn from one specific type of network—incident response networks convened in reaction to large-scale wildfire events. As such, the present study is limited in its ability to draw conclusions concerning how much pluralism exists in other network contexts. Rather, we are content to assert only that pluralism *can* be significant in disaster response networks and therefore should be considered in any network performance measurement endeavor.

In addition, since our findings suggest that pluralism may not be explained away easily based on obvious stakeholder groupings and differences, this raises additional issues about what might explain within-incident variation. One explanation is that variation is clustered within alternative subgroups that we did not test. Alternative groupings such as local versus extra-local participants, professional affiliations (e.g., public information, law enforcement, fire management roles; see Nowell & Steelman, 2015), or individuals with greater or less experience with large wildfires may help us understand variation at the subgroup or individual level. An additional alternative explanation is that evaluation of network performance on many disasters is largely an individual-level phenomenon. Conceptually, we understand that within networks, different constituencies will have different experiences and will interpret these experiences through different filters and with different points of reference. Differences in how these criteria are experienced can lead to different evaluations of performance effectiveness. Unpacking these biases will be important to further conceptualizing how we better understand pluralism in evaluations of network performance.

We see two promising paths forward for future research. Operationalizing high performance in terms of the level of agreement among network members is one possibility. For example, Mandell and Keast (2008) have argued that high-performing networks develop shared identity, goals, and understandings across members that would, presumably, result in higher agreement in evaluating the network's performance. In this approach, it is the standard deviation rather than the average that is considered. The theory here is that, if network members do not agree, the network did not function well to create a shared point of reference and therefore would be evaluated as a lower-performing network. This offers a different approach compared to the functionalist performance approach that we embraced in this study.

Another potentially promising methodological path forward is to focus on extreme case comparison methodologies. Just because there was not substantial variation across incidents when all 21 events were considered does not mean that there were not a few incidents that distinguished themselves from the pack. In fact, it was clear from our data that most of the observed between-network variation was being driven by a few key problematic incidents in which there were markedly lower scores on performance across responders. A qualitative comparison of these extreme cases against cases characterized by corresponding high scores could lead to key insights about what factors distinguish higher- and lower-performing incidents.

With this research, we have just scratched the surface of what we need to know to evaluate network performance. In this chapter, we have identified a way forward in terms of approaching how we might view network performance measurement, some of the assumptions about plurality and its consequences for network measurement and demonstrated how this kind of analysis can be accomplished. Next steps include further efforts to overcome the challenges identified in our own work and looking at other disaster networks to see how and if pluralism matters in other domains. More research is needed to understand the determinants of within-incident variation and build practical guidelines for developing measures of network performance that avoid the pitfalls of the past and account for the plurality that is integral to complex problem domains. Without this progress, we will never realize the full value of what networks can accomplish or how to understand them.

ACKNOWLEDGMENTS

Data for this study were supported by funds from the National Science Foundation, the Joint Fire Science Program, and the USFS Northern Research Station. It was collected as part of a research initiative led by Dr. Branda Nowell (North Carolina State University) and Dr. Toddi Steelman (University of Saskatchewan). See firechasers.ncsu.edu for more information. Special thanks to members of the Fire Chasers Project team and all the fire and emergency personnel who contributed their time and insights in support of this project. The views and conclusions contained in this document are those of the authors and should not be interpreted as representing the opinions or policies of the US government. Mention of trade names or commercial products does not constitute their endorsement by the US government.

INTERORGANIZATIONAL RESILIENCE: NETWORKED COLLABORATIONS IN COMMUNITIES AFTER SUPERSTORM SANDY

Jack L. Harris, Marya L. Doerfel

Rutgers University, New Brunswick, NJ, United States

CHAPTER OUTLINE

INTRODUCTION

Disaster response and recovery rely upon networked collaborations that cross multiple geographic and organizational boundaries. The National Response Framework (Department of Homeland Security, 2008, 2013a) details a wide variety of public, private, and nonprofit organizations and agencies as participants in emergency response and long-term recovery, while extant disaster-related literature highlights the roles played by formal and informal organizations, volunteers, and emergent groups over

Social Network Analysis of Disaster Response, Recovery, and Adaptation. http://dx.doi.org/10.1016/B978-0-12-805196-2.00006-6

time (Drabek & McEntire, 2003; Majchrzak, Jarvenpaa, & Hollingshead, 2007; Kendra & Wachtendorf, 2003; Simo & Bies, 2007). Disaster response and recovery are phased processes, with different organizational activities taking precedence at different points in time (Doerfel & Haseki, 2013; Doerfel, Lai, & Chewning, 2010). Effective emergency response and long-term recovery from disaster require connected public agencies, private sector businesses, and nonprofit organizations that enable communication and coordination among multiple organizations with different goals, missions, information, and resources (Nowell & Steelman 2013). While the role of bystanders, disaster tourism, unaffected volunteers, and emergent groups is well documented in the disaster literature (Drabek & McEntire, 2003; Helslott & Ruitenberg, 2004; Majchrzak et al., 2007), formal disaster planning and response tends to overlook the important preexisting community connections and grassroots relationships in disaster-struck communities.

Prior community relationships activate information flows and material assistance necessary for survival and recovery via interorganizational collaboration. Collaboration in disaster-struck communities occurs at grassroots or neighborhood levels as well as at the level of formal institutional response in cities, towns, counties, and states. Indeed, such grassroots organizing is seen as foundational to civil society (Taylor & Doerfel, 2011) and can be viewed as the social structure that undergirds resilient communities (Doerfel, 2016). This chapter examines community resilience as a social phenomenon involving networked organizations that collaborate after a natural disaster.

We present the case of Oceanport, a small coastal town in New Jersey that was struck by Superstorm Sandy in 2012. Some 1200 of the community's 2100 residences were damaged with 700 of those having more than 50% of the given home significantly damaged or destroyed. Power outages ranged from 5 to 13 days throughout the town. The community's town hall and police station were destroyed, and the community's volunteer first aid squad was inoperable. Oceanport is home to state and federal installations with large amounts of property that were used to stage state and national response efforts to the impacted region. In addition to providing support for impacted residents in Oceanport, the Oceanport community had to coordinate flows of people and materials into town to service other impacted communities in the region.

ORGANIZATIONAL NETWORKS IN THE FABRIC OF COMMUNITIES FACING DISASTERS

SOCIAL RESILIENCE PROCESSES

Resilience processes are a mix of discursive and organizing activities that integrate improvisation and problem recognition with solution building. Social resilience is accomplished through community partnerships that enable organizations to access tangible and intangible resources to prepare for and manage threats. In disaster contexts, networked partnerships become a critical resource for responding to and rebuilding after disasters (Doerfel, Chewning, & Lai, 2013; Doerfel et al., 2010). Resilient systems can withstand stresses without degradation or loss of function, identify problems, establish priorities, and mobilize resources during disruptions, have redundant/substitutable elements, and meet priorities and achieve goals in a timely manner (Bruneau et al., 2003; Chewning, Lai, & Doerfel, 2013; Doerfel et al., 2013; Kendra & Wachtendorf, 2003). Social resilience is by its very nature communicative, that is, it relies on interaction between people and organizations in order to make sense of their surroundings, problem solve, and achieve specific outcomes (Buzzanell, 2010; Doerfel & Harris, 2016; Weick, Sutcliffe, & Obstfeld, 2005).

COLLABORATION AND INTERORGANIZATIONAL RELATIONSHIPS

Collaboration is designed through communication activities that involve both talk and action (Aakhus, 2007) and, in disaster-struck communities, collaboration is heavily dependent upon preexisting relationships (Doerfel et al., 2010). Collaboration is a matter of social structure, process, and relationships, and is reflected in the interorganizational linkages that comprise communities, organizations, and organizational sectors (Diani, 2015). Collaboration can be created through the dyadic and group actions of individuals in partnering organizations or imposed by external structures that require collaboration in some way. Collaborative activities can even be suggested through the use of frameworks, governing documents, and other management and funder documents. The United States National Response Framework (Department of Homeland Security, 2008, 2013a) clearly identifies how public, private, and nonprofit organizations are expected to come together during crisis to deliver services and resources to impacted communities.

In disaster-struck communities the ever-shifting reality of disrupted physical and social environments impacts organizational relationships as communication patterns shift during transitions from emergency to transitional to recovery phases. A key question for understanding disaster response and recovery is the role that community-based interorganizational collaboration plays in: (1) constituting new patterns of networked stakeholder relations in disrupted social environments, and (2) building interorganizational resilience.

NETWORKED MODELS OF STAKEHOLDER RELATIONSHIPS

Rather than focusing on the specific ties of specific organizations to other organizations, a networked model of stakeholders accounts for the dynamic, multiple ties among different organizations populating the plural, public, and private sectors that give a community its social structure (Diani, 2015; Mintzberg, 2015; Rowley, 1997). Key features of the networked stakeholder relationships we explore include organizations' attributes, tie strength, and the role of centrality in broader interorganizational networks. Organizational populations consist of a combination of organizational attributes and relational ties that interact to shape the broader network of which they are a part (Monge, Heiss, & Margolin, 2008). The composition and function of these networks are understood through the organizational attributes, relational ties, and the strength or weakness of those relational ties.

Within a network structure, nodes represent the actors (in this study, organizations) with each actor composed of particular attributes. Network analysis measures the types of relationships between the nodes, in addition to the characteristics or nature of the nodes themselves. Yet, much of the interorganizational analyses of disaster response and recovery focus on the formal institutions of response rather than on the broader community network of stakeholders with their multiplicity of relational ties and organizational attributes (Carlson et al., 2016; Kapucu & Hu, 2014). We extend postdisaster research by considering the larger community stakeholder network, which serves to cement the community together as they engage in collaborative public activities (Diani, 2015).

Relational ties may carry informational, financial, material, or reputational resources that are the building blocks of interorganizational relationships (Doerfel et al., 2013). Within geographic communities, local organizations connect with one another through relational ties that involve both personal and organizational relationships that give shape to the broader community (Diani, 2015; Mische, 2003). Relational ties connect organizations with different attributes within a community by tying together their associated stakeholder networks. Weak ties become salient after significant community disruption by enabling organizational resiliency and lessen dependence on any one part of the network (Doerfel et al., 2010; Harris & Doerfel, 2016; Taylor & Doerfel, 2011). Strong ties may help to create dense or embedded networks that facilitate collaboration and resource sharing and consist of links between organizations with similar goals or missions. Latent ties become activated after disruption and may help to create or increase redundancy in a disaster-struck organizational or community network when resources are stretched thin and organizational systems are stretched to the breaking point (Doerfel et al., 2010).

Organizations with more ties to other organizations have high degree centrality and tend to play a more prominent role in the network (Doerfel & Taylor, 2004). While knowing the most nodes in a network is one way to increase access to and control over information, an organization can also occupy a brokerage role. Brokers connect organizations, people, and resources from distant or disconnected parts of a network. In disaster response and recovery, brokers often coordinate voluntary and institutional organizing efforts and mediate between emergent or converged response and the formal response efforts of established organizations. Researchers have identified brokers using different algorithms including betweenness centrality (Borgatti, 2005) and structural holes (Burt, 1992). Taken together, aspects of centrality reveal which organizations are leaders in terms of organizing and maintaining the network through bridging, gatekeeping, and coordination activities that bring disparate parts of the network together.

It is important to distinguish between centrality and centralization in understanding how networks operate. Centrality measures reveal how well connected any one organization is within a network based on the number of incoming and outgoing ties of a particular organization. Centralization measures reveal the overall level of connection or disconnection within a network. Centralization focuses on whether a network is highly centralized and organized around a particular focal organization or organizations, or whether it is decentralized, such that information flows in and around a network with no particular organization controlling the flow. Paired with density, which measures the general level of cohesion (extent to which members of the network are connected with each other member), centralization can provide a view of network organization (Borgatti, 2005; Borgatti & Everett, 2006). Key network questions we consider include: Does centralization change as the stakeholder network moves from tasks associated with emergency response to tasks associated with long-term recovery? Are local or national organizations more central in stakeholder networks in a disaster-struck community? What implications do these network questions have for response and recovery activities and public policies that foster disaster preparedness, emergency response, and community recovery?

POSTDISASTER CASE STUDY: OCEANPORT, NEW JERSEY

We conducted field research in Oceanport, New Jersey, and along the New Jersey coast after Superstorm Sandy with the goal of assessing interorganizational interactions used to recover at both organization and community levels. Over the course of two years, the first author volunteered as a leader at the Oceanport microshelters, in the formation of Oceanport Cares, as part of the Monmouth County Long-term Recovery Group's (MCLTRG) volunteers committee, and observed disrupted community and organizational networks at both community and regional levels through these activities. This chapter focuses on the three-month period from October 29, 2012, to December 24, 2012. This time period ranges from Sandy's landfall to the official incorporation of Oceanport Cares as an outcome of the volunteer activities conducted in Oceanport during this period of disaster response and the subsequent transition into recovery activities.

Using longitudinal field research and document analysis, we reconstructed network relationships with the use of communication categories that represented network activities. The network was constructed using data obtained from participant observation, purposive sampling of active organizations, meeting agendas, and rosters obtained through the first author's volunteer roles. Network ties were defined as interaction between organizations in the data set as observed in the field and through joint activities. This interorganizational interaction was defined as one or more contacts between organizations active in the recovery using a variety of face-to-face and mediated communication channels (Doerfel, & Haseki 2013; Doerfel et al., 2013; Doerfel, Lai, & Chewning, 2013). The network analyzed in this study is comprised of community and regional organizations ($n = 26$) active in Oceanport during this time period. Observations began casually on October 30, 2012, the day after the storm hit. More comprehensive mixed methods, including additional observation and participant observation, unstructured and semistructured interviews, small-scale surveys, and document analysis were used to supplement initial observations and notes over the longer course of the study from 2012 to 2015.

Community organizations in the network included Oceanport groups such as the Boy Scouts and Girl Scouts, the Oceanport Sports Foundation, and a local pizzeria. The Oceanport Sports Foundation acted as a pass-through organization for monetary donations until Oceanport Cares was formalized. Enzo's local pizzeria provided meals when operations relocated to the volunteer fire station and helped provide food to the senior citizen apartment building in town. Oceanport Girl Scouts provided babysitting and group activities for children at the middle school and often worked with mental health responders developing systems of social and psychological support. Oceanport Boy Scouts organized and distributed supplies delivered by the Monmouth County Park System. The local teachers union worked with the local parent–teacher organization to collect, organize, and distribute clothing in the middle school library, while school food service personnel worked with citizen volunteers to prepare meals and snacks for impacted residents. These were ad hoc activities organized by citizen volunteers as conditions warranted. Operations then shifted to more formal organizations such as faith-based organizations, formalized local organizations (eg, Oceanport Cares; Sea Bright Rising) the Occupy Sandy network, government agencies, and coordinating groups such as the MCLTRG.

Oceanport operations at the middle school also served as a staging area for nearby communities and the severely impacted Bayshore region, directly south of Staten Island alongside Raritan Bay. In the days after Sandy impacted the area, Monmouth County Park System warehouses were overwhelmed with donations. These donations were then sent out to those communities that could organize and distribute them. Since the New Jersey National Guard and Federal Emergency Management Administration (FEMA) Corps were staging at the Oceanport middle school, and Oceanport's Office of Emergency Management had two-and-a-half-ton trucks that could navigate through flooded beach and bay towns, supplies and relief operations were routed through Oceanport for distribution to Oceanport and nearby communities even while state and federal response operations were underway across town. The New Jersey National Guard provided security and search-and-rescue operations in flooded areas of Oceanport, Monmouth Beach, and Sea Bright, while FEMA Corps volunteers provided information about federal insurance, assistance, and claims processes as well as application assistance for residents of Oceanport, Monmouth Beach, Sea Bright, and Long Branch. Oceanport's Office of Emergency Management made a direct request for a Corps presence in the community.

STUDY VARIABLES

Time. We bracketed the time frames of this study into four periods: Time 1 and Time 2, which cover the poststorm emergency response from October 30 to November 11, 2012; Time 3, which represents post-storm emergency response and short-term recovery activities from November 12 to December 24, 2012; and Time 4, which starts with December 24, 2012, representing the start of the long-term recovery period. Times 1 and 2 were combined for analyses of degree centrality and betweenness centrality.

Qualitatively, Times 1 and 2 are different phases because a transfer of activities from a local middle school to a local volunteer fire station occurred sequentially, marking the transition from emergency response to short-term community recovery. The spatial location of activities changed sharply between Time 1 and Time 2, while organizational and network activities slowly transitioned from a state of emergency response to a state of recovery. Times 1 and 2 are combined for this chapter to clearly reflect those differences between spatial and organizational activities. The analysis in this chapter focuses on the temporality of organizational activities rather than the spatial location of those activities.

Time 3 started with the initial organizing meeting of the MCLTRG in the second week of December 2014. Time 4 is represented by the incorporation of the emergent disaster relief organization, Ocean-port Cares, within the State of New Jersey on December 24, 2012, and represents a turn from emergent to more formal interorganizational relationships within the community. Following Doerfel and colleagues (2010), Times 1 and 2 represent the emergency response and short-term recovery phases. Emergency response and short-term recovery phases often overlap in the first few weeks after disaster. Times 3 and 4 represent the transition from short-term recovery to long-term recovery activities in the community.

Organizational Network. Communication is two-way and multichannel between the organizations active in this network during the time period analyzed. *Degree centrality* identifies organizations that helped activate and maintain Oceanport's stakeholder network after Sandy made landfall. All centrality calculations were made with UCINET (Borgatti & Freeman, 2002). We used Freeman's betweenness measure in UCINET; though betweenness and degree centrality measures are similar measures, high betweenness does not require high degree centrality. If few organizations in the overall network have high degree centrality, then they also fall between otherwise unconnected organizations. UCINET's *network centralization* represents the degree to which information flows were reliant on a few key organizations or dispersed across the network. High-centralization scores represent a more centralized network in which information flows through relatively fewer organizations. Low scores represent a relatively decentralized network in which information flows in and around members where no one or very few organizations act as gatekeepers/brokers of the information (Borgatti & Freeman, 2002).

LOCAL STAKEHOLDER NETWORKS IN OCEANPORT, NEW JERSEY

This case's research questions considered how interorganizational relationships built new patterns of stakeholder relationships and created resilience through networked activities and whether centrality and centralization changed as the network transitioned from emergency response toward long-term recovery activities. The findings showed that patterns of interaction shift over time through interorganizational collaborations both within the local Oceanport stakeholder's network and with larger regional, and national organizations. Two local organizations—one that was part of the formal institutional response and a second that emerged as a new organization created by citizen volunteers—held primary roles throughout the four phases of response and recovery.

To answer these questions, we first describe the networks over time. Fig. 6.1 depicts Times 1 and 2 and shows two distinct clusters joined by the communication activities of the Oceanport Office of Emergency Management and the nascent Oceanport Cares, a group of unaffiliated volunteers loosely organized by Oceanport's Office of Emergency Management. These citizen volunteers and emergent organizing processes morphed into a formal 501c3 nonprofit in the year following the storm, with state incorporation occurring on December 24, 2012 (start of Time 4) and 501c3 status achieved in January 2014. Both the Oceanport Office of Emergency Management and Oceanport Cares connected a local network organized around issues of community support with a larger network of local, regional, state, and federal agencies providing more traditional emergency response and law enforcement activities.

As shown in Figures 6.2, 6.3, and 6.4, the Oceanport Office of Emergency Management and Oceanport Cares were consistently among the organizations ($n = 26$) with the highest degree and betweenness centrality across each of the time periods measured. This is consistent with their role as key brokers of resources and information throughout the three-month period analyzed after Sandy made landfall. A high degree of relational ties to other organizations would be expected given the coordinating role that emergency management offices are expected to play as part of the National Response Framework and New Jersey's disaster response protocols.

The stakeholder network began to shift from the emergency response to the short-term recovery phase when operations changed locations from the middle school to the volunteer firehouse during the middle of Times 1 and 2. Toward the end of Times 1 and 2, the Oceanport Office of Emergency Management evolved into a broker for local organizations expecting reimbursement from FEMA for their supplies and labor, while Oceanport Cares continued to connect directly with organizations providing resources and social support to impacted residents and businesses. National institutions like FEMA and the American Red Cross were in peripheral roles in the local communication network at this time.[1]

Stakeholder network changes during Times 1 and 2 were complicated by the material impacts of Sandy on Oceanport. The municipal building was destroyed by the storm, resulting in moving regular municipal operations to the borough's emergency operations center on the second floor of the Oceanport Hook and Ladder Company. Floodwaters disabled the operation of the Oceanport First Aid Squad's headquarters and disabled their ambulances. After short-term recovery operations were moved to the firehouse from the middle school, the first aid squad essentially became disconnected from the network as they focused on repair and recovery of their equipment and buildings.

Time 3 included the launch of the MCLTRG, which took over the coordinating role in disaster response from local Offices of Emergency Management as organizing the rebuilding included both social and physical reconstruction. The social elements of reconstruction included case management services, volunteer coordination, interorganizational coordination, arranging temporary housing for displaced homeowners, financial assistance through state, federal, foundation, and nonprofit sources, and spiritual and emotional support services. Physical elements of reconstruction included demolition and remediation, replacement and rebuilding of destroyed homes, repairing and renovating impacted homes that could be saved, elevating and lifting homes to be compliant with new FEMA flood insurance regulations, warehousing donations for impacted homeowners, and the local permitting processes associated with demolition and reconstruction. This network is depicted in Fig. 6.5.

---

[1] The Federal Emergency Management Agency continued to focus on financial recovery from Sandy as it does after every disaster (in conjunction with the Small Business Administration and the Department of Housing and Urban Development) but financial assistance from the federal government is administered by a byzantine structure of agencies, nonprofits, state and municipal governments (in the case of large cities like New York City), programs, and administrators, discussion of which is beyond the scope of this chapter.

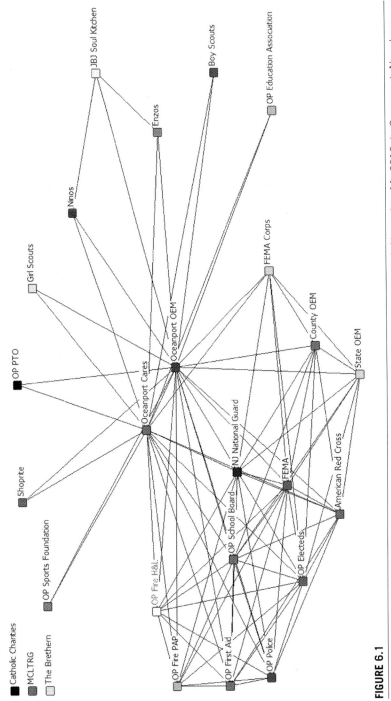

Catholic Charities

MCLTRG

The Brethern

JBJ Soul Kitchen

Enzos

Boy Scouts

OP Education Association

Ninos

Girl Scouts

FEMA Corps

Oceanport OEM

OP PTO

County OEM

Oceanport Cares

State OEM

Shoprite

NJ National Guard

OP Sports Foundation

FEMA

OP School Board

American Red Cross

OP Fire H&L

OP Electeds

OP Fire PAP

OP First Aid

OP Police

FIGURE 6.1

Times 1 and 2. Emergency response and short-term recovery stakeholder networks through November 11, 2012, in Oceanport, New Jersey.

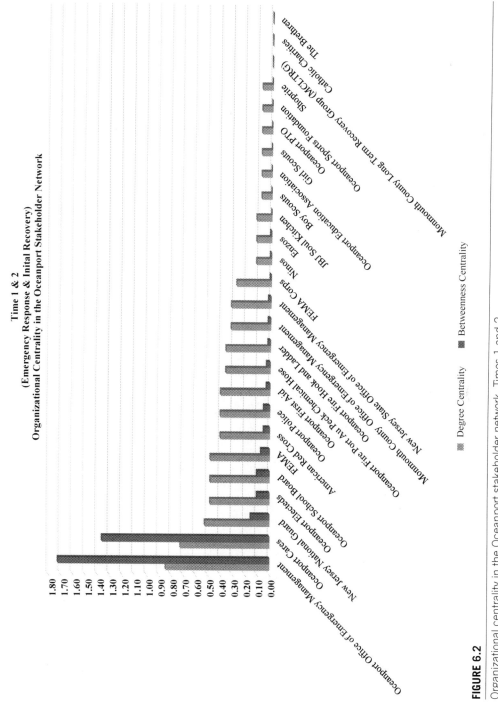

FIGURE 6.2

Organizational centrality in the Oceanport stakeholder network, Times 1 and 2.

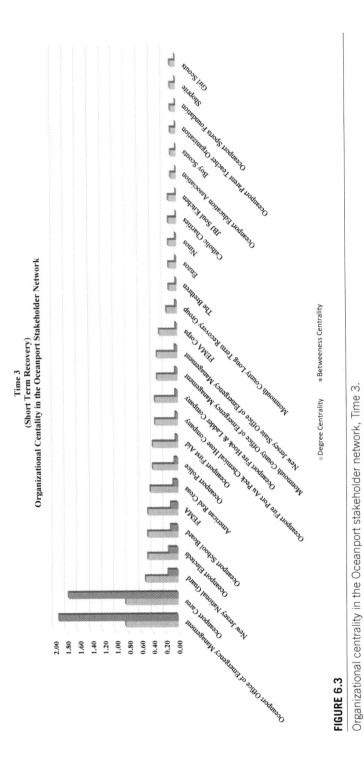

FIGURE 6.3

Organizational centrality in the Oceanport stakeholder network, Time 3.

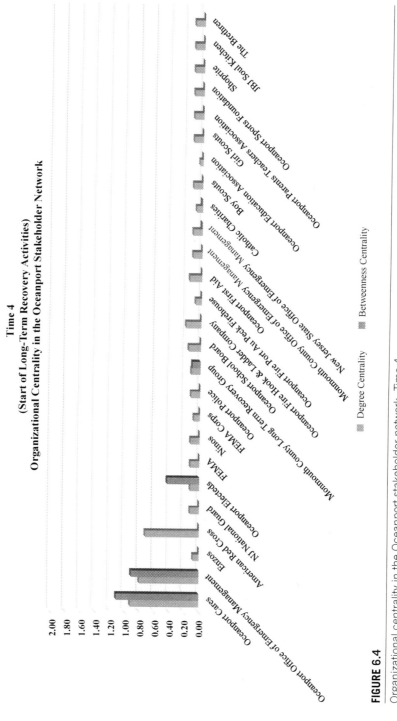

Time 4
(Start of Long-Term Recovery Activities)
Organizational Centrality in the Oceanport Stakeholder Network

■ Degree Centrality ■ Betweenness Centrality

FIGURE 6.4

Organizational centrality in the Oceanport stakeholder network, Time 4.

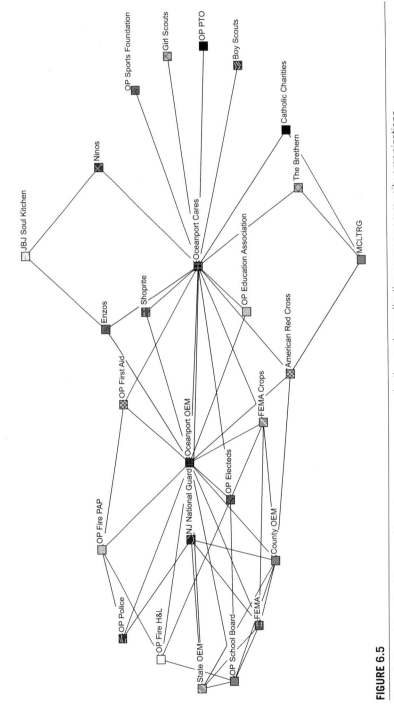

FIGURE 6.5

Time 3. Organizing short-term and long-term recovery: reorganization and coordination among community organizations.

At this point, regional organizations (MCLTRG and Catholic Charities) and national faith-based organizations (The Brethren, Lutheran Social Ministries, United Methodist Committee on Relief) became primary partners in recovery with local and other regional organizations. Resource deployment and management became more predictable and routinized at this phase of the recovery. As depicted in Fig. 6.6, the Oceanport Cares played a significant role in communication with nongovernmental organizations and distributed confusing and sometimes conflicting information from state and federal agencies and national nonprofits to local organizations and residents. Similar to Time 1, coordination activities were essentially split into two different networks, one comprising local and regional businesses and nonprofits coordinated by Oceanport Cares, and a second set of government agencies and elected officials coordinated by the Office of Emergency Management.

Oceanport's stakeholder network evolved in response to both exogenous events (a physical move in locations, the formalization of citizen volunteer efforts) and as organizations entered and exited the network. By Time 4, Oceanport's stakeholder network became smaller and more centralized as local organizations exited the network, relying on the Oceanport Office of Emergency Management and Oceanport Cares as information and resource brokers. Overall network centralization increased from 61% in Time 1 to 78% in Time 4. Centralization dipped to 59% in Time 3 as the network transitioned from emergency response and short-term recovery operations into long-term recovery activities. Other organizations also began to play more central roles as roles and tasks changed. The MCLTRG, organized in Time 3, became a broker for national organizations entering to engage in long-term recovery efforts. The mix of local organizations declined from 87% to 73% between Time 1 and Time 4 as

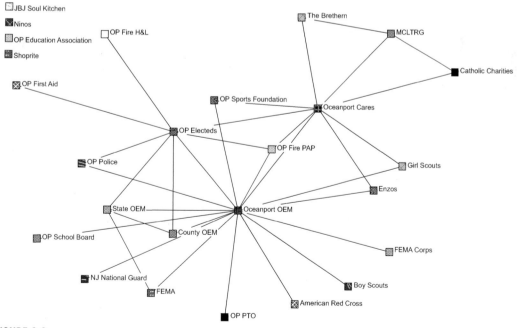

FIGURE 6.6

Time 4. Transitioning to long-term recovery: professionalizing the network.

national organizations increased their share of the network from 13% to 18% during the same time period. As the overall network shrank, betweenness centrality scores decreased for organizations engaged in the Oceanport stakeholder network.

NETWORK TRANSITIONS FROM SHORT-TERM TO LONG-TERM RECOVERY

Shifts that occurred in the network from Times 1 and 2 to Times 3 and 4 included the departure of the FEMA Corps and the New Jersey National Guard during Time 3 and the assumption of social support services, volunteer coordination, warehousing of food and supplies, and case management services by more formally organized nonprofit organizations. This move to more formal nonprofit interorganizational relationships included organizing efforts by local citizen volunteers who created formal local nonprofit organizations as part of their long-term recovery efforts. These activities created local nonprofits such as Oceanport Cares and Sea Bright Rising, which participated in the MCLTRG.

THE NATIONAL RESPONSE FRAMEWORK IN ACTION

Formal emergency response operations in the Oceanport stakeholder network were similar to those outlined in the National Response Framework. The National Response Framework developed by the Federal Emergency Management Agency calls for coordination of emergency and community resources through an incident command system (Department of Homeland Security, 2013a). In New Jersey, incident command after a disaster runs through the Office of Emergency Management, which is a nested set of command structures running from the state emergency operations center in the state's capital city, through county command centers, and into the local municipalities. Communication is two-way with local emergency management offices reporting their community's status, situation, and needs through the county emergency command center run by the director of the county office of emergency management, who then updates the state emergency command center. The state office of emergency management shares information about the current status of communities and emergency operations, and the on-the-ground situation, through the FEMA's incident commanders and command centers. As emergency response operations end, it is assumed by disaster planning frameworks and public officials that nonprofit and voluntary organizations will begin to fill key roles in impacted communities by locating temporary housing for residents who couldn't return home, providing food, clothing, and other assistance, and providing assistance with insurance claims, funding applications, overall case management, social support, and mental health services.

BUILDING INTERORGANIZATIONAL RESILIENCE THROUGH NETWORKED COLLABORATION

This case makes visible (1) the interaction of private, public, and plural sector organizations in response and recovery in one small coastal community, (2) the role of centrally located, local organizations in disaster response and recovery, and (3) the limitations of existing policy frameworks in fully accounting for the breadth of local organizational activities involved in disaster response and recovery.

By connecting local stakeholder networks with larger-scale regional and national response efforts, information and resource exchange occurred at the point of impact. Two local organizations, an existing formal government agency, and an emergent organization staffed by citizen volunteers were central to the network throughout all four phases of response and recovery. These two organizations brokered information between local, regional, and national organizations during the personal emergency, transitional, and recovery phases of the disaster.

CROSS-SECTOR COMMUNICATION AND THE PLURAL SECTOR

The networked activities of response in Oceanport after Superstorm Sandy reflect Henry Mintzberg's (2015) "plural sector," which is comprised of those organizations that are neither fully public nor fully private sector organizations. In Oceanport, this plural sector took shape as unaffiliated citizen volunteers organized in the days after the storm by creating a system of work routines and strategic interactions designed to assist the community's recovery. These daily work routines enacted a series of interorganizational relationships involving local businesses, local associations, local membership organizations, and local unions that were connected with, and ran parallel to, the larger institutional interorganizational response network. These activities connected organizations with different characteristics, missions, and identities into a singular stakeholder network centered on emergency response and long-term recovery. Citizen volunteer activities in conjunction with the Oceanport Office of Emergency Management balanced Oceanport's different organizational sectors in pursuit of a common goal. These private/public/plural sector partnerships were dynamic relationships socially constructed through stakeholder network members' collaborations. Interorganizational relationships reorganized over time as operations transitioned from emergency response to short-term recovery and then into long-term recovery activities. The Oceanport case is among the first empirical applications of the plural sector framework for community research in disaster-struck communities.

COORDINATING RESILIENCE THROUGH BROKERS

Within the post-Sandy Oceanport stakeholder network, the central role of two key organizations emerged by brokering two separate networks, a network of social support based on informal relationships comprised of strong, weak, and latent ties coordinated by the citizen volunteer group—which later became Oceanport Cares—and the other, an emergency response network based on formal relationships between the Oceanport Office of Emergency Management, regional, and national organizations. The Oceanport network was comprised of both strong and weak ties and is reflective of the overall National Response Framework and existing emergency response protocols in New Jersey. The Oceanport Office of Emergency Management also connected the two stakeholder networks into a larger community-wide social infrastructure that coordinated information flows and resource exchanges. Organizations with high degrees of centrality play a key role in information and resource exchange activities within a network and are able to influence communication patterns and partnering activities within the network. The Oceanport Office of Emergency Management, by virtue of its role and authority in the network, brought citizen volunteers, plural, and private sector organizations into the emergency response network to provide resources that were unavailable or inaccessible from larger regional and national organizations that were participating in the interorganizational response network.

Brokerage and coordination activities created an opportunity for a new local organization to emerge and created a new resource cluster much closer to the point of impact. While this new resource cluster may seem redundant, it actually played a key role in the resilience of the network and lessened the community's dependence on other parts of the network. Redundant resource clusters are created through multiple communication activities by centrally located, local organizations that broker information and resources within the stakeholder network in order to fill in the gaps within the larger interorganizational emergency response network. The emergence of these two centrally located organizations within the local stakeholder network in Oceanport adds to existing literature on the role of network redundancy and multiple network ties in survival, recovery, and social resilience after a disaster by demonstrating how community-level interorganizational relationships supplement formal emergency response and disaster recovery activities (Doerfel et al., 2010; Kapucu & Garayev, 2014; Kapucu & Hu, 2014; Kendra & Wachtendorf, 2003; Nowell & Steelman, 2013).

EXPANDING POLICY FRAMEWORKS TO ACCOUNT FOR LOCAL STAKEHOLDER NETWORKS

The Oceanport Office of Emergency Management played an important role in bringing together private, public, and plural sector organizations as would be expected under the National Response Framework. However, given their responsibility for leading local community response along with coordinating the influx of resources designated for the regional response, Oceanport Office of Emergency Management relied upon unaffiliated volunteers and existing organizations within the community to execute and operate a microshelter, community support activities, and to transition rescue and response operations into short- and long-term recovery activities. While the National Response Framework emphasizes the need for local emergency management offices to coordinate all resources and all types of organizations within a community, Oceanport Office of Emergency Management expanded this framework in a real-time, on-the-ground response to material conditions and interorganizational response efforts after Sandy.

   Despite Oceanport's role in the formal state and national response as a physical site for regional relief efforts, local needs were not met by national and state response agencies. Instead, relationships with other local government agencies were activated, and local businesses and organizations were accessed for food, clothes, and additional volunteers as the Oceanport operation expanded to include neighboring communities. Local organizations in an impacted region may play more significant roles during emergency response and short-term recovery than do national organizations, but these contributions may not be recognized within formal planning frameworks or in the actual operation of response and recovery efforts. While the national response framework calls for private, public, and plural sector partnerships (Department of Homeland Security, 2008, 2013a), in practice unaffiliated volunteers and new organizations are often seen as ancillary parts of the network and as not critically important to formal response and recovery operations (Drabek & McEntire, 2003; Helslott & Ruitenberg, 2004; Majchrzak et al., 2007). Findings from the Oceanport case, however, underscore the relevance of the local stakeholder environment and community resources. This case illustrates a community that coordinated private, public, and plural sector resources and volunteers in an informal social support network, providing redundant resources at the point of impact and in concert with the existing formal emergency response network. Such informal activities contribute to broader community recovery and social resilience and need to be accounted for in disaster planning and public policy.

IMPLICATIONS

Local organizations are located within stakeholder networks at the point of impact and despite being victims, themselves, are able to broker resources and information more efficiently due to their physical location within the community and to their organizational position within the network (Doerfel et al., 2010, 2013; Doerfel & Haseki, 2013). The Oceanport case points to the need for a new look at the National Response Framework to ensure that community stakeholder networks are supported and incorporated in broader planning and operational activities. Our research indicates that it is a mix of formal and informal relational ties that contributes to the activation and expansion of a stakeholder network in a disaster-struck community.

As seen in myriad news, blogging, and social media critiques of large relief organizations like the Red Cross and FEMA (Elliot & Sullivan, 2015; Feuer, 2012; Mayton & Kazem, 2015), there is a general perception that regional and national institutional response efforts should meet all community and organizational needs after a disaster and make communities whole, but that these formal institutional responses often fall short. This case suggests that it is integrated local-, regional-, and national-level interorganizational interactional activities that create new partnering opportunities. These emergent interorganizational relationships in turn contribute to processes of social resilience. In the case of Oceanport, we contend that it was the mobilization of the two networks (policy-based interorganizational response networks and local stakeholder networks) through two locally based central organizations brokering information flows that facilitated resilience processes. The findings in this study add to a growing call for an understanding of social networks in disaster by policy planners and emergency responders (Feuer, 2012; Department of Homeland Security, 2013b).

LIMITATIONS

This is a site-specific case study of a small, middle class community during the three-month period following Sandy's impact. Larger communities may have multiple, competing grassroots efforts, which may make it harder for emergency management efforts to coordinate them. Unlike other examinations of emergent volunteer efforts in disaster response (Drabek & McEntire, 2003; Helslott & Ruitenberg, 2004), Oceanport's emergency management office played a role in getting citizen volunteer efforts in Oceanport organized with adjoining communities and delegated responsibilities to these citizen volunteers. This is consistent with existing disaster response frameworks and protocols in the United States, but in stark contrast to parallel response efforts like Occupy Sandy (Department of Homeland Security, 2013b; Feuer, 2012) that organized and distributed information and resources outside of formal disaster institutions and networks.

CONCLUSIONS

The role of local stakeholder networks is a crucial component of community resilience. Community resilience in turn is dependent upon interorganizational resilience, which rests upon a strong foundation of social and organizational relations between private, public, and plural sector organizations. However, network composition changes over time and reflects the dynamics of the interorganizational response represented within these embedded networks. Organizations become active in postdisaster networks based on stakeholder needs and whether or not they hold resources that could meet stakeholder needs. However, resource holders are not exclusively outside organizations with a large resource base and clearly identified reputations. Resource brokerage, like in the case of Oceanport, can take place as different organizations and organizational populations connect with one another within a disaster-struck community. Finally, the role of the plural sector as a networked set of relationships among organizations with very different goals, missions, and reward structures is vital to resilience.

SHIFTING ATTENTION: MODELING FOLLOWER RELATIONSHIP DYNAMICS AMONG US EMERGENCY MANAGEMENT-RELATED ORGANIZATIONS DURING A COLORADO WILDFIRE

Zack W. Almquist[1,2], Emma S. Spiro[2], Carter T. Butts[3]

[1]*University of Minnesota, Minneapolis, MN, United States;* [2]*University of Washington, Seattle, WA, United States;*
[3]*University of California, Irvine, CA, United States*

CHAPTER OUTLINE

INTRODUCTION

Emergency management organizations rarely act in a vacuum: both governmental and nongovernmental organizations look to one another for topical information and practical guidance in both routine and nonroutine settings (Comfort et al., 2004; Drabek & McEntire, 2002). The most basic manifestation of this interaction is *attention*; that is, a systematic effort on the part of one organization to observe the actions of and/or to receive information from another. Attentional relationships may be particularly important as conduits for information diffusion during crisis events, and for the diffusion of organizational routines, practices, and standards during periods of routine operations (DiMaggio & Powell, 1983; Reeder et al., 2014). When visible to third parties, attentional relationships may also have a signaling function, indicating a form of affiliation between attender and attendee to third parties (Kwak et al., 2010). For example, if a federal agency begins to follow a local agency, then it may be perceived by the populace that the local agency has *federal* support or affiliation.

In contrast to realized communication relationships, attentional relationships reflect the *potential* for information transmission and hence are potentially useful for probing the structure of opportunities for interorganizational information flow net of sender behavior; likewise, they can indicate pathways through which information may flow to a receiving organization without a deliberate effort on the part of the sending organization to target it (e.g., by the attender observing actions taken or generalized announcements made by the attendee).

Despite their importance, attentional relationships have been historically difficult to study. This has been due in large part to the difficulty of measuring who is attending to whom, as retrospective surveys of organizational informants and archival materials can provide only limited evidence regarding attentional relationships. The expansion of organizational activities into the online domain provides a remarkable opportunity to study these otherwise elusive networks, due to the fact that certain computer-mediated communication systems record attentional relationships as a side effect of their operation. Records of relationships derived from these sources provide a unique window into the process by which organizations form and dissolve attentional relationships over time, at least within particular settings.

One such setting in which organizations directly and publicly articulate attentional relationships is Twitter, a popular microblogging service that has seen increasing emergency management presence and utilization during both routine conditions and crisis events. Prior work exploring the use of social media, specifically Twitter, during emergency contexts has primarily focused on its facilitation of rapid information dissemination and transmission, as well as its capacity to support collective sense-making processes and rumoring (e.g., Spiro et al., 2012; Sutton, 2010; Sutton et al., 2014). Other work has also looked at its affordances for digital volunteerism and situational awareness enhancement (e.g., Starbird & Palen, 2011; Vieweg et al., 2010). However, little of this work has investigated the activities of emergency response organizations themselves in this context (some exceptions being Hughes et al., 2014; Reeder et al., 2014). Even less research has sought to understand the evolving landscape of attentional relationships among organizations as they seek to access and provide information to each other and to the public at large (an early example being Sutton et al., 2012). To date, no work exists that examines online attentional relationships and resulting information flow among emergency management organizations over an extended period of time, and that systematically probes the mechanisms governing tie formation and dissolution.

We address this gap by employing a dynamic network logistic regression (DNR) modeling approach to uncover the mechanisms that govern the evolving follower (i.e., subscription) relationships among a set of United States (US) emergency management-related organizations (federal and state levels) on Twitter over an extended period. DNR allows us the ability to directly model the temporal relationship of historical interactions—e.g., preferential attachment—to understand and predict future interactions; for complete details see Almquist and Butts (2013, 2014b). Here, we relate features of organizations' temporally evolving structural positions within a social network to their public information exchange patterns and directly estimate the effect of disaster events on regional Twitter feeds. Our analysis provides a first look at the factors that drive the allocation of attention among emergency management organizations in the online domain.

Below, we review past work on the use of Twitter during crisis situations. We then discuss the data and methods employed in our analysis, followed by a summary of our empirical findings. Finally, we conclude with a discussion of the implications of this work for disaster management.

LITERATURE REVIEW

Disasters serve as extreme and dangerous disruptions of commonplace life. These disruptions can, among other things, inhibit the functioning of existing support networks, communication networks, and other infrastructures. This has led to a large and growing literature on the importance of social network theory and methods to the application of disaster research. This work includes the study of communication networks during disaster (e.g., Butts, 2008a; Smith & Simpson, 2009; Sutton et al., 2012, 2014); research into the effects and importance of social support during and after disasters (e.g., Jones et al., 2013; Mathbor, 2007); and this research area has further explored the importance and multifaceted effects of social networks on disaster relief and management (e.g., Hamra et al., 2012; Marcum et al., 2012). In this chapter we focus on the importance of attentional dynamics in a disaster setting by response organizations and explore how modern technology can enhance the current state of the art in research in this area of study.

Social media and the online environment have radically changed the ways in which public officials, organizations, and individuals engage in conveying warnings, alerts, and other information, and in coordinating task performance during both routine periods and disaster events (Kavanaugh et al., 2011; Palen, Vieweg, & Sutton, 2007). While it is true that traditional communications methods such as television and radio remain essential for engaging the public at large, these channels restrict communication to be largely broadcast in nature and are controlled by a small number of major private media outlets. Online communications methods, on the other hand, allow for more flexible dissemination strategies; these channels often allow for two-way communication in addition to relatively low-cost broadcast dissemination, give the sending organization direct control over message timing and content, and can be updated in near real time (Bruns et al., 2011). Moreover, in the case of social media, retransmission mechanisms are typically built directly into the communication infrastructure. These encourage individuals to repost content to other users, facilitating information diffusion via the underlying social network (Sutton et al., 2014, 2015a,b). As a result, social media and other Internet-based communication channels offer an increasingly attractive option for reaching at-risk populations before, during, and after emergencies or disaster related events (Sutton et al., 2012).

Emergency management organizations have recognized the potential of these new communication platforms and now actively use these tools during crisis events (Hughes et al., 2014, e.g., the Federal Emergency Management Agency (FEMA) follows or retweets a hazard event and then local agencies notice the signal and begin following or retweeting current messages). Despite this recognition, however, governmental emergency management organizations are still learning how to best make use of these new channels for crisis communication and disaster response. Research on the use of social media platforms during crisis has argued that most organizational engagement with social media to date can be viewed as falling into two broad categories: first, these platforms can be used to disseminate information; and second, online communication infrastructure can be used as a management tool itself (e.g., to receive victim requests for assistance) (Lindsay, 2011). This division omits a third role of social media platforms as tools for information *collection* and improved situational awareness (Mehrotra et al., 2004). The attentional side of social media use also extends to organizations employing feedback from users to alter their own communication patterns. In recent work documenting the 2012 Hurricane Sandy response online, for instance, researchers found that some emergency responders adapted their social media protocols over the course of the event—altering their behavior from a purely broadcast paradigm to acknowledging requests for help, while simultaneously trying to reinforce the use of official channels for aid requests (Hughes et al., 2014). As the Sandy case indicates, prior work suggests a mismatch between use of social media platforms by public officials and the expectations for their use held by the general public. Government command-and-control protocols rarely integrate seamlessly with social media, leading to legal barriers, insufficient use of resources, and lack of training by communicators—all of which can prevent emergency responders from effectively engaging with those in need via social media (Hughes et al., 2014).

Use of the microblogging platform Twitter exemplifies many of the advantages and limitations of using social media during disaster preparedness, response, and recovery. These tools can help facilitate information sharing at a scale and speed previously unattainable, enhancing dissemination of emergency information, early warning systems, and coordination of relief efforts (Kryvasheyeu et al., 2016). Eyewitness reports from disaster survivors are readily available, increasing situational awareness. However, such platforms also have important drawbacks. Misinformation and disinformation are prevalent; indeed, emergency responders are quick to point to this as one of the primary factors contributing to their reluctance to use social media as an information source (Hiltz et al., 2014). It can also be difficult to identify credible or relevant disaster-related content within the larger stream of posts.

Twitter has become a widely utilized platform by the general public as well as emergency responders during crisis contexts (Vieweg et al., 2010). As such, it has attracted the attention of scholars from a variety of disciplines interested in better understanding informal communication during crisis situations, as well as in the distributed coordination tasks that take place in these new venues. Sutton (2010), for example, demonstrated that the public utilizes social media to fill gaps that occur when official sources are slow or nonexistent. Others have similarly looked at these collective sense-making processes and the proliferation of rumors as individuals attempt to understand uncertain events as they unfold (Spiro et al., 2012; Sutton et al., 2013; Arif et al., 2016; Starbird et al., 2016). This research has provided good evidence that Twitter provides access to rapid exchange of up-to-date information about a given situation (Sutton et al., 2014); however, it is less clear on the role of emergency responders within the online information ecosystem.

There is a notable lack of empirical evidence about how and why government agencies use social media to communicate emergency-related information and whether this information is effective in reaching vulnerable, diverse populations (Hughes et al., 2014; Reeder et al., 2014). Even less work has studied how complex social relationship patterns evolve in varying disaster contexts within Twitter communities, especially in the context of US emergency management-related organizations. To fill this gap, and to further our understanding of the importance of communication and social relations (in this case governmental interaction) for information passing and gathering, we explore the dynamic social interactions of US emergency management-related organizations over a more than 100-day period. The rest of this chapter presents the data, social mechanisms, and analysis necessary to explore such phenomena.

MECHANISMS OF ATTENTIONAL INTERACTION DURING A DISASTER

We propose a series of basic social mechanisms for determining whether a given Twitter account is likely to follow another account. We consider a broad array of potential inertial terms and social inertial effects. An inertial model is particularly appealing in this context because following relationships on Twitter, like many social media platforms, have a relatively low initial cost—ties can typically be formed at the click of a button. Ties can become costly if social contacts post frequently, leading to subscriptions with high volume. In one sense, following relationships are more costly to remove because they require the user to go through the additional step of finding and deleting the subscription relationship. These properties suggest that ties will have strong inertial effects.

We propose that there will be some baseline probability of any two accounts having a relationship, and also a general tendency for a tie to persist over a given k days; we expect the former effect to be very small; that is, we expect ties to form at random at very low rates. Next, we propose a series of homophily driven mechanisms based on location (e.g., same state) and FEMA designation (e.g., whether they are a FEMA affiliate or not) to account for similarity between organizations on such features, allowing us to explore the extent to which similar organizations are more likely to be tied. Further, we expected that these networks would be governed by dependence mechanisms associated with key network features.

In this chapter, we propose nine core network-based hypotheses, focusing on the mechanisms of an attentional nature that we parameterize in our model. First, we posit that organizations' attentional relationships do not change instantaneously but are instead subject to "inertia" (Almquist & Butts, 2013, 2014b). Next, we hypothesize that organizations will be more likely to attend to others that are similarly situated institutionally (McPherson, et al., 2001). Specifically, we focus on the "homophily" effects of node match on same state, and node match for being a FEMA subunit. Organizations that follow "shared partners" are posited to come to follow one another. This mechanism is captured by an effect that, for focal pair (i,j), counts the number of organizations k such that (i,k), (j,k) at the specified time lag (see Butts, 2008a). In addition to the shared partner effect, we posit that attentional networks will be driven toward transitive closure (Wasserman & Faust, 1994); if, during a given time point, organization i follows organization j and organization j follows organization k, then i will be more likely to initiate or persist following k at a future time point. Organizations that are "active"—by following many others—or that are "popular"—by being followed by many others—are likely to become especially salient targets for attention. This has been previously found to be important for dynamic networks in the online context by Almquist and Butts (2013). Since following is to some extent a public statement of importance—and since one's own followers are especially salient targets for attention—we expect organizations' follower ties to be biased in the direction of past reciprocity (Almquist & Butts, 2013; Wasserman & Faust, 1994). Last, we hypothesize that disasters in a given region result in increased saliency of local Twitter accounts (motivated by Sutton et al., 2008).

DATA

The data used here consists of a large, dynamic network of following (i.e., subscription) relationships among a set of US government emergency management-related organizations (federal and state levels) obtained from Twitter. Network data were collected daily from June 24, 2010 to February 27, 2012 by the Hazards, Emergency Response, and Online Informal Communication (HEROIC) Project (Butts et al., 2011).[1]

Building a complete set of all government, emergency-related Twitter accounts is a difficult task. Twitter is an ever-changing environment and, further, lacks a centralized database of such organizational accounts. Project HEROIC researchers (Butts et al., 2011) identified and enumerated a set of 213 actors over this time period by searching all known state and federal emergency management Web pages and checking to see if the page contained a link to a Twitter account. These targeted accounts were identified because they represent the population of public officials at the state and federal levels who were serving in a public-safety capacity and who were actively advertising their Twitter account. To find these accounts required a researcher to review all known Websites and check for a link to a given Twitter account (for example, see Fig. 7.1).

Once Twitter handles (or user names) were identified, the Twitter REST application programming interface (API) was used to collect information about which organizations were "following" each other on the platform. Following relationships on Twitter indicate that one actor subscribes to another's messages; messages (i.e., tweets) are automatically delivered to an actor's followers. Following thus represents a directly measurable and publicly visible attentional relationship. These social ties were sampled daily over the observation period; however, due to restrictions in data access there are missing links over some of the period. Here, we utilize a 125-day period from April 28, 2011 to August 30, 2011, where data is complete for the collection of the network.

THE DYNAMIC NETWORK

This interorganizational network is comprised of 213 nodes, each representing one Twitter account. The nodes are comprised of Twitter accounts for organizations based in all 50 states, Washington DC, and Puerto Rico. Many accounts represent state entities, but regional and federal-level organizations are also present. For example, there are 13 accounts that are part of the federal organizational entity;

---

[1] http://heroicproject.org/.

FIGURE 7.1

Image of the Federal Emergency Management Agency (FEMA) homepage with link to the FEMA handle highlighted.

these correspond to the eight FEMA regions, the director of FEMA, the FEMA regional office in Louisiana, and a general FEMA account (for a full tabulation by state, see Table 7.1.). For this social network the average density over the period covered is 0.0574, indicating that the data is relatively sparse in terms of observed social ties. Put another way, each of the 213 accounts attends to (i.e., follows) an average of approximately 13 other accounts at any given time.[2]

To provide a sense of the global structure of the follower network, Fig. 7.2 shows the set of all organizations present within the sample, with adjacencies determined by the average duration of connection over time (i.e., the fraction of time points for which a tie is present). The network appears to have a strong core-periphery structure, which we verified by fitting a confirmatory block model (Wasserman & Faust, 1994) to the dichotomized data (dichotomizing ties at the median strength). For a directed graph, there are four canonical core/periphery models: a mutual core with an effectively isolated periphery; a mutual core with a receiving periphery; a mutual core with a sending

---

[2]For comparison, Twitter (http://www.twitter.com) reports the average number of followers per Twitter user as 208 in December 2015. Others have likewise explored the degree to which politicians follow each other (e.g., see http://nymag.com/daily/intelligencer/2013/08/who-do-members-of-congress-follow-on-twitter.html).

Table 7.1 Number of Twitter Accounts by State or Federal Affiliation

State	Freq	State	Freq	State	Freq	State	Freq	State	Freq	State	Freq
AK	4	FL	4	MA	6	NE	3	RI	4	WI	3
AL	6	GA	5	MD	4	NH	2	SC	4	WV	2
AR	3	HI	2	ME	5	NJ	4	SD	1	WY	1
AZ	2	IA	4	MI	5	NM	1	TN	3		
CA	7	ID	3	MN	5	NY	9	TX	4		
CO	10	IL	2	MO	3	OH	6	UT	2		
CT	4	IN	3	MS	2	OK	3	VA	6		
DC	3	KS	6	MT	2	OR	3	VI	3		
DE	5	KY	4	NC	2	PA	1	VT	1		
Federal	22	LA	5	ND	2	PR	1	WA	6		

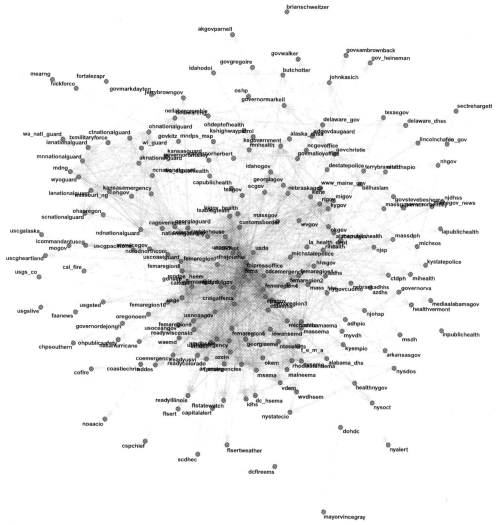

FIGURE 7.2

Time-averaged network of attentional ties for all 213 United States emergency management-related organizations. Edges are shaded based on attention level over the observed time period (darker ties indicate longer duration following).

periphery; and an "expansive" core with no effective periphery. Here we find that the best-fitting block model is one where there is very little interaction between the core and periphery. The organizations found to be in the resulting core can be found in Table 7.2, sorted into four subgroups by affiliation. We see that the federal agencies (FEMA, Centers for Disease Control and Prevention, Department of Homeland Security, etc.) are among the most important players in this network,

Table 7.2 The Core Members of the Twitter Follower Network, Categorized by Organizational Affiliation (FEMA, National Guard, State and Federal Agencies)

FEMA	National Guard	State	Federal
craigatfema	georgiaguard	alabamaema	cdcemergency
fema	ndnationalguard	ar_emergencies	customsborder
femaregion1	uscoastguard	azein	dhsjournal
femaregion2		calema	fbipressoffice
femaregion3		emdsc	readydotgov
femaregion4		georgiaema	statedept
femaregion5		gohsep	tsablogteam
femaregion6		Mass_hhs	usagov
femaregion7		massema	usgs
femaregion8		massgov	usnoaagov
femaregion9		michemhs	usoceangov
femaregion10		mndps_hsem	whitehouse
		msema	
		nmdhsem	
		nysemo	
		okem	
		readycolorado	
		t_e_m_a	
		utahemergency	

although we do see a number of very active state accounts (e.g., readycolorado). We further note that there are a few organizations that are very active in general; for example, *fema* has the highest degree across all time points.

This broad core-periphery structure reflects the long-term average behavior of a complex dynamic process within which organizations make decisions to start or cease following other organizations within their field. To better understand the mechanisms governing such attentional dynamics in a disaster setting, we model the time evolution of the Twitter follower network during the period of a major event—the Colorado Duckett fire. In addition to allowing us to capture general factors affecting attentional dynamics, this case also allows us to contrast effects associated with Colorado-based Twitter accounts, as compared to all other accounts in the network.

DISASTER CASE STUDY: COLORADO DUCKETT FIRE

Wildfires are of major concern to many regions within the United States. Major forest fires can cause more than a billion dollars in damage and burn more than 100,000 acres of land during any one event (Fischetti, 2011). These fires occur most often during the summer months, as a result of dry weather conditions, and often require federal assistance for management and recovery. In June 2011, a fire broke out in southern Colorado in the San Isabel National Forest. FEMA declared this fire a "disaster" from June 15, 2011, to June 24, 2011, and allocated federal funds to the recovery effort.[3] During the event, an estimated 650 individuals were required to help fight the fire. Approximately 130 homes in Custer and Fremont counties were evacuated (a photograph from the fire can be seen in Fig. 7.4). Overall damages incurred as a result of this fire are estimated to cost at least 4 million dollars (Associated Press, 2011). The Colorado Duckett fire was a significant regional event, one we explore in our analysis of interorganizational interaction. In Fig. 7.3, we show the location of Colorado-based emergency management Twitter accounts in relation to the location of this event.

METHODS

To analyze the evolving nature of attentional relationships among US emergency management organizations on Twitter, we employ DNR (for complete details, see Almquist & Butts, 2013, 2014b). DNR is a simple, scalable special case of the family of temporal exponential family random graph models and is a useful and robust starting point for dynamic network modeling. We employ standard network notation and structural definitions in describing our modeling framework, to clarify the connection between the formal techniques employed and the substantive phenomena they represent. This section begins first with this basic notation and then reviews some important details of DNR, ending with a discussion of the computational methods employed to apply DNR to the Twitter data set.

NOTATION

Social networks are often represented as *graphs*, mathematical objects that are defined by two sets: a vertex set V (e.g., organizations) and edge set E (e.g., collaboration). A graph may be represented as $G = (V,E)$. In graphs for which no meaningful distinction can be made between senders and receivers of edges (*undirected* graphs), edges correspond to unordered vertex pairs (i.e., $\{i,j\}$, for vertices i,j in V); in cases where the sending and receipt of an edge are distinct (*directed* graphs or *digraphs*), edges correspond to ordered vertex pairs (i.e., (i,j)). The Twitter follower network studied here is most naturally represented as a *simple digraph*, i.e., a graph in which (1) edges are directed, (2) no vertex may send an edge to itself, and (3) no vertex can send more than one edge to another vertex at any given time. This representation follows immediately from the way that follower operations operate: following is distinct from being followed (Friended, in the language of Twitter API); accounts cannot meaningfully follow themselves; and each account either does or does not follow another at a given time (i.e., one account cannot partially follow another, and obviously one cannot follow the same account multiple times in the same instant).

Although the set theoretic formulation described herein is often used in theoretical development, it is more common to use equivalent matrix-based representations in statistical settings (see Wasserman & Faust, 1994). The common matrix form used here is referred to as an *adjacency matrix*. An adjacency matrix, Y, is an N by N matrix (where N is the number of entities, e.g., individuals or organizations, in the network) such that each Y_{ij} cell is either a "1" (if i sends a tie to j) or a "0" (if i does not send a tie to j). This can be extended to the temporal case with the addition of a t index to indicate time period (e.g., days),

---

[3]FEMA disaster declaration dates, see https://www.fema.gov/disaster/2923.

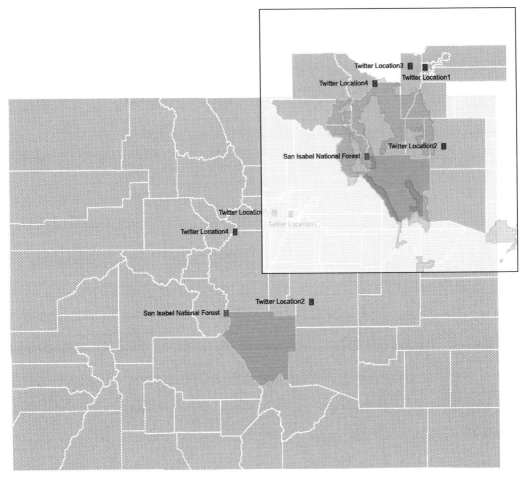

FIGURE 7.3

Colorado state with county borders projected in latitude/longitude space. *Blue dots* represent Twitter handle geocodes and the *red dot* represents San Isabel National Forest location, which was the start location of the Duckett Fire. The affected counties (Fremont and Custer) are labeled in *blue*, and the national forest boundaries are shaded in *red*.

FIGURE 7.4

Type 1 helicopter flying over Duckett fire captured by the US Forest Service and posted on June 15, 2011 10:38 p.m. (http://inciweb.nwcg.gov/incident/photograph/2306/17/18449/).

i.e., Y_t encodes the state of the network at time t, with $Y_{tij}=1$ if i sends a tie to j at time t (otherwise $Y_{tij}=0$). To make this concrete, consider a collaborative tie between two organizations existing on day 1 and dissolving on day 2, this would correspond to $Y_{112}=1$ (on day 1) and $Y_{212}=0$ (on day 2).

DYNAMIC NETWORK REGRESSION

We begin with a series of cross-sectional temporal networks $(\ldots,Y_{y-1},Y_t,Y_{t+1},\ldots$ with $Y_t \in 0, 1^{N_t \times N_t}$, where N_t is the number of nodes at time t) and look to model Y_t conditioned on the past, $Y_{t-1}^{t-k}=Y_{t-1},\ldots,Y_{t-k}$ (Almquist & Butts, 2014b). The most general form of these models is the so-called temporal exponential random graph model family (Hanneke et al., 2010). Although the general temporal exponential family random graph models—as they are referred to in the literature—are extremely flexible, this flexibility comes at a cost; if not careful, the analyst can easily create models with very unrealistic behavior (for a recent review in the exponential family random graph models context, see Schweinberger, 2011). However, much work has demonstrated that while these challenges exist in the general temporal exponential family random graph models case, one can under certain regularity conditions stabilize the models through the temporal structure (see Cranmer & Desmarais, 2011; Hanneke & Xing, 2007; Hanneke et al., 2010). The best known of these frameworks is DNR. Almquist and Butts (2013, 2014b) have shown that this framework is predictively powerful and provides intuitive and interpretable results as either a logistic choice process or as a behavioral model. DNR is so named because it can be rewritten as a logistic regression, with edge states at each point in time as the dependent variables; it therefore inherits the usual inferential and computational properties of logistic regression, and DNR model coefficients can be interpreted in a manner analogous to logistic regression parameters. Formally, the DNR likelihood for a given time slice is

$$Pr\left(Y_t \mid Y_{t-1}^{t-k}, X_t\right) = \prod_{(i,j) \in V_t \times V_t} B\left(Y_{ijt} \mid \log it^{-1}\left(\theta^T u\left(i,j,Y_{i-1}^{i-k},X_t\right)\right)\right), \qquad (7.1)$$

where B is the Bernoulli pmf, X is a covariate set, u are sufficient statistics for the edge set, and θ an edge parameter vector. Due to the (temporal) causal structure of the model, the joint likelihood of an entire time series is just the product of likelihoods for each time slice (Eq. (7.1)) conditional on the ones before it. As noted before, the DNR model family is equivalent to a logistic regression with lagged predictors, allowing for a conditional odds interpretation of the results. Because the DNR family is also a special case of the temporal exponential family random graph models, any general result or technique usable on the latter is applicable to the former. Bayesian extensions of this model have been explored by Almquist and Butts (2014a) and are generally advised.[4]

MECHANISMS OF ATTENTIONAL DYNAMICS

We operationalize each of the mechanisms in section "Mechanisms of Attentional Interaction During a Disaster" by translating them into lagged graph statistics for the DNR model. Further, we engage in model selection as suggested by Almquist and Butts (2014b), where we will build a series of models from simplest to most complex and use this method as an initial stage of hypothesis testing—if a parameter is not included in the criterion-selected model, then we take this as evidence that the

---

[4]All models used here were computed in the R statistical computing environment (R Core Team, 2015) based on the original code used in Almquist and Butts (2013, 2014b) and updated in Yang and Almquist (2015) using custom code that builds on sna (Butts, 2008b), ergm (Hunter et al., 2008), and arm (Gelman et al., 2009) packages.

associated mechanism does not demonstrably contribute to explaining interaction dynamics (net of other effects). We employ the Bayesian information criterion (BIC), a popular model selection metric for exponential family models (Kass & Wasserman, 1995), for model selection. Each mechanism is discussed substantively in the section "Mechanisms of Attentional Interaction During a Disaster," and a description for how each mechanism is parameterized and necessary baseline parameters included in the model follows in the next section.

Base Rate of Attending

We parameterize the model with an intercept or edge effect that captures the baseline proclivity of accounts to follow others (Wasserman & Faust, 1994); this baseline is then modified by other factors to determine the actualized rate at which organizations attend to one another.

Inertia

To evaluate inertia in follower behavior, we parameterize the model with 1–7 (daily) lag terms to incorporate differential hazards of tie dissolution over a week-long period (Almquist & Butts, 2013, 2014b).

Homophily

We parameterize the model with two homophily effects: (1) node match on same state, and (2) node match for being a FEMA subunit.

Shared Partners

Shared partners are captured by an effect that, for focal pair (i,j), counts the number of organizations k such that (i,k), (j,k) at the specified time lag. Where the associated parameter is positive, organizations following the same third-party organizations have an enhanced probability of initiating or sustaining following relations with each other.

Transitivity

We parameterize transitivity by employing, for each (i,j) pair, the number of (i,k), (k,j) two-paths at the specified lag as a predictor for the i,j edge variable.

Popularity and Activity

We capture popularity and activity effects by respectively adding model terms for lagged indegree (followers) and outdegree (others followed). Specifically, the popularity effect takes the lagged indegree of j as a predictor for the (i,j) edge variable, while the activity effect takes the lagged outdegree of j as the analogous predictor.

Reciprocity

We parameterize reciprocity as lagged mutuality term, i.e., a single statistic for the number of times an organization i is in a mutual relation with an organization j for a given lag. We expect the number of past reciprocal relations to increase the likelihood of a tie in the future with a given organization. Further, we expect this to be magnified if the organization has been in mutual relationship over multiple lag periods.

Seasonality

Organizations are not expected to be equally active at all times of the week; past network studies have found strong seasonal components in online activity and turnover rate in tie formation (Almquist & Butts, 2013; Butts & Cross, 2009), and we expect that here as well. We parameterize seasonality via a simple indicator for weekend versus weekday and observe the change in relationship to lag terms under consideration.

Hazard Event Effects

To test the hypothesis that disasters in a given region result in an increased degree of saliency of local Twitter accounts, we parameterize the model with an indicator for Colorado Twitter accounts and the Duckett fire period as defined by FEMA.

RESULTS

To test the hypotheses outlined in section "Mechanisms of Attentional Interaction During a Disaster", we begin with a parsimonious model with fundamental controls: a term for density; two terms for homophily; a seasonality effect; and an inertial term for a single lag. We then steadily add inertial effects up to an entire week. Our decision theoretic framework (Bayesian information criterion)[5] selected the 7-lag model, we begin with this case and then follow up by adding in the next set of network features. We then add Mutuality, Popularity (indegree) and Outdegree, Shared Partner, and finally Transitivity effects. At this stage we add an effect for Duckett fire (CO Fire; the period effect for Colorado Twitter accounts). Here again we employ, as discussed earlier, the Bayesian information criterion method for selecting whether to include a parameter in the model or not. This procedure can also be considered as a first stage of hypothesis testing, allowing us to directly test whether each proposed mechanism adds sufficient explanatory power to justify inclusion. Based on the Bayesian information criterion metric, we find that all the hypothesized mechanisms for interaction appear to be present. The final model can be seen in Table 7.3. As a point of reference, we also designate coefficients as "significant" in a Bayesian sense if the central 95% posterior interval about the estimate does not include 0— all estimates are posterior modes under independent standard Cauchy parameter priors.

INTERPRETATION

Our analysis focuses on the complete model, since it has the lowest Bayesian information criterion value, and thus was chosen under our criterion as the best fitting model (see Table 7.3). We find that at baseline there is very low probability of interaction, as we would expect in the dynamic network regression setting (see Almquist & Butts, 2014b, for details). There is very strong evidence of the effect of inertia in this system, as this term was strong in all models (i.e., large and significant). In particular, in the model under consideration, the lag term is quite important. Consider an i,j dyad with no lag effect, i.e., organization i did not interact with organization j in the past seven days, then we would expect the log-odds of a tie (when considering only the intercept and lag effects) to be large and negative. However,

---

[5]The Bayesian information criterion avoids overfitting by introducing a penalty term for the number of parameters in the model, such that if the model is not sufficiently improved, we reject the addition of the parameter.

Table 7.3 Dynamic Network Logistic Regression Table of Bayesian Posterior Mode Estimates and Parameter Standard Deviations (Under a Cauchy Prior) for the Best Fit Model Under Bayesian Information Criterion Decision Criterion for Modeling the Attentional Network

	Parameters	SD		Parameters	SD
δ	−6.6860*	0.0548	–	–	–
FEMA	−0.0201	0.0382	–	–	–
STATE	0.0623	0.0772	–	–	–
I{Weekend}	0.7646*	0.0356	–	–	–
I{COFire}	1.4819*	0.1042	–	–	–
$\text{Trans}(Y_{t-1})$	−0.0237*	0.0041	$\text{SP}(Y_{t-1})$	−0.1237*	0.0156
$\text{Trans}(Y_{t-2})$	0.0010	0.0016	$\text{SP}(Y_{t-2})$	−0.1195*	0.0153
$\text{Trans}(Y_{t-3})$	0.0569*	0.0106	$\text{SP}(Y_{t-3})$	0.1159	0.1707
$\text{Trans}(Y_{t-4})$	0.0696*	0.0104	$\text{SP}(Y_{t-4})$	0.0207*	0.0064
$\text{Trans}(Y_{t-5})$	0.0897	0.1190	$\text{SP}(Y_{t-5})$	−0.0036	0.0025
$\text{Trans}(Y_{t-6})$	0.0241*	0.0060	$\text{SP}(Y_{t-6})$	−0.0047	0.0166
$\text{Trans}(Y_{t-7})$	−0.0002	0.0023	$\text{SP}(Y_{t-7})$	−0.0072	0.0160
$\text{InDeg}(Y_{t-1})$	−0.3907*	0.1837	$\text{OutDeg}(Y_{t-1})$	−0.0018	0.0023
$\text{InDeg}(Y_{t-2})$	−0.0406*	0.0055	$\text{OutDeg}(Y_{t-2})$	−0.1985*	0.0159
$\text{InDeg}(Y_{t-3})$	0.0034	0.0021	$\text{OutDeg}(Y_{t-3})$	−0.1924*	0.0152
$\text{InDeg}(Y_{t-4})$	0.1087*	0.0139	$\text{OutDeg}(Y_{t-4})$	−0.4868*	0.1728
$\text{InDeg}(Y_{t-5})$	0.0967*	0.0135	$\text{OutDeg}(Y_{t-5})$	0.0363*	0.0058
$\text{InDeg}(Y_{t-6})$	0.3616*	0.1557	$\text{OutDeg}(Y_{t-6})$	−0.0044	0.0023
$\text{InDeg}(Y_{t-7})$	0.0606*	0.0060	$\text{OutDeg}(Y_{t-7})$	−0.1297*	0.0154
$\text{Mut}(Y_{t-1})$	−0.0897*	0.0145	Y_{t-1}	2.8834*	0.0589
$\text{Mut}(Y_{t-2})$	−0.0595	0.1654	Y_{t-2}	−1.0183*	0.1187
$\text{Mut}(Y_{t-3})$	−0.0338*	0.0038	Y_{t-3}	1.5953*	0.1174
$\text{Mut}(Y_{t-4})$	−0.0028	0.0016	Y_{t-4}	3.5918*	0.0761
$\text{Mut}(Y_{t-5})$	0.1021*	0.0107	Y_{t-5}	−2.6662*	0.0926
$\text{Mut}(Y_{t-6})$	0.0677*	0.0102	Y_{t-6}	3.2747*	0.0771
$\text{Mut}(Y_{t-7})$	0.2947*	0.1168	Y_{t-7}	6.0267*	0.0475
BIC	71,312.96				

A Bayesian analog of "significance" at $p < .05$ based on 95% central posterior intervals is denoted with ''.*

if the organization i interacted with organization j for the entire time period, we would have a log-odds of forming an attentional relationship, which is large and positive (a massive increase). Note that if organization i interacted with organization j only in the immediate past (e.g., the first lag), it would still increase the likelihood of interaction on a log-odds scale substantially—in this case a nearly twentyfold increase in terms of the odds of a tie.

We find little evidence of a strong state-based or FEMA-based homophily effect, neither being significant. There is some evidence of a moderate lagged effect of transitivity, with 2-paths at 3–6 days having a positive effect on following and recent 2-paths (e.g., lag 1) having a negative effect; taken together, the net impact of these terms is for long-running 2-paths to engender transitive closure while ephemeral ones inhibit it, suggesting a substantive difference between enduring and transient targets of attention. Shared partners show a similar qualitative pattern, though the negative short-term effects are dominant; thus, organizations tend to avoid following others that are attending to the same third parties, all other things being equal. Mutuality behaves like 2-path embeddedness, with long-running mutual relationships tending to enhance following and ephemeral mutuality tending to promote relationship decay. We also see a similar pattern for indegree, with negative effects at short lags and positive effects at longer lags whose total magnitude is larger than the negative short-lag effects. In this case, the interpretation is that organizations that are transiently popular (e.g., for less than approximately one week) tend to lose followers, while those with enduring popularity tend to gain them. Outdegree tends to be more consistently negative, suggesting that, ceteris paribus, organizations that follow many others tend to receive fewer followers themselves. (It should be noted that this effect, integrated across lags, is stronger than mutuality, so following many others appears to be a losing proposition overall.)

We note that while many of these coefficients appear to be small in magnitude, their net effects can become large for organizations that are embedded in dense groups. The seasonality term for weekend days (versus weekdays) is large and positive, disconfirming our hypothesis that these Twitter accounts would be largely focused on the standard work week. On the contrary, it would seem that substantially more attentional activity occurs on weekends (perhaps suggesting that account maintenance is being offloaded to nonstandard work hours during this period).

Finally, there is clear and strong evidence that these Twitter accounts become highly active in attending to others when a disaster occurs in their local region. Specifically, the COFire effect is very large, positive, and significant. We can see this effect visually by looking at the induced egocentric network (Almquist, 2012; Wasserman & Faust, 1994) for just the Colorado-based Twitter accounts on the day FEMA "Fire Management Assistance Declaration" was declared.[6] This occurs on June 16, 2011, where the Colorado subgraph is composed entirely of isolates, and on June 17, 2011, it is almost a completely connected graph (Fig. 7.5). These Twitter accounts, furthermore, are active with a number of other accounts in the network during this period. We visualize the strength of this effect with Fig. 7.6, which traces the incoming and outgoing ties for the CO Twitter accounts and the rest of the network compared to the mean degree within the CO Twitter account group.

DISCUSSION

The shifting attentional relationships among US emergency management–related organizations in the context of Twitter demonstrates a number of basic social mechanisms and activities that individuals and organizations employ to improve both information flow and situational awareness. We see that as emergent activities occur, the attentional dynamics of the system shift so that the organizations in the immediate impact area become more focal and engaged both internally and externally. These attentional shifts take place against a background of dynamics that overall favors switching ties away from more active organizations, and those with whom one has shared partners; that is, we often see that local organizations step into the spotlight during crisis events, while large, national organizations do not.

---

[6]https://www.fema.gov/disaster/2923/designated-areas.

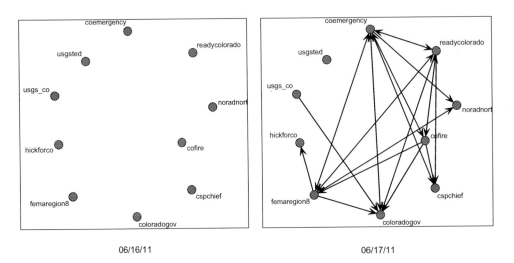

06/16/11   06/17/11

FIGURE 7.5

Induced attentional network of Colorado-based US emergency management–related organizations over June 16 and June 17, 2011. Note the clear surge in attentional relationships associated with the onset of the disaster.

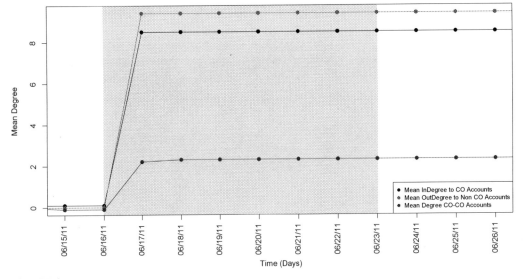

FIGURE 7.6

Mean degree within CO Twitter accounts and mean in- and outdegree of the CO Twitter account groups. Federal Emergency Management Agency (FEMA) disaster declaration dates for the CO fire period is highlighted in light red.

This behavior is rather different from what has been found in other contexts, where social dynamics tend to preserve existing group and positional characteristics. We do see an overall tendency toward inertia within particular attentional relationships, however, as well as toward reciprocity for relationships that have been in place for an extended period. Indeed, there is a clear asymmetry (manifesting via several distinct mechanisms) between long-lasting and ephemeral structures, with dyadic embeddedness in the former being more conducive of tie formation and the latter being inhibitory. The organizations in this population thus appear to treat ongoing attentional relationships differently from short-duration relationships, with the former being inertial and the latter being actively cycled. We conjecture that in the case of the Colorado Duckett fire, organizations were employing these tactics to maximize their situational awareness under constraints of time and focus of a given organization.

Our findings suggest that having a large number of followers over an extended period increases the likelihood of being the object of attention, while attending to many others makes one appear unimportant and decreases the attention paid to that account. This result could mean that followers represent a strong signal of being worthy of attention, and following a large number of other organizations sends a strong signal of being "out of the loop." However, it appears that a major disruption to the system (e.g., a disaster) can alter these relations significantly, leading to a rapid reorganization of the attentional network.

Incorporating attentional dynamics into models of informal communication online is vital for understanding social processes that occur via these social pathways. Processes such as collective sense-making and organizational learning all occur in these online platforms and can be impacted by attentional dynamics. The results presented here build on prior studies of rumoring behavior—natural social processes likely to occur in disaster contexts—and demonstrate how attentional dynamics can structure the transmission of information in these settings. For example, our analysis reveals strong inertia within the system—inertia that can be overcome but at the cost of highly disruptive shocks (i.e., major disaster events). If attentional relationships become "stuck" in preevent states, they may be unable to adapt to the highly dynamic response environment; as a result, emergency responders may lose critical opportunities to send and receive event-related information, increase situational awareness, and clearly communicate event-related information to the general public. In extreme cases, this may imply that attentional relationships are determined at the time the account was created and rarely adapt to changing circumstances. In both cases, response and management of the event could be impacted.

SUMMARY

This chapter has provided a first look at the shifting attentional network of US emergency management–related organizations in the context of Twitter. Using statistical network models, we have been able to test for the interaction of multiple distinct mechanisms driving social interaction. Our findings show a fairly complex pattern of historical dependence in attentional ties, with effects that vary substantially over a 7-day lag period. In addition to an unusual tendency toward switching away from especially active nodes, we find a moderate tendency for triadic closure between organizations and a modest hierarchy effect. Ties that endure over a 5–7 day period have different effects than those that are ephemeral, the former generally being inertia and closure enhancing and the latter being inhibitory. Further, we find that Twitter activity and engagement are increased dramatically for the organizations directly affected by disaster, at least in the case of the Colorado Duckett fire. The ability to probe these complex mechanisms while also controlling for the influence of external events highlights the potential of tools like DNR for understanding network dynamics during unfolding events. We believe that there is considerable potential for further development in this area, both for attentional networks and for other sorts of interorganizational interactions.

ACKNOWLEDGMENTS

This work was supported in part by Army Research Office awards #W911NF-14-1-0577 (YIP), #W911NF-15-1-0270 (YIP), and #W911NF-14-1-0552, and by National Science Foundation awards CMMI-1031853, CMMI-1031779, and CMMI-1536319.

THE EFFECT OF HURRICANE IKE ON PERSONAL NETWORK TIE ACTIVATION AS RESPONSE AND RECOVERY UNFOLDED

Christopher Steven Marcum[1], Anna V. Wilkinson[2], Laura M. Koehly[1]

[1]*National Institutes of Health, Bethesda, MD, United States;* [2]*University of Texas Health Science Center at Houston, Houston, TX, United States*

CHAPTER OUTLINE

INTRODUCTION

It is well known that disasters pose opportune settings to study the ways in which social structure is affected by disruption to normative processes that unfold over time (Drabek, 1986). For an affected population, those shocks may reverberate through channels along social networks as individuals respond to their sudden need for assistance by reaching out and activating ties that are normally held dormant (Smith & McCarty, 1996). While past research on population responses to disasters has revealed important findings on displacement patterns (Drabek & Boggs, 1968; Gray & Mueller, 2012), changes in demographic composition of affected areas (Donner & Rodríguez, 2008; Smith & McCarty, 1996), the timing of evacuations (Huang, Lindell, Prater, Wu, & Siebeneck, 2012; Ng, Diaz, & Behr, 2015), and adaptation through social support (Norris & Kaniasty, 1996), relatively little is known about the impact that disasters have on personal networks within families. Indeed, most research on disaster networks of individuals—whether studied as whole networks or personal networks—implicates the networks as independent variables rather than as outcomes.

Social Network Analysis of Disaster Response, Recovery, and Adaptation. http://dx.doi.org/10.1016/B978-0-12-805196-2.00008-X

In this chapter, we present findings from a natural experiment—an empirical scenario by which subjects are exposed to experimental and control conditions determined by nature (Dunning, 2012). In the summer and fall of 2008, our study team was in the process of collecting longitudinal multiple-informant family-centric networks from Mexican-origin families residing in the Greater Houston, Texas, area. During our recruitment, Hurricane Ike carved its destructive path through the region. We hypothesized that the disaster response activated latent network ties as individuals engaged both their local and distant network members. Testing this hypothesis at the level of personal networks would normally be very challenging given the unpredictable nature of most natural hazards. Hurricanes, due to their seasonal nature (similar to wildfires and floods), are one scenario where network study designs can benefit from moderate predictability (Butts, Acton, & Marcum, 2012). Alternatively, such disasters may strike in the middle of an unrelated study by happenstance, as our case highlights.

As a result of the unanticipated intersection between our study and Hurricane Ike, we are able to evaluate the impact of the disaster on the types of networks activated in periods before, during, and after the response. In this study, we examine the extent to which disasters affect interactions with members of family networks within the Mexican-origin families participating in the main study. To address this, we first examined differences between the numbers of network members each respondent listed on their family network assessment in three different time periods. Our network enumeration elicited first- and second-degree relatives and up to 20 other important people that participants felt "played a significant role in [their] life during the past year." We hypothesized that, on average, networks would be larger during the response period (H1). We then examined differences in the frequency distributions of types of relationships enumerated in each period. Our hypothesis here was that during the response period there would be relatively more distant relatives than first-degree relatives and nonrelatives enumerated (H2). Finally, we explored differences between periods with respect to reports of sending and receiving of favors among our respondents and their enumerated alters. We hypothesized that the response period would realize a greater number of reciprocal favors than expected by chance, while the predisaster and recovery periods would not deviate from expectation (H3). By testing these hypotheses, we shed considerable light on how disasters affect family networks in the immediate aftermath of a storm like Ike.

The balance of this chapter is outlined as follows. First, we briefly review the natural disaster event and discuss the family context of responding to a hurricane that motivates our hypotheses. Then, we move to a description of our data and methods, touching on sampling and recruitment in the context of the original study. Finally, we summarize our results and discuss our findings in the context of research on family network processes that may unfold during disasters.

THE EVENT IN FAMILY CONTEXT

Early in the morning of September 13, 2008, Hurricane Ike—by then a Category 2 hurricane—made landfall on Galveston Island before arcing northeast toward Houston, Texas (Emmett, 2013). In its wake, Ike left a path of destruction that displaced some 1 million people and left at least 3500 families immediately homeless in Galveston and Harris Counties. To date, at nearly $38 billion in damages, Hurricane Ike remains one of the costliest Atlantic hurricanes to strike the US coastline, ranking third behind Hurricanes Katrina (2005, at $125B) and Sandy (2012, at $71B), respectively. The cost in human life was modest, with 195 US deaths attributed to Ike, including 74 in the Galveston–Houston area (Zane et al., 2011).

Despite evacuation orders from Texas Governor Rick Perry, National Weather Service media messages predicting that Ike would bring "certain death" (Morss & Hayden, 2010) along with a rare presidential emergency declaration—more than 100,000 people chose to remain and shelter in place (Cutter & Smith, 2009). Variability in family decisions on whether, and where, to evacuate strained immediate response strategies (McGraw, 2008). Past work has shown that individuals considered a number of social and ecological factors unrelated to expected disaster intensity—including risk assessments based on home ownership, built environment, prior Hurricane survival history, housing tenure, and family structure—when forming their immediate response decisions (Adger et al., 2005; Riad, Norris, & Ruback, 1999). There is some evidence that social relationships are important to family decisions to evacuate. For example, coming from a network perspective, early work by Thomas Drabek and Keith Boggs (1968) showed that having fewer social ties on which to rely decreased the likelihood of evacuation. Moreover, their work showed that the preponderance of network relations tapped during the response period were family members. This previous research on evacuation patterns provides prior evidence that informs our hypothesis that affected people activate extended family network ties during an immediate response phase.

Even while families had to make more-or-less immediate response decisions during Hurricane Ike, the official response period took months to unfold. This observation is supported by the fact that the United States Federal Emergency Management Agency's (FEMA) housing recovery program, an indicator of transition from response to recovery periods, did not launch until November (Jadacki, 2011). Non-immediate or medium-term family-level responses to the hurricane often differ depending on whether families decided to evacuate or shelter in place. During this recovery period, affected individuals may experience increased awareness of their broader family networks since making such response decisions likely involved reflecting on their emergency contacts or otherwise trying to reunite with previously disconnected members (Blake & Stevenson, 2009). We expect that such processes heighten the salience of family, which in turn differentiates how individuals utilize their network resources before, during, and after disasters.

SOURCES OF SUPPORT FOLLOWING DISASTERS
FAMILY SUPPORT

Norms of family composition, function, and form vary within and between societies and over time (Thornton, 2013). One of the primary functions of family conserved across societies is to provide protection and support to its members (Lowenstein & Daatland, 2006). Perhaps this role is nowhere more evident than during disasters, where family members are suddenly under collective threat and stress. Enrico Quarantelli (1960) made the case that immediate family members (those living together) are the true first responders. He argued that family members assess each other, provide aid, and respond as a collective during disasters. Family members are often those most proximal to each other and therefore the people most likely to be available to provide aid during a response (Sampson, 1991; Bales & Parsons, 1956). Indeed, in much of the early sociological work on disasters, families were a social context that was studied in an effort to understand the mechanisms associated with preservation of society following crisis (Fogleman & Parenton, 1959; Form et al., 1956; Killian, 1952; Litwak & Szelenyi, 1969). Historically, this focus shed light on how families evacuate and reunite (Drabek, 1969) and economically adapt (Kunreuther, 1967) by members relying on one another for support and resources.

OTHER KINDS OF RELATIONSHIPS

More recently, this line of research has shifted beyond the immediate family to emphasize the contributions of extended kin, friends, neighbors, and other members of personal networks to individuals in need during crises. Quarantelli (1960) pointed out that ties to personal network members—including family and friends—often provide a second line. Indeed, work by Thomas Drabek and colleagues (1975), and later by Joanne Nigg (1995), showed that ties to immediate family members are more active during the response period while extended kin may play important roles during the longer-term recovery period. Along with providing housing to their affected family members, extended kin help with child care, meals, and even financial support. Likewise, neighbors and friends with strong social bonds may also provide such instrumental support to one another through the provision of help and favors during disasters as both Casagrande, McIlvaine-Newsad, and Jones (2015) and Kaniasty and Norris (2000) have shown. In Israel, Yossi Shavit, Claude S. Fischer, and Yael Koresh (1994) found that friends filled greater psychological support roles than did immediate family members during the missile attacks on Haifa in First Gulf War. These relationships are particularly important in the recovery process as communities are rebuilt and restored. Following the Kobe earthquake, for instance, Yuko Nakagawa and Rajib Shaw (2004) found that individuals residing in a community with strong social bonds between residents (measured as individual perceptions of a gestalt rather than through neighborly closeness) rated their satisfaction with recovery and rebuilding efforts higher than those who lived in a community with weaker social bonds.

TIES ACROSS SPACE

One reason that tie activation during disaster response and recovery—through the seeking and provision of support—varies between different types of relationships may have to do with geographic propinquity of alters. That is, immediate families are often geographically proximal, and extended family members may be more geographically distant. Distant family members and other relations may not be as salient in normal day-to-day life as those close by, but become activated during times of crisis because they represent a place to go to escape the disrupted local environment. Supporting this notion, Xin Lu, Linus Bengtsson, and Petter Holme (2012) found that residents of Port au Prince, Haiti, likely experienced reduced local network capacity in the first three days after the 2010 earthquake, which was remedied by a large outmigration to seek help from family and friends living in unaffected areas. Therefore, close-by friends and neighbors, while likely salient during normal life, may fall off as significant social resources as they become occupied by their own disaster response efforts. Indeed, Valeri A. Haines, Jeanne S. Hurlbert, and John J. Beggs (1996) found that residential proximity to others was not predictive of providing support during the early period of Hurricane Andrew but was positively associated with providing support during the recovery period. Thus, the increasingly dispersed family structure in the United States may have a protective effect during disasters—both on individuals who may need evacuation destinations and on communities that may have limited number of temporary shelters.

For some underserved and underresourced populations, however, increased geographic distance between themselves and their extended family members may not be protective, or even relevant, when disaster strikes since family disaster preparedness and response strategies vary by socioeconomic status (Elliott & Pais, 2006; Fothergill, Maestas, & Darlington, 1999; Fothergill & Peek, 2004). As Kathleen Tierney (2006) puts it, the "axes of inequality" affect one's "ability to engage in self-protective activities across all phases of the hazard cycle" (p. 113) including the ability to access wells of social capital.

This is as true for poor individuals as it is for poor families. Members of recent immigrant families, for instance, may be separated by legal and geopolitical boundaries that constrain their ability to respond to each other's needs during a crisis (Foner & Dreby, 2011; Treas, 2008). The risk rising from an inability to activate their ties is compounded by the very real possibility of deportation for undocumented individuals who seek out assistance from authorities in lieu of being able to tap family resources during disaster response periods—such as happened to dozens of undocumented Latino shelter seekers during Hurricane Andrew (Phillips, Garza, & Neal, 1994). This may be one reason that Hispanic families have been found to disproportionately rely on extended kinship ties when responding to disasters. During the Denver flood of 1965, for example, Thomas Drabek and Keith Boggs (1968) found that the majority of Hispanic families (63%) evacuated to a relative's home rather than relying on friends (13%) or an official shelter (24%), which were used more frequently by non-Hispanic families. Alternatively, such a reliance on kinship ties for support may offset families' needs—or desires—to seek assistance from formal support institutions. Indeed, Jeanne S. Hurlbert, John J. Beggs, and Valerie A. Haines (2001) showed an inverse relationship between reliance on kin networks and reliance on formal institutions for help during Hurricane Andrew.

NETWORKS AND COPING

The availability of extended kinship ties may offer families a wellspring of social, as well as instrumental, support that can be accessed during disasters. Indeed, social support represents available interpersonal resources that can be tapped when coping with a given stressor (Lin, Woelfel, & Light, 1985; Schwarzer & Luszczynska, 2012), such as that encountered within the context of a disaster. There is evidence that suggests having access to such social resources is associated with successful coping with disasters, as Ralf Schwarzer, Rosemarie M. Bowler, and Cone (2014) found with people recovering from posttraumatic distress after the September 11, 2001, terrorist attacks in the United States and as Filip K. Arnberg, Christina M. Hultman, Per-Olof Michel, and Tom Lundin (2012) found among victims of the 2004 Indian Ocean tsunami. Of course, stressors emerge and change as different challenges unfold throughout the various stages of disaster response and recovery. Here, we posit that the structure underlying resource exchange in these contexts is important to coping with the stressor and changing social support needs over time. Therefore, we expect variability in the exchange of psychosocial and instrumental support resources across the predisaster, response, and recovery phases. This expectation is largely guided by communal coping theory, in which members of a social unit (e.g., family or community) view a stressful event as a common problem that will be addressed through cooperative coping actions (Lyons, Mickelson, Sullivan, & Coyne, 1998).

Activation of cooperative coping strategies may be embodied in the network structure through reciprocal exchange processes, particularly when each group member is impacted by the stressful event. Thus, we hypothesize that families recruited into our study during the disaster response and recovery phase will be more likely to engage in reciprocal exchange of psychosocial and instrumental resources as compared to those recruited in the predisaster phase. Reciprocal exchange is thought to drive cooperation and social organization (Gouldner, 1960; Molm, 1994). However, we note that Valerie A. Haines and colleagues (1996) found that reciprocity of support exchange was not associated with support provision in either preparation or recovery phases of Hurricane Andrew once additional network and background factors were taken into account.

Moreover, such long-term mutual assistance may be particularly meaningful to the Mexican-origin families participating in our study. Latinos have been shown to have greater expectations of reciprocity from their personal network members than other ethnic groups (Dilworth-Anderson & Marshall, 1996); at the same time, they have been found to experience accelerated declines in access to social support resources across phases of disaster recovery during Hurricane Andrew (Kaniasty & Norris, 2000). The nature of reciprocity in Hispanic families during the immediate aftermath of disasters may also be shaped by socioeconomic factors: in disaster-affected areas in Mexico and Ecuador, for example, Eric C. Jones and colleagues (2015) found evidence of inequity in reciprocal exchanges of material and instrumental support reported between classes, but equitable levels of reciprocity within classes.

In summary, past research on how families respond to disasters predominantly focused on evacuation strategies and the provision of instrumental and social support. Those aspects of disaster response are unevenly distributed across families by economic and social inequality. In what follows, we describe our sample, which taps directly into a vulnerable immigrant population. For this group, reliance on family members may be especially important for survival when responding to a disaster.

DATA AND METHODS

The Project Risk Assessment for Mexican Americans was a community-based study designed to understand how family health history might be used as an intervention to spark health communication about the risk, prevention, and maintenance of heritable diseases within families of Mexican heritage. Participants from Project Risk Assessment for Mexican Americans were recruited from the Mano-a-Mano Mexican American Cohort Study, a population-based cohort of Mexican American households launched in 2001 by the Department of Epidemiology at the University of Texas M.D. Anderson Cancer Center (Chow et al., 2015; Wilkinson et al., 2005). Households in the cohort study were initially recruited via probability random-digit dialing, door-to-door recruitment, intercepts, and networking approaches. All were of Mexican heritage, having at least one first-generation family member, and resided in underserved communities in the greater Houston area.

Detailed descriptions of the primary objectives, recruitment, and sampling design of Project Risk Assessment for Mexican Americans have been previously published elsewhere (Koehly et al., 2011; Marcum & Koehly, 2015). Between three and four family members ($N=497$) from 162 families were interviewed in home at baseline in English or Spanish and then completed two follow-up phone surveys approximately 3 and 10 months later. Baseline assessments required all participating family members to be available in the home at the same time for assessments; participants were separated for these assessments to limit information exchange across informants. Each participating family member received a $20 cash incentive (totaling either $60 or $80 per family) per assessment. Our staff trained community-embedded recruiters and interviewers who administered the survey on the ground in Houston. The study enjoyed a 93% retention rate over the follow-ups, owing much to the efforts of our community-embedded interviewers and recruiters.

We conducted a multiple-informant family network assessment as part of this study. Respondents listed their first- and second-degree relatives, and up to 20 additional people they considered to be important members of their personal networks; these enumerated network members are referred to as alters. Here, we only consider network members who were alive at the time of interview. A host of information was collected on these alters from the perspective of each reporting respondent, including key health factors, and relationships between respondent and alter. The baseline network assessment is the focus of this chapter.

Of particular importance is that our network study recruitment was conducted between the summer of 2008 and the spring of 2009—Hurricane Ike struck the area during this time and, as evidenced in our recruitment tracking records and personal correspondences, both participating families and our community-embedded recruiters were affected by the natural disaster.

We divided up our sample recruitment into three time periods: predisaster, response, and recovery. With respect to our respondent recruitment, the predisaster period started on June 6, 2008, and ended on September 10, 2008, which was the day prior to evacuations that were ordered by Texas' governor (Mount, 2008). The response period likewise ranged from the evacuation order on September 11 through the following two months—approximately the length of time it took FEMA to start its housing recovery operations (Jadacki, 2011). We did not recruit during the first two weeks immediately following Ike, so this period actually represents about six weeks of recruiting. Finally, our recruitment during the recovery period lasted from mid-November until March 13, 2009. Of course, the actual recovery period is ongoing even seven years postdisaster. Due to our requirement that all participating family members must be present at the time of interview, we have no families where interviews are split across time periods.

For social interactions occurring within these family networks, we focus on reports of any favors exchanged between respondents and their enumerated alters. Apart from being one of the only relational items from our network assessment questionnaire that is relevant to this scenario, favors tap directly into social interactions that are likely to transpire during a disaster response. Specifically, respondents indicated for whom among their alters they did favors, and also from whom they received favors. We combined these dichotomous items to characterize the four possible dyadic configurations: no favors, incoming favors only, outgoing favors only, and mutual favors (which is a traditional network measure of reciprocity).

To address our research question, we use contingency tests to determine whether the background characteristics of the respondents and families differed by time period to rule out the possibility that Hurricane Ike led different types of individuals to participate during different periods. We use linear regression to evaluate the hypothesis that the average number of enumerated alters differs by time period (H1). To evaluate period differences in enumerating particular types of relations in each family we use t-tests (H2). Finally, we use a dyadic log-linear model to examine period differences in favor exchange configurations (H3).

RESULTS

Table 8.1 reports the descriptive statistics of the informants across the three time periods. Where appropriate, we report p-values from omnibus chi-squared tests of contingency tables. We successfully recruited 35 families during predisaster, 45 during response, and an additional 82 during recovery periods. Owing to our three to four informants per family study design, this recruitment translated to totals of 107, 137, and 253 interviewed family members, respectively. Likewise, those informants provided relational information on 1732, 2470, and 4133 additional dyads in each respective recruitment period.

As reported in Table 8.1, there were no significant differences in background characteristics of the informants across time periods with respect to average age, gender composition, marital status, place of birth, and socioeconomic status (as measured by joint home and car ownership). These results

Table 8.1 Descriptive Statistics of the Informants and Families Recruited in RAMA by Hurricane Ike Period

	Predisaster	Response	Recovery	$Pr(>|x|)$
No. informants	107	137	253	–
No. families	35	45	82	–
No. dyads	1732	2470	4133	–
Pct. female	54%	54%	55%	0.96^{α}
Mean age	41.4	40.5	40.7	0.89^{β}
Pct. foreign born	69%	68%	69%	0.98^{α}
Marital status				0.97^{α}
Divorced	1%	1%	2%	–
Married	71%	69%	70%	–
Never married	25%	28%	25%	–
Separated	3%	2%	2%	–
Widowed	1%	1%	1%	–
Owns house & car	62%	65%	66%	0.85^{α}

p-values based on contingency from $^{\alpha}$Chi-square test, or $^{\beta}$F-test.

provide us with evidence that Hurricane Ike did not disrupt the sampling design of the study. Concurrently, these results also suggest any differences found in our subsequent analysis comparing network size and composition across time periods are not likely to be attributable to differences in the informant sample drawn in each period.

Table 8.2 reports the regression results that test for differences in the average number of enumerated network members between time periods. As previously mentioned, these networks consist of all first- and second-degree relatives and up to an additional 20 friends[1] and family that respondents felt played a significant role in their life during the past year. On average, respondents interviewed during the response period enumerated 1.8 more network members than those interviewed during predisaster and recovery periods. At the pooled family level, this translates into an additional 2.5 unique family members, on average, captured as a result of the immediate response to Hurricane Ike. Thus, we find evidence in support of H1—more network members were enumerated by participants recruited during the response phase.

Now that we have identified that there are, on average, more network members enumerated by families recruited during the response phase of Hurricane Ike, we turn to the question of who are those additional members? Fig. 8.1 reports the results of independent pairwise t-tests for the difference in the family-level average number of enumerated relationship types between time periods. The set of relationships on the left side of the figure represent mean differences in family member counts between response and predisaster periods, while the relationships on the right side represent differences between recovery and predisaster periods. There is a significant increase in the number of aunts and uncles

---

[1]"Friends" represent the plurality of relationships enumerated in this category, which also included neighbors and coworkers, but there were too few of those latter relations to separate them out.

Table 8.2 Linear Regression of Number of Network
Members Enumerated on Time Period

Constant	16.19*** (0.53)
Time period (ref = Predisaster)	
Response phase	1.84** (0.71)
Recovery phase	0.15 (0.64)
Observations	497
R^2	0.020
Adjusted R^2	0.016
Residual std. error	5.51 (*df*=494)
F-Statistic	4.96** (*df*=2; 494)

**p < .01; ***p < .001.

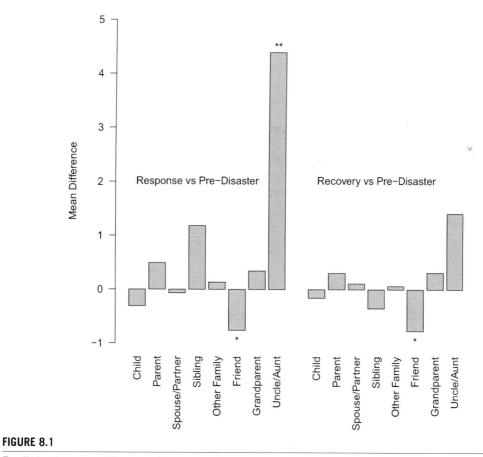

FIGURE 8.1

Family-level average differences in number of relations enumerated comparing pre-disaster with response and recovery phases.

Table 8.3 Standardized Pearson Residuals from the Log-Linear Model of Dyadic Favor Exchanges on Time Period

	No Favors	Received Only	Sent Only		Reciprocated Favors	
Predisaster	−0.619	0.766	−5.697	***	3.828	**
Response	−0.776	0.282	1.353		−0.087	
Recovery	0.995	0.710	2.627	*	−2.398	*
	$\chi^2 = 64.7482$; $df = 6$; $p < 4.856e^{-12}$					

$^*p < .05$; $^{**}p < .01$; $^{***}p < .001$.

enumerated during the response period ($t = 2.439$, $df = 75^2$, $p < .017$). Additionally, there is a significant decrease in the number of friends enumerated in both post-Ike periods ($t = −2.086$, $df = 51^2$, $p < .042$, and $t = −2.200$, $df = 46^2$, $p < .033$, respectively). No other count of a particular relationship differed significantly between time periods.

These results suggest partial support of H2; there were more second-degree relatives, i.e., aunts and uncles but not grandparents, enumerated during the response period. Part of this result can be attributed to the enumeration questionnaire; parents and grandparents were elicited by design as parents and grandparents were explicitly asked about by the enumeration instrument. However, enumeration of others was unaffected by this design, which has the implication that we can only find partial support of this hypothesis. We also observe a significant drop in the number of enumerated friends[1] that persists over time. This suggests that there is a short-term trade-off preferring extended family members over friends as a result of the disaster; but the likelihood of enumerating friends continues to be low even when the propensity to nominate second-degree relatives returns to baseline. While providing only partial support of our hypothesis, this result is consistent with the conclusion by Jari Saramäki and colleagues (2014) that individuals must drop some ties in favor of others during periods of high communication in their network, and that there is considerable persistence of that signature over time.

Finally, we evaluate whether the perceptions of sending, receiving, reciprocating, or not exchanging favors differs across response period. Table 8.3, reports the standardized Pearson residuals and chi-square contingency test from the log-linear model of dyadic favor exchanges on time period. These values represent relatively how much the observed counts of favors differ from expected counts of favors based on the total number of favors reported (i.e., from chance). In the predisaster period, we observe significantly fewer reports of favors sent (but not received) and more reports of reciprocated favors than expected by the margins of the log-linear model. The table also shows that these results represent a trade-off in the recovery period, where we observe more favors sent and fewer favors reciprocated than expected. Interestingly, we observe the expected frequencies of each dyadic favor exchange configuration during the response period. Thus, we find no evidence in support of the hypothesis that the response period is a time of greater than expected favor exchanges (H3).

---

[2]Degrees-of-freedom were rounded to the nearest whole number.

DISCUSSION: FAMILY NETWORKS AND DISASTERS

Hurricane Ike's disruptions to the original research project—how ways of providing family disease risk information influenced how families approach health—may warrant modification of analytic designs to account for potential impacts on social mechanisms that are integral to intervention outcomes. Fortunately, in our diagnostics, Hurricane Ike does not appear to have affected the results on the key health communication outcomes, even when larger networks were obtained during the response period, presenting potentially more communication pathways (Dunning, 2012). Moreover, while current US federal guidelines for human subject research require storing records for at least three years [45 CFR 46.115(b)], we found that permanently archiving our study records in digital form to be an invaluable resource in both reconstructing exactly when participants were recruited from participant tracking records and contextualizing study efforts from correspondence sent during the disaster. We advise future researchers planning to conduct studies in disaster-prone areas to consider the longevity of their record keeping in the development of their protocols.

As reported earlier, the increases in the number of enumerated family network members were largely due to increases in the number of aunts and uncles enumerated. Our study design largely engaged parents and their adult children; thus, the siblings of our participating parents were the aunts and uncles of our participating adult children. Examination of the enumeration data during the predisaster period showed that adult children left out some of their aunts and uncles in their network enumeration, as ascertained from their parents' enumeration of siblings. Hurricane Ike likely increased communication ties between parents and their siblings, which in turn influenced the salience of such relationships for their adult children. This may be a particularly relevant point for immigrant families, where younger generations may have limited interaction with their relatives living in older generations' country of origin (Foner & Dreby, 2011; Treas, 2008).

Additionally, in some cases, the younger participants recruited during the response period may have identified and enumerated their *compadres* as aunts and uncles. Compadres play a special role in Mexican heritage families. Typically, they are not blood relatives, but are rather close friends of the parents, who are relied on as if they were members of the family and remain in the lives of children indefinitely (Kana'Iaupuni, Donato, Thompson-Colon, & Stainback, 2005). The relation is akin to *godparent* in English, but would have been translated as aunt/uncle by our study confederates. Thus, it is plausible that some of the excess in enumerated aunts and uncles during the response period—a time of heightened need and familial salience—reflect such special relationships.

Our results also provide insights into how families respond to natural disasters and the potential psychological implications of that response. Families that live in the Gulf Coast region are in the hurricane zone and many may have developed a certain resilience to storm damage; these families are diverse, however, and have unequal abilities to anticipate, prevent, mitigate, respond to, and recover from disaster (Tobin, Whiteford, Jones, & Murphy, 2007). Tellingly, when considering our results with regard to the exchange of favors, we find that during the recovery period, participants perceived inequality in exchange. There were statistically fewer than expected perceived reciprocal exchanges in favors with an excess in perceived provision of favors. This is in stark contrast to the predisaster period, where reciprocity exceeded expectations, and to the response period, where expectations were met.

Research suggests that reciprocal exchange is a social norm that drives cooperation and social organization and is ultimately related to survival (Gouldner, 1960; Molm, 1994). Here, we consider perceptions of social exchanges. What is particularly interesting is the perception of fewer than expected reciprocal ties. Thus, participants recruited during the recovery period did not perceive equity in exchange. From an equity theory perspective (Hatfield, Walster, & Berscheid, 1978), this inequity in the exchange of favors may lead to increased distress and ambiguity in the relationship (Väänänen, Buunk, Kivimäki, Pentti, & Vahtera, 2005). In particular, those who provide more favors than expected without receipt of favors in return—as is the case here—feel frustrated, angry or hurt because they received less than they gave (Väänänen et al., 2005; Whatley, Webster, Smith, & Rhodes, 1999). Future work will evaluate the extent to which absence of reciprocity is associated with having more conflict between actors, and whether this effect holds under the uncertain environs of disaster response.

Another explanation, which is particularly relevant for family systems, suggests that this inequity in exchange may reflect an adaptive psychological response in which willingness to help those close to others, in absence of reciprocal exchange, is an important communal process. This idea builds upon findings that suggest the norm of reciprocity in families is grounded in more generalized exchange processes based on members' established history—and expected future continuance—of support (Lawrence & Schigelone, 2002; Rook, 1987) especially in Hispanic families during disasters (Kaniasty & Norris, 2000). Thus, the perceived outflow of favors by our participants may reflect a perception of altruistic behavior in supporting their family members through the recovery process, without the expectation of reciprocal exchange; in turn, such altruistic responses can result in increased well-being and more cohesive social relationships (Liang, Krause, & Bennett, 2001). Indeed, there is some evidence to suggest that this type of perceived inequity positively impacts health outcomes, particularly for women (Väänänen et al., 2005). Consistent with this interpretation, Froma Walsh (2007) found evidence that such communal coping in the face of disaster strengthens resilience among families and communities.

This explanation is also consistent with the past behavior of Hispanic families during disasters. During Hurricanes Hugo and Andrew, Jasmin K. Riad and colleagues (1999) found that Hispanic families and those with higher perceived social support were among those most likely to evacuate. While we cannot know for certain where all of the families participating in our study evacuated to, we suspect that some of the increase in family ties we observed during the response period reflects such a family network response. As we discussed, this may be due to younger participants' heightened awareness of aunts and uncles by observing their parents' increased communication with their siblings who were not actively engaged in their predisaster lives.

In conclusion, our results suggest that individuals in this population activate ties to their extended kin, deactivate friendship ties, and perceive imbalance in the extent to which the favors they send to members in their networks are reciprocated during the long-term disaster recovery. This suggests that the protective role of family extends beyond the nucleus of the home and outward to extended kin, but that exchanges between such members may be unevenly distributed during disaster response. Future disaster planning and response programs might promote extended kin as a well of instrumental and social support on which affected individuals and their immediate families might rely for communally coping with crises.

NETWORKS IN DISASTER RECOVERY

3

THE FAMILY'S BURDEN: PERCEIVED SOCIAL NETWORK RESOURCES FOR INDIVIDUAL DISASTER ASSISTANCE IN HAZARD-PRONE FLORIDA

Michelle Meyer

Louisiana State University, Baton Rouge, LA, United States

CHAPTER OUTLINE

INTRODUCTION

Friends, family, coworkers, and acquaintances comprise the networks of individuals with whom one shares confidences, provides and receives social support, and discusses problems and important topics (Paxton, 1999). Research in disasters and discussed in this book shows that family, friends, and neighbors are often the real first responders when disaster strikes. Members of one's social networks provide vital disaster assistance including the communication of warning messages and recovery information; provision of supplies; search and rescue; shelter during evacuation and rebuilding; social, emotional, and financial support; and help with debris removal, rebuilding, and repairs (Elliott, Haney, & Sams-Abiodun, 2010; Hawkins & Maurer, 2010; Hurlbert, Haines, & Beggs, 2000). Thus, social networks contribute to the ability of a person to respond and cope with extreme events.

One way to understand the effect of social networks in disaster settings is by understanding how size and composition of these networks determine support received. Individuals with *larger* social networks receive more tangible (e.g., debris removal), informational (e.g., directions to formal aid resources), and emotional (e.g., encouragement) assistance following a disaster (Kaniasty & Norris, 1995). Research has found that individuals receive informal support from 1.5 to 5 individuals on

Social Network Analysis of Disaster Response, Recovery, and Adaptation. http://dx.doi.org/10.1016/B978-0-12-805196-2.00009-1

average (Hurlbert et al., 2000; Jones et al., 2015). Further, the types of people—or composition—in those networks also affects the disaster assistance one receives. Research from Hurricane Andrew in 1992 found that individuals whose networks are comprised of more men, younger people, and family members are more likely to receive informal support (e.g., rebuilding assistance, child care assistance) from their social networks (Hurlbert et al., 2000).

Family members, in particular, are central to what I refer to as disaster support networks (Drabek & Boggs, 1968; Garrison & Sasser, 2009; Haines, Hurlbert, & Beggs, 1996). Kin embeddedness, or how connected one is with family members, has been shown to be an important mechanism for individuals to receive needed assistance, especially as community-wide structures and institutions that normally provide social support are fractured during a disaster (Bolin, 1976; Erikson, 1976). Disaster studies have found that family members provide evacuation and material assistance during disaster (Casagrande, McIlvaine-Newsad, & Jones, 2015; Pettigrew, 2011; Sprinkle, 2012). These findings in disaster situations coincide with research on the role of family in non-disaster "everyday" emergency situations, such as borrowing money or helping with emergency child care (Chua, Madej, & Wellman, 2011; Longino & Lipman, 1981). Family is often who individuals will turn to for assistance that is more burdensome, such as financial loans and long-term housing (Furstenberg & Kaplan, 2004).

In addition to size and composition, the types and amount of resources available through the network are key to understanding how social networks affect individuals' life trajectories. This line of research in non-disaster situations has found that individual demographics—especially socioeconomic status, race, and gender—create variation in how individuals access and use social networks for particular resources (Lin & Dumin, 1986). For example, individuals with access to high-status persons are able to get information about high-status jobs (Lin, 2000). At the same time, disadvantaged persons often share resources and favors within their networks to meet daily survival needs (Stack, 1975). In disasters, these inequalities in existing social network resources can be exacerbated, meaning that those with fewer ties to individuals who can provide the required resources often have reduced capacity to cope and bounce back from the impact (Messias, Barrington, & Lacy, 2012).

There are few studies that clearly distinguish between the types of social ties and resource availability in disaster. James Elliott and colleagues (2010) studied two neighborhoods in New Orleans: the Lower Ninth Ward, a poor, majority African American neighborhood; and Lakeview, an affluent, majority white neighborhood. They found that Ninth Ward residents relied on close personal relationships (family and close friends) for informal support during Hurricane Katrina, and that they received less support overall—including less sheltering assistance and less contact with neighbors in the year following the event, compared to Lakeview residents. The authors concluded that affluent, white Lakeview residents had more relationships with persons outside of the area who were unaffected by the hurricane and a greater variety of resources available through these networks to support their recovery. Robert Hawkins and Katherine Maurer (2010) also found that minority and low-income individuals had limited access to aid information from their social networks following Hurricane Katrina. Earlier research following Hurricane Andrew showed that individuals living in areas of higher poverty provided less support to others than those living in higher income areas (Haines et al., 1996). Thus, networks are not created equal in their ability to increase individual resilience.

To build on our understanding of networks in disaster, in this chapter I describe data on individuals' perceptions of their social networks as providers of disaster assistance. Using a common social network measure—the name generator—and in-person interviews, I asked participants to whom in their social networks would they turn for assistance during and following a disaster. Studies of social networks in disaster are commonly post-event and rely on recall of the survivors about their use of social networks during response and recovery. This study differs from previous research by gathering pre-disaster data and asking participants about their *perceptions* of how they would use their social networks in a future disaster event. How accurate these perceptions are is unknown, but these perceptions provide a novel take on networks in disasters and contribute to our understanding of disaster social networks in several ways.

First, perceptions of social support have a history of research across a variety of contexts (Furman & Buhrmester, 1992; Pierce, Sarason, & Sarason, 1991; Procidano & Heller, 1983). Perceptions of social support, whether accurate or not, independently predict mental health outcomes in stressful events (Haden, Scarpa, Jones, & Ollendick, 2007; Zimet, Dahlem, Zimet, & Farley, 1988). Further, perceived support mediates the role of actualized support on disaster outcomes (Norris & Kaniasty, 1996). To date, data on perceived support in disasters has been collected post-event. Thus, the literature includes no data that would allow scholars to understand how much these perceptions coincide with post-event disaster assistance or test any assumptions of accuracy of pre-event perceptions. Second, these data help us understand how individuals view their social networks' ability to provide disaster assistance. This information indicates resource depth within a social network and whether mobilization of support from outside their usual social networks is assumed or will be necessary. Both resource depth and mobilization are important to understanding trajectories following disaster (Fothergill & Peek, 2015), and individual views on their networks provides one method of collecting information about the resource depth available in their social world. Third, perceptions affect the actual support received. If one does not believe that those in their network could provide evacuation shelter, for example, they may be less likely to ask for that assistance. Fourth, as much disaster literature shows, community members come together, whether they know each other or not, to offer support after an event (Quarantelli & Dynes, 1977).

Many of the participants in this research, as described in the following discussion, have experienced previous disasters, and often used those situations to develop their perceptions. Thus, while not based on a specific disaster, individuals are often basing their perceptions on cumulative disaster experience. This fact makes these results on perceptions interesting, in that individuals—many of whom are disaster-experienced—were unlikely to list church members, organizations, or other loose community ties as sources of assistance that are commonly highlighted as part of the therapeutic community. This result indicates how family, as one participant states, "comes first" and thus these perceptions provide information on the first line of support that individuals would seek in a disaster, and any remaining needs would be met by the, hopefully, therapeutic community.

Based on this data, I discuss the size and composition of individuals' perceived disaster support networks. Perceptions of the resources needed and resources available from social networks highlight how the type of resource affects perceptions of disaster support networks. Specifically, individuals describe financial assistance as limited to a small number of network ties. Further, my results point to the centrality of family as a perceived source of disaster assistance and variations in perceptions based on indicators of social vulnerability. I conclude with the implications of these results for theory and practice on social networks in disaster.

METHODS

Data for this chapter are drawn from a mixed-method study in two Florida counties. Florida has an annual hurricane season lasting from June through October along with a regular threat of tornadoes, floods, and wildfires. The two counties selected—Leon and Dixie—are exposed to similar natural

Table 9.1 Survey Name Generator for Disaster Support Networks

Person	Relationship (parent, sibling, friend, neighbor, etc.)	Age	Gender	Race	Location (City, State)	Help they could provide (check all that apply)	Have they helped you in past hurricanes?	Have you helped them in past hurricanes?
1			☐ Female ☐ Male	☐ White ☐ Latino/a ☐ Black ☐ Asian ☐ Other:_____		☐ Financial ☐ Non-financial	☐ Financial ☐ Non-financial ☐ No	☐ Financial ☐ Non-financial ☐ No

hazards but vary in economic, social, and built environments and provide data from both suburban/ urban populations as well as rural residents. I collected mail surveys and conducted in-person interviews with a subsample of the survey participants. The interviews provided in-depth information on how individuals define the role of their social relationships in disaster contexts (Schensul, LeCompte, Cromley, & Singer, 1999).

SOCIAL NETWORK MEASURES

The survey included a name generator to collect social network data (Knoke & Yang, 2008, p. 21). This data collection technique provides information on "ego networks" or the individuals' relationships with other persons. To apply the name generator to disaster support, respondents were asked to list up to eight people to whom they would turn to for assistance after a hurricane.[1]

Respondents also provided the following information about each person listed: relationship, age, location, race, gender, type of disaster assistance that they could provide (financial or nonfinancial), and previous assistance supplied to the interviewee and provided from the interviewee (Table 9.1). Financial assistance in this study was defined as any type of monetary assistance, such as helping with evacuation or rebuilding costs, while nonfinancial assistance was other forms of assistance such as offering labor, child care, or emotional support.

I calculated network size as the total number of social ties each survey respondent listed in the name generator and, based on type of resource indicated, I calculated the size of financial and nonfinancial networks separately. Then, I used relationship to describe network composition. Survey respondents listed a variety of relationships, which I categorized into three groups: family, friends, and neighbors. Family included parents, children, siblings, grandparents, grandchildren, aunts/uncles, and cousins, as well as family through marriage (in-laws) and blended family members (step-parents and step-siblings). Friends included friends and coworkers. Neighbors were only those individuals listed as neighbors.

The semistructured interviews confirmed and expanded on the information provided in the survey. Gathering social network data through these open-ended questions allowed respondents to use stories

---

[1] An upper limit was set for practicality and to avoid overburdening respondents (Marsden, 1990) as well as based on results from previous literature. Previous research has shown that core discussion networks include approximately two to three persons (e.g., McPherson, Smith-Lovin, & Brashears, 2006). Previous disaster research found network size averaged near four persons (e.g., Haines et al., 1996). Further, this upper limit of eight persons was tested for validity through two open-ended questions that asked respondents the total number of people that they could rely on for financial and nonfinancial assistance, respectively. Median number of people listed for financial assistance was two persons and four persons for nonfinancial assistance. Thus, the name generator may slightly underrepresent the size of these networks.

and allowed me to understand how individuals determine which network ties are important in disaster contexts, such as what particular traits of individuals led to their inclusion in the support network.

SAMPLING AND DATA COLLECTION

Study participants were selected via mailing addresses. Because the two counties differed greatly in population density, I used different sampling strategies. In Dixie County, the low population density allowed for a random sample of 300 mailing addresses from the entire county. In Leon County, which has a higher population density, I performed stratified cluster sampling—first selecting six census tracts and then 50 households within each tract for a total of 300 households.[2] This process helped me gather a completed set of surveys that were generally representative of Leon County (see Table 9.2).

After undeliverable addresses were removed, 529 valid households remained (275 from Dixie County and 254 from Leon County). Twenty-two households declined to participate in the survey. The response rate was 27% with 138 completed surveys. Survey respondents were mostly representative of the population of each county, except that respondents were older and more likely to have a person with a disability in the household than was the general population.

The final question on the mail survey asked respondents about their interest in an in-person interview and offered a $10 incentive for interview participation. I interviewed 25 individuals, nine from Dixie County and 16 from Leon County; 52% of interviewees were female and 76% were white. The interviews ranged from 20 minutes to 2 hours and were digitally recorded and transcribed verbatim for analysis.

DISASTER SUPPORT NETWORKS: SIZE AND RESOURCES

Glen and his wife, Jane, live in an established suburban neighborhood of Tallahassee. He nervously wrung his hands as we talked about assistance from friends and family in a disaster. As he explained, he had one local friend who could provide nonfinancial assistance in a disaster, but there is no one that he would call on for financial assistance. Glen elaborated during our interview:

> I guess [George and I] have been friends for 20 years or so. He lives pretty close to our house. We've assisted each other on home projects and stuff so it's just a friend who would, you know, we've shared labor back and forth, off and on.... I guess we haven't really had a lot of discussion [about disasters], maybe he wouldn't even be here for that matter.... Uh, well I don't really have somewhere else to turn [for financial assistance]. I suppose I'm kind of on my own so... I don't quite know what else I would do, you know? I mean, I have what I have, and that would carry us as far as it would carry us.

Many respondents were similar to Glen—they listed few social ties available for disaster assistance, and when divided by type of assistance (financial versus nonfinancial), they listed even fewer social ties

---

[2]It was expected that households with higher social vulnerability would respond at a lower rate than others. Because I could not determine which addresses contained more or less socially vulnerable individuals, I oversampled at the mailing stage from census tracts with higher populations of socially vulnerable individuals to compensate for this concern. I used eight variables common in economic and demographic analysis of social vulnerability to identify census tracts with potentially more or less vulnerable populations: percent poverty, median income, percent racial minority, household size, percent female-headed households, percent renters, percent elderly, and percent children. I purposefully selected two tracts with the highest vulnerability on these factors, two tracts with the lowest, and two tracts in the middle.

Table 9.2 Participant and County Demographics

	Dixie		Leon		Total	
	Survey (n=75)	Population 2010	Survey (n=63)	Population 2010	Survey (n=138)	Interview (n=25)
Median age (in years)	62	45	57	30	60	60
Reported disability	28%	34%	10%	9%	20%	36%
Education						
High school degree or higher	90%	73%	94%	91%	92%	83%
College degree or higher	25%	6%	57%	41%	40%	42%
Female	51%	46%	54%	52%	53%	52%
Race						
White	92%	89%	67%	63%	81%	76%
Black/African American	1%	8%	31%	30%	15%	20%
Asian	0%	0%	0%	3%	0%	0%
Other	7%	3%	2%	4%	4%	4%
Median years living on coast	34	n/a	29	n/a	31	35
Median hurricanes experienced	3	n/a	2	n/a	3	2
Household income		Median income: $32,312		Median income: $44,490		
Less than $15,000	23%		21%		22%	32%
$15,000–30,000	25%		14%		20%	18%
$30,000–45,000	22%		13%		19%	14%
$45,000–60,000	9%		11%		10%	0%
$60,000–75,000	8%		11%		9%	9%
$75,000–130,000	9%		25%		16%	23%
Greater than $130,000	3%		2%		4%	4.6%

that would provide financial assistance. Fig. 9.1 shows the percentage of survey respondents who reported each size of network, from zero to eight persons. The mean and standard deviation of network size are included. The overall network column identifies the number of all individuals listed in the name generator regardless of type of assistance. Network sizes are also specified by type of resource: financial or nonfinancial.

Fig. 9.1 shows that there is variation in size of reported disaster support networks. Nearly 10% of all survey respondents listed no individuals in their disaster support network, irrespective of resource type. At the other end of the spectrum, 18% of respondents reported eight individuals in their disaster support network, which is the maximum possible from the name generator.

When the type of resource is included, the number of people that respondents believe could provide financial resources in a disaster was smaller than the number of people they felt could provide nonfinancial support. Specifically, nearly 40% of all respondents reported *no* friends, family, or neighbors that could or would provide financial assistance to them during a disaster, whereas only 17% reported the same for nonfinancial assistance. Average network sizes also varied by resource, with 3.7 persons listed on average regardless of resource, 3.1 persons listed on average for nonfinancial support, and only 1.7 persons reported on average for financial support.

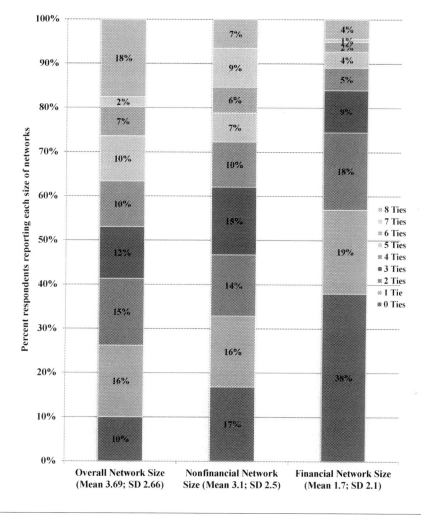

FIGURE 9.1

Percent survey respondents by disaster support network size (0–8) and resource available.

   The in-depth interviews from each county confirmed the general pattern about disaster support network size gleaned from the survey. Many interviewees discussed very few other persons that they could rely on for support in disaster, while other interviewees identified numerous others that they could and would receive assistance from as needed. As the following quote from Stan, a middle-aged man who worked for a local construction company, emphasizes, some interviewees had trouble naming each individual person that could help them and instead used phrases such as "many" or "a lot of" people. Stan remarked, "Oh, I think we'd be OK, depending on what kind of hit my house took, you know…. We've got family and friends all around to help. I wouldn't have a problem findin' shelter with someone close by."

   Marty and Adelle provide further examples of each end of the spectrum. Marty, a married father of two young children living in poverty in Dixie County, stated that he would rely on his parents who live a couple miles away as needed during a disaster: "My parents, my kids, pretty much is what I think about all the time. If I need something, I talk to my mom and dad, see if they can help. A lot of times I don't ask my mom, I pretty much get myself out of everything." At the other end of the spectrum, Adelle, a 90-year-old lifelong, middle-class resident of Dixie County, discussed how her four children and their spouses, a weekly housekeeper, two neighbors, and several fellow church members could all support her if she needed them.

These findings indicate that nonfinancial assistance is perceived as more readily available in a disaster than is financial assistance. But if individuals need money to complete a repair or to pay for a stay in a hotel during evacuation, there are fewer individuals who they think they would or could ask for assistance, with nearly half of the respondents listing no one. This result indicated that individuals will rely on savings or insurance, if they have it, governmental disaster aid or aid from other organizations when available, or go without financial assistance after an event.

DISASTER SUPPORT NETWORKS: COMPOSITION

Jill is a young, white woman who recently relocated with her husband from Michigan to Tallahassee. Even though she had only lived in Florida for a few short months, she and her husband had made friends through their new church. She felt these new, local friends could provide shelter in case of a hurricane, but she was uncomfortable asking them for financial assistance, as she states here:

> There are some people in our church, we know that they would have the money to help. We've been to people's houses, to know they would have an extra room, if we were to call and need to stay somewhere. Then in that church atmosphere, we know that some of them are pretty giving. They probably wouldn't say no, but definitely nothing that I've ever spoken to them about. If it was any kind of monetary help that we were seeking, we would go to our parents first as opposed to someone who was local.

Interviewees' responses were affected by respondents' perceptions of the availability of certain resources (financial or nonfinancial) from different ties and the social acceptability of asking those ties for each resource. In other words, respondents weighed their decisions about whom to include in these network lists between whom they knew had the resource and how appropriate it was to ask that person for that type of assistance based on their relationship. Jill's decision to go to parents for financial assistance was a common response and brings forward the next issue of disaster support networks—composition.

Moving from size to composition, I analyzed relationship type for the ties listed in the name generator and again used the interview data to provide context. The findings confirm that family is central to disaster support networks as respondents overwhelmingly believe they will rely on kin to support them both financially and nonfinancially in a disaster. While friends, neighbors, and others were included in many survey respondents' networks, family members dominated these networks.

Fig. 9.2 shows the average composition of disaster support networks by type of relationship between the respondent and the listed social tie. Recall that 10% of all survey respondents indicated no disaster social network ties. On average for respondents reporting at least one network tie, 59% of these respondents' overall disaster networks were comprised of immediate or extended family. As Meryl, an

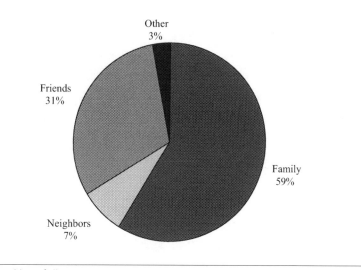

Other
3%

Friends
31%

Family
59%

Neighbors
7%

FIGURE 9.2

Mean percent composition of disaster support networks.

interviewee from Dixie County, explained, a disaster is "when family have to take care of each other. When they have to step up to the bat. If they would still do it. I don't have a mother or father no more.... My kids, they're in Tennessee, so I'd have to move up there, in with them." On average, neighbors comprised only 7% of respondents' networks. In contrast to neighbors, the average percent of friends in the networks was higher at nearly 30% overall. If a person reported six ties, this average network indicates three or four of those ties would be family, two would be friends, and at most one would be a neighbor.

To provide more detail, Fig. 9.3 shows the percent of respondents who included at least one of the common relationships reported (any family member, parent, sibling, child, extended family, grandchild, grandparent, neighbor, and friend). Of people who named at least one person in response to the name generator, over 85% included at least one family member in their disaster support networks. Parent was the most common familial relationship type, with 45% of respondents including at least one parent. Sibling and child were the second and third most common familial relationship listed. While the average percent of neighbors in networks was small, approximately 56% of respondents listed at least one neighbor in their disaster support networks. Fewer respondents reported at least one friend in their networks (24%).

When describing the ability of neighbors or friends to help in disasters, interview respondents often recalled routine emergencies as examples that foretold the type of assistance these nonfamilial ties could provide. These stories included shared car rides, electrical outages and shared generators, and help moving, among others. Savannah, a 25-year-old white woman, lived in an apartment in Leon County with her boyfriend. She focused on nearby friends as potential assistance based on previous experiences helping each other with routine emergencies:

> We help friends out with moving all the time and helping patch up walls and painting and stuff like that. So if something happened to their house or if they needed a place to stay with their animal or needed food or something, definitely I would lend a helping hand to those in need and I feel the same, that they would do that for me as well.

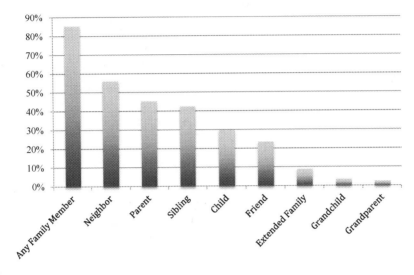

FIGURE 9.3

Percent respondents who listed at least one of the relationship types in disaster support networks.

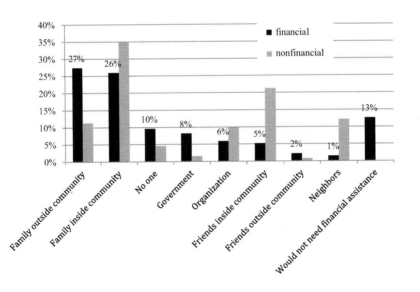

FIGURE 9.4

Anticipated first source of financial and nonfinancial disaster assistance.

Nathan, a white man from Leon County, explained that he would call on friends for assistance *if* local family were unavailable: "We also know we have friends who are just houses away from us, and I'm sure that they would do the same thing [as family would] and let us sleep over."

While friends and neighbors were described by interviewees as potential sources of nonfinancial disaster assistance, family was viewed as the most common source of financial assistance. Fig. 9.4

shows the differences between relationship and type of resource one expects to receive during a disaster. Overwhelmingly, respondents indicate that family is their primary source for financial assistance compared to neighbors, friends, organizations, and government. Fifty-three percent of respondents selected either local or extralocal family as their first source of financial assistance. While relationship seems to drive the selection of ties for financial assistance, location drives the selection of ties for nonfinancial assistance with local family, friends, and neighbors being the most common first sources of nonfinancial assistance.

PERCEPTIONS OF SOCIAL NETWORKS AND SOCIOECONOMIC STATUS

During disasters, as in everyday situations, social network resources often fill the gaps in individual resources. For example, poor individuals may need to ask for financial assistance from social ties more so than others who have financial savings or insurance coverage. Thus, it is important to understand who does and does not have these informal ties, and how that interacts with other social vulnerabilities during disaster.

To address this issue, I analyzed the effect of different indicators of financial capital and social vulnerability on the odds of reporting at least one financial tie (Table 9.3 shows the odds ratios for each independent variable). Recall that financial resources are perceived as more difficult or less appropriate to access than nonfinancial resources through network ties, and that nearly 40% of respondents across the study lacked a social tie to provide monetary assistance. Due to the small sample size, I present the regression results in variable groups of interest: demographics, housing status, and race.

Table 9.3 Odds Ratios of Reporting at Least One Network Tie That Could Provide Financial Support in a Disaster

Variable Odds Ratio (SE)	County	County and Demographics	County and Housing Status	County and Race
Leon county	3.20*** (1.20)	3.27*** (1.49)	2.21* (1.05)	3.38*** (1.35)
Female	–	0.81 (0.38)	–	–
At or below 150% poverty	–	0.84 (0.41)	–	–
Elderly	–	0.31*** (0.14)	–	
Disability	–	0.45 (0.25)	–	–
Mobile home	–	–	0.82 (0.38)	–
Homeowner	–	–	0.32** (0.18)	–
Minority	–	–	–	0.95 (0.59)
Constant	1.00 (0.23)	2.51** (1.13)	3.15* (1.98)	0.92 (0.22)
Pseudo R^2	0.056	0.163	0.085	0.060
Observations	137	119	134	132

*p < 0.10, **p < 0.05, ***p < 0.01.
Dixie County is reference category.

The strongest relationship between demographics and odds of having at least one network source of financial assistance is the difference between residents of Leon and Dixie Counties. Recall from the previous demographic information that Leon County is more suburban/urban and has a higher median income overall. Thus, it appears from this comparison of just two counties that population density and/ or median income at the county level may relate to perceptions of the availability of financial assistance through social ties. This relationship of county context to individual perceptions of network resources needs further research.

Controlling for county of residence, these results indicate that in this sample being poor, elderly, disabled, female, a homeowner, or living in a mobile home reduced the likelihood of reporting a social tie that could provide financial disaster assistance. Being elderly and being a homeowner were both statistically significant and had a large substantial effect on the odds of reporting this network resource. Elderly respondents and those owning their homes were one-third as likely to report a financial tie as were nonelderly respondents or nonhomeowners. The demographic variables together explain the most variance in the odds of reporting a financial tie ($R^2 = 0.16$). Together, these data indicate that socially vulnerable respondents are also more vulnerable in terms of network resources and will be required to mobilize new sources of resources in the aftermath of an event.

The interviews were able to add insight into these results. Specifically, interviewees evaluated (1) the resources the tie would have to offer, (2) their individual socioeconomic status and potential needs, and (3) the strength of the tie when determining how they would activate the available social networks in a disaster. First, social ties were often viewed as having a specific role during disasters; and that role was more quickly associated with nonfinancial resources, such as evacuation sheltering, help during debris removal, and information sharing. Interviewees commonly recalled a previous disaster event or routine everyday emergency to explain who would help them in nonfinancial ways. For example, the following quote describes sheltering assistance received by Barbara from Leon County and is a representative example of nonfinancial assistance from social networks: "I went with my friends over to their apartment building and of course we stayed up all night. You can hear the wind blowing and everything like that, and there was maybe about six couples in the apartment 'cause it was a sturdy building."

Next, while interviewees of varying socioeconomic status described the importance of social ties for nonfinancial assistance, those with higher socioeconomic status discussed social capital use in disasters *only* as it related to nonfinancial assistance. For example, Debra, a middle-aged woman from Leon County, clearly described the type of resources she expected social ties to provide:

> One is the financial aspect. We have insurance for the house, always have, always will. Financially, we have the savings that we would need to recover from anything that would happen, and then the immediate support system of friends and family and neighbors for any annoyance kinds of things. You know, the house wrecked is a financial thing. A broken window and water on the carpet is, you know, an annoyance kind of thing that you would call family and friends to assist.

Other higher income individuals described family members they could ask for financial assistance, but emphasized that situation would be incredibly unlikely. Thus, they were accounting for three concerns: the type of resource, what ties had the ability to provide the resource, and whether they would even need that resource. As Neil, a white respondent from Leon County explained, he knew which family members would have the resources to share, but still expected to not have to ask for them:

> We have family that actually would help us but we're not in the position where we would have to ask for financial help except maybe in the short term. Even then we probably wouldn't do it. But we do have family in town who are wealthier than we are, so they actually could afford it if we needed them to.

On the other hand, interviewees from lower socioeconomic strata more easily described whom they could ask for financial assistance during a disaster. Instead of indicating this type of assistance was unsuitable for social ties to provide, low-income individuals evaluated who in their networks had the resources to share. This result mimics research on resource sharing among low-income populations to meet daily needs (e.g., Stack, 1975). For low-income individuals, many of the same people that could provide nonfinancial assistance were viewed as unable or unsuitable for providing financial assistance. As Sheila from Dixie County described, her local friends and neighbors could assist with debris removal but were too poor to provide financial assistance:

> The ones close by don't have two nickels to rub together, so if we needed money, we would probably be calling Chicago, you know, to relatives. If it's assistance moving trees or whatever that are down, I've got friends down here.

Sheila's quote was typical of low-income interviewees, who indicated that they knew who in their social network may have financial resources to share. Further, they indicated that this resource, money, was almost exclusively a burden on family, no matter the socioeconomic status of the respondent. This result indicates that network size alone cannot fully predict how social capital resources will be used in a disaster, but perceptions of the relationship, the resources available in the whole network, and individual needs all affect reporting of social capital ties for disaster.

DISCUSSION

This research, conducted outside of a disaster situation, shows how individuals perceive the role of their social networks in providing support during a future disaster. In summary, the results show that perceived disaster support networks are relatively small—less than four people per respondent—and that financial support is limited to only one or two social ties, if any. The use of a name generator and corresponding interviews to triangulate the data showed how individuals perceive the use of social ties for nonfinancial and financial resources, indicating that financial resources are less commonly associated with social networks, and is dependent upon perceptions of the relationship, capacity of the social tie to provide money, and self-perceived potential need.

Compared to previous research on networks in disasters, my results are similar to findings post-disaster. Related to network size, Jeanne Hurlbert and colleagues (2000) found that respondents reported on average 1.6 persons supported them during preparation for Hurricane Andrew and 2.5 persons supported them during the recovery, which is slightly smaller than the perceived 3.1 persons found in my study. Eric Jones and colleagues (2015) also found that individuals received and provided support to between three and five members of their social networks. The comparability of results between this pre-event data and post-event research indicates that pre-event perceptions may be close to accurate, at least in the size of support networks.

Next, my results point to the need to understand social networks in light of other forms of capital individuals may access, especially financial capital through savings or insurance. My results echo those from Deanne Hilfinger Messias et al. (2012) and James Elliot and colleagues (2010) that marginalized groups may be less able to access resources necessary for recovery through their social ties, limiting the effect of social networks among already vulnerable populations. This issue raises questions about the interconnection of personal networks and resilience with community-level social capital (Aldrich & Meyer, 2015). Future research needs to tease out the use of social networks within the context of various other assets or capacities that individuals have access to in a disaster. For example, how do social networks allocate financial resources during a disaster? Is this type of support limited to family ties, or do other individuals emerge and provide support these respondents would not predict?

In addition, this study underscores the importance of the family unit for perceived disaster assistance, thus supporting the importance of kin embeddedness for individual resilience (Bolin, 1985). These data indicate that disaster social capital is heavily dependent upon family ties, and thus is more the "strength of strong ties." Immediate family, extended family, and family through legal changes (e.g., in-laws, divorce, marriage) are all perceived as the main members of disaster support networks. The centrality of family in my results mimic findings from David Casagrande and colleagues (2015), Olivia Pettigrew (2011), and Nicholson Sprinkle (2012) who all found that family ties provided more material support than did nonfamily ties. When disaster occurs, it appears that individuals expect to and do consolidate their social networks to the strongest ties. As one respondent, Vera, explained, family is first priority for getting and giving disaster assistance, and she expected others to do the same.

> [During an evacuation,] I know my family would have to meet somewhere in the middle of central Florida, because my mother, sister, and brother are the Fort Myers area, and that's not good. They're always being hit [by hurricanes]. That's about it, really. Friends, you know, I suppose they would do the same thing, probably. The family, we always kind of stick together a little bit.

The centrality of family to understanding social networks in disaster implies the concern for those isolated individuals in a community with few family ties (Swanson, Forgette, Van Boening, Kinnell, & Holley, 2007). As family dynamics change in America, such as smaller family sizes, what will be the outcomes for disaster support? For those without family, or lacking family that have resources to share, the effect of individual social networks in disaster situations will be limited. Are these individuals able to find support through other persons in their social networks, or do they turn to formal institutional philanthropic or governmental support? Or, will they, like the elderly isolated persons in the Chicago Heat Wave of 1995, perish at a disproportionate rate when disaster strikes (Klinenberg, 2002)?

This research has a variety of limitations. The small sample size and focus on only two counties in Florida limits the generalizability in the statistical sense of these results. The results did show patterns that would need to be assessed with larger and more diverse samples. Also, the name generator as a data collection tool may have been burdensome to respondents and thus limited the number of ties that they listed. In this case, these results are downwardly biased considering network size. Further, the design of the pre-event survey highlighted interesting points about perceptions on social networks' utility in disasters. Future research that incorporates pre- and post-event measures, although difficult to gather, would provide the most fruitful method of testing the accuracy of perceptions as well as the effect of perceptions on disaster outcomes.

For practitioners looking for ways to incorporate measures of social networks in disaster assessments and to predict emergency needs within their communities, these results show that measuring family ties can help highlight vulnerable populations, along with other traditional measures of social vulnerability. For populations known to have fewer family ties, nonfinancial and financial institutional assistance should be targeted and nonfamilial network mechanisms should be supported. As disasters increase in frequency and impact, policies that address both individuals and community social capital and foster network support will only grow in importance, especially for socially vulnerable groups.

INTERORGANIZATIONAL NETWORK DYNAMICS IN THE WENCHUAN EARTHQUAKE RECOVERY

10

Jia Lu[1,2]

[1]*California State University-Los Angeles, Los Angeles, CA, United States;* [2]*University of Southern California, Los Angeles, CA, United States*

CHAPTER OUTLINE

INTRODUCTION

Disaster scholars examine various aspects of interactions within the public sector, the private sector, and emergent citizen groups (Stallings & Quarantelli, 1985) at different phases of a disaster cycle, such as preparedness, emergency response, recovery, and mitigation (Neal, 1997). More recently, the cross-sector interactions, particularly the proactive role of civil society actors during disaster recovery, have been cited as critical for areas of risk reduction and long-term development (Arnold, 2006). In her examination of World Bank disaster projects implemented in a cross-country context, Margaret Arnold (2006) concluded that empowerment and capacity building at the local community level are keys to effective risk management when seeking to support economic development. After investigating the long-term recovery and redevelopment efforts of eight catastrophic events from the attacks of

Social Network Analysis of Disaster Response, Recovery, and Adaptation. http://dx.doi.org/10.1016/B978-0-12-805196-2.00010-8

September 11, 2001, in Washington, DC and New York City, to Hurricane Katrina in 2005 and the Haiti earthquake in 2010, Jeffrey Garnett and Melinda Moore (2010, p. 1) found that both the bottom-up approach involving local people and the top-down approach incorporating a long-term development economics vision from the state are all needed for disaster recovery. They cited local empowerment, organization, leadership, and sustainability planning as integral aspects of recovery planning and risk management. These principles of participation, empowerment, and collaboration were also important in disaster recovery at the local community level for countries recovering from the 2004 Indian Ocean tsunami (Rowlands & Tan, 2008).

The sociopolitical conditions under which the responses of civil society occur vary from one country to another. For the case of the Golcuk earthquake in Turkey in 1999, the relationship between society and the state was being renegotiated (Pelling & Dill, 2010) because of a shattered role of the state (Ganapati, 2005). The role of civil society—in the form of the emergence of civic networks—arose out of the context of a weak state. An active role of civil society in promoting disaster mitigation and prevention strategies was also being noted in the recovery process from Hurricane Mitch in El Salvador in 1998 (Wisner, 2001). A dogmatic neoliberal state accompanied by a lack of capacity for local municipal governments to rebuild after the disaster was found to serve as the sociopolitical backdrop in the El Salvadorian case. Ben Wisner (2001) found that the municipal government's lack of capacity included problems in the areas of planning, program delivery, creating budgets, general management, and litigation. The investigation of Japanese disaster recovery showed another side of the sociopolitical landscape. Archival research reviewing Japanese disaster recovery processes showed greater role of the state in building up the physical infrastructures as compared to the cases investigated by Wisner (2001) and Nazife Ganapati (2005). While the state government paid less attention to social infrastructure reconstruction, civil society was active in bringing residents back into the disaster-damaged cities (Aldrich, 2008).

However, little is known about the institutional processes through which players in civil society actively develop their roles while contributing to the social adaptive capacity following perturbances. In other words, there is a lack of studies that adopt a procedural-action approach to understanding the transition from emergency response to long-term recovery from the perspective of civil society actors. The case in which I employ this process-oriented approach to understanding the action domain of civil society is the recovery of China's Wenchuan earthquake, which struck 92 km northwest of the Sichuan provincial capital, Chengdu, on May 12, 2008, with a magnitude of M8.0 on the Richter scale (United Nations Development Programme, 2010). It was the strongest earthquake to occur since the 1950 Chayu earthquake (M8.5), and the deadliest since the 1976 Tangshan earthquake (240,000 + deaths).[1] The Wenchuan earthquake affected more than 100,000 square miles and about 30 million people; 69,226 deaths were attributed to the disaster (EERI, 2008). Shortly after the disaster, large numbers of citizen volunteers flooded into the region to assist with the response and recovery activities. Three years after the earthquake, my qualitative research shows that the grassroots participation in disaster recovery remained active through both formal and informal organizational networks (Lu, 2013). However, the process through which such civil society action has evolved and is maintained remains underexplored and will constitute the main focus of this study.

---

[1]China's deadliest quake in Shaanxi (east-central China) in 1556 claimed over 800,000 lives, the most in any known earthquake in human history.

CIVIL SOCIETY THEORY IN THE CHINESE CONTEXT

From a concept that was central to the public sphere (Habermas, 1989) in late seventeenth- and eighteenth-century Europe, to the current debate of its nature given the rise of the market economies, civil society has lacked a consensual definition (Edwards & Gaventa, 2001). The term has always been in flux with social realities. One approach emphasizes the civil society development process through the changes in institutional structures, and thus includes the possibilities of further theoretical adjustments to the term itself. From this dynamic perspective, civil society becomes "a sphere of social interaction between economy and state, composed above all of the intimate sphere (especially the family), the sphere of associations (especially voluntary associations), social movements, and forms of public communication" (Cohen & Arato, 1992, p. ix).

The concept of civil society was first applied to liberal Western democracies, and there is reason for caution in employing the concept for the analysis of organizations and associations in Chinese society. Due to important historical and sociopolitical differences, civil society in China is in many ways distinct from civil society in the West (Wang, 2009; Wu & Gong, 2008). From the founding of the People's Republic of China in 1949 through the late 1970s, the boundaries between the state and the society were hard to differentiate as the two established a near-unity relationship with one's functional territory almost fully overlapping with the other (Han, 2002; Tang, 1996; Tao, 2009). The market reform and opening-up policy implemented in 1978 and onward is a direct driving force that facilitated China's social diversification and the rise of a distinct Chinese civil society (Wang, 2009; Zhang, 2009).

In contrast to the Western experience, China's process is originally promoted by the nation-state, nurtured by the emergence of a socialist market economy, and thus develops under the interaction between the state and the market institutions (Wu & Gong, 2008). When treating civil society as a distinct domain developed alongside the state and the market system, few have systematically explored the interactive nature of civil society organizations from a dynamic point of view. By saying dynamic, I mean tracing and projecting a set of rules governing the actions of civil society organizations over time.

Earlier research on Chinese civil society paid significant attention to qualitatively depicting the characteristics and functions of nongovernmental organizations in postreform China, especially within the context of evolving state–society relations (Lu, 2009). However, no research has looked into the patterns of interactions among a set of nongovernmental organizations, given the institutional environment in the Chinese context. Such examination will shed light to understanding the evolution of civil society in a sociopolitical context where the state continues to play an important part in social development.

On the one hand, the party-state established a large number of officially controlled government-organized nongovernmental organizations, such as the All-China Women's Federation and the All-China Federation of Trade Unions. The government also has begun to encourage the development of certain types of nongovernmental organizations such as social welfare organizations, trade associations, and the like to shoulder some of the responsibilities for social services since 1978 (Ma, 2006; Yu, 2009). On the other hand, the party-state has learned a lot about the potential political risks of self-governing nongovernmental organizations from the democratic movement in 1989. Thus, it restricts the development of self-governing civic organizations in China through policies regulating nongovernmental organizations.

According to the basic characteristics distinguished by Yu Keping (2009), Chinese civil society organizations can be formally divided into nine categories: (1) trade organizations such as those professional and management associations connected to various industries; (2) charitable organizations such as disaster and poverty relief charities and foundations like the Red Cross and disabled persons' federations; (3) academic groups composed of scholars with similar interests; (4) political groups that safeguard citizens' political rights, some of which include the villagers' committees and neighborhood committees; (5) community organizations such as homeowner associations and community welfare centers; (6) social service organizations that provide social and public welfare services such as environmental protection, culture, education, and health; (7) citizens' mutual assistance organizations such as rural agricultural associations, and urban and rural mutual assistance associations; (8) common-interest organizations based on citizens having common leisure, career, and sports interests; and (9) nonprofit consulting service organizations including privately operated noncommercial entities.

One type of organization that experienced dramatic emergence after the Wenchuan earthquake was the social service organization, particularly in Sichuan province. During the emergency response stage, Chinese citizens utilized social media such as online messenger tools to virtually connect with each other and later meet up in person to organize various kinds of collective support for the earthquake-impacted areas (Lu, 2013). Groups or organizations would gradually form as more people joined the effort to help with disaster recovery. In fact, the adoption of social media technologies not only can provide opportunities for public participation (Sutton, Palen, & Shklovski, 2008) but it was also found to facilitate communication and collaboration for both local and distributed networks (Sutton, 2010). During my interview with some of these group and organizational leaders in 2011, some of the most important motivations for them to participate and eventually establish their grassroots organizations involve recognizing their social responsibilities, an awakened sense of giving to others, and participating in a larger cause. However, over time, especially starting with the long-term recovery stage, many of these grassroots nongovernmental organizations suffered from shortages of funding mainly due to difficulties in receiving legal status as a registered organization (Lu, 2013).

Historically, one major problem for obtaining legal registration status is the difficulty for nongovernmental organizations to get approval from professional supervisory agencies. For the professional supervisory agencies, approving a nongovernmental organization means being held responsible for the activities of that organization. The political risks that nongovernmental organizations may bring make qualified agencies reluctant to serve as the supervisory agencies for nongovernmental organizations. In order to overcome the registration barrier, many nongovernmental organizations have to register as for-profit business firms (Ma, 2006). Meanwhile, a lot of nongovernmental organizations operate without registration, thus bearing the risks of being illegal organizations. It is estimated that the number of these nonregistered nongovernmental organizations is about 10 times that of registered nongovernmental organizations in China (Wang, 2007). Wang (2007) also argues that the dual management mechanism actually puts nongovernmental organizations into a conflicting relationship vis-à-vis the government because the major goal of government agencies is to avoid the risks that nongovernmental organizations may cause, rather than to promote the development of the nongovernmental organizations.

The other institutional barriers that Chinese nongovernmental organizations face include the minimum membership requirements—no less than 50 individuals or 30 legal entities—and the noncompetition requirement that no new nongovernmental organizations can be established if a nongovernmental organization already exists in the same administrative field and in the same region (Ma, 2006). These high bars for establishing a nongovernmental organization is one of the most important policy tools for China's government for controlling nongovernmental organizations, particularly at the grass roots (Ma, 2006; Yu, 2009). Though the party-state encourages the development of certain types of nongovernmental organizations that are nominally apolitical, it is keenly aware of the potential challenges that nongovernmental organizations may pose to its rule if they are strong enough. The massive involvement of citizen-led civil society organizations after the 2008 Wenchuan earthquake generated an opportunity to closely examine the development processes of self-governing civic organizations. Theoretically speaking, such examination also provides insight into the dynamic aspect of civil society within the context of contemporary China.

SOCIAL NETWORKS AND DISASTERS

I use social network analysis to understand the role of civil society in disaster response, recovery, and mitigation from a process-oriented perspective. Social networks have several advantages investigating a process of change. Networks allow investigators to emphasize structural environments that civil society actors enact and that constrain them. Different types of structures, either at the whole network level or at the substructure level, reveal variation in cohesiveness vs. disintegratedness of interorganizational relations. Basic network structural properties such as reciprocity, transitivity, and clustering can provide the key for examining how actors choose to situate themselves across different time stages of the disaster recovery process. For example, a communication network that lacks reciprocity during the emergency response stage of a disaster means that actors are not engaged in general information sharing with each other. One's act of reaching out for information does not lead to responses by the other party. This kind of behavior would eventually lead to more isolated actors carrying out duties on their own at a time when sharing information about each other's needs and resources is most important. When a communication network is characterized by a pattern of transitivity, it can be expected that if actor A directs communication to actor B, and actor B directs communication to actor C, then A also directs communication to C or vice versa. During the disaster recovery stage, such structure not only enhances information flow to more members of the network but also promotes group formation, clustering, and possibly future collaborations among actors.

Recent development in network methodology allows theories to be formulated based on tracing the rules governing institutional change process, thus exemplifying their predictive power in longitudinal models. Through the dynamic network modeling process, both network structural changes and factors related to actor attributes can be specified to explain the longitudinal development of civil society organizations. For example, the tendency for organizations to respond to each other's information requests signifies that communication network is more likely to be driven by reciprocal relationships between pairs of actors. When organizations have a tendency to only work with those that have similar areas of expertise, the network will be more likely to be segregated by different clusters of actors with rare communication lines across clusters. The main research question of this study is designed to have two components: (1) What are the dynamics of the interorganizational structures in civil society in the disaster recovery context? (2) Are there rules of change that can help explain interorganizational behaviors short term and long term after the disaster?

RESEARCH DESIGN
NETWORK BOUNDARY CLARIFICATION

The level of focus for this study is a whole network. Each civil society organization active in the Wenchuan recovery is referred to as a network actor. With the permission of a key civil society organizational informant, I was able to gain access to a list of the civil society actors with whom the informant and their organization had contacts throughout the earthquake response and recovery settings. The determining factor for inclusion is that the actors participated in the emergency response stage and sustained their actions into the long-term recovery period. Some actors, upon my first contact with them through email or phone conversations, confirmed their participation or engagement long term after the disaster, and in these instances, the network boundary clarification was an ongoing process throughout the data collection stage.

NETWORK CONTENT SPECIFICATION

I explore communication content to capture the process of information exchange after the earthquake. During both the emergency response and the recovery periods, information seeking and exchange became an important activity for Chinese civil society actors. Organizations communicated with each other in order to locate the most updated information regarding the types of assistance needed across the earthquake-affected region. As local grassroots organizations are better informed of the location and of the degree of help needed for the earthquake-damaged areas, they became the key sources of information for other out-of-province civil society organizations seeking to provide support. As Edward Laumann and David Knoke (1987) noted, "the greater the variety of information and the more diverse the sources that a consequential actor can tap, the better situated the actor is to anticipate and respond to policy events that can affect its interests" (p. 13).

THE SURVEY QUESTIONNAIRE METHOD

I adopted a survey questionnaire technique as part of my data collection procedure. The first section of the survey questionnaire was designed to collect the attribute data regarding an actor's date of establishment as an organization, whether they have been registered as a formal nongovernmental entity with the state, and the types of activities the actor engaged during the disaster recovery stage. The second section of the survey compiles a roster list of all the actors with whom the respondent interacted in any way during response or recovery. The means of communication included emails, phone calls, online messengers, and in-person contacts.

The response formats utilized the binary judgments (Wasserman, Scott, & Carrington, 2005) design within which respondents specify whether the organizations that they represented had and/or were having a communication relationship with each actor on the roster. Communication was further defined as those activities related to information exchange. The respondents were asked to name their communication activities for the following time phases: (1) before the earthquake, (2) during emergency response period, and (3) during the disaster recovery period. The main unit of analysis of this study is at the organizational level, while both informal and formal ones were counted. Thus, each respondent represented the organization that they were working for at the time of data collection.

DATA TREATMENT

A total of 138 actors comprised the roster list, 136 of which were civil society actors. Among them, nonregistered informal social groups constituted 21.3%. Sichuan-based local grassroots groups and nongovernmental organizations comprised 75.7%. This category also included the civil society organizations that established long-term field offices for disaster recovery in Sichuan after the 2008 earthquake. The state actor is designed to be one node representing all the state agencies. The market actor is another node in the network representing all the private enterprises. The purpose of including these two general actors in the study is to take into account the cross-sector institutional environments within which the communication networks developed. The survey questionnaires were distributed to a total of 136 civil society organizational informants. A total of 63 questionnaires were returned—each from a different civil society organization. The response rate was 46.32%. The theoretical concern for this level of missingness in network data is the loss of information in terms of how organizations connect to each other, thus creating problems in fitting models to the data and maybe leading to difficulties in reaching good convergence.

Considering the current available tools in treating the incompleteness of responses of longitudinal network analysis, I first ran the categorical core-periphery analysis implemented in UCINET program (using CORR algorithm) on all three periods of the communication networks. A total of 70 actors eventually composed the core of communication networks across all three time periods. These 70 actors performed various types of activities during the recovery stage of the Wenchuan earthquake. They engaged in housing reconstruction, assisting special groups such as elderly, disabled, women, and children, providing psychological support for people impacted by the disaster and supporting the livelihood development for urban and rural areas. Aside from these six categories, I added one "other" category to account for those actors that did not target their services for a particularly group of people or region. Major activities in this category include community service provision and community capacity building.

I then utilized the composition change function in simulation investigation for empirical network analysis (SIENA) to further reduce the impact of nonresponsiveness in the data. This means the entry time of each actor into the communication network was being documented through the data coding process and also specified a composition change file, which was being modeled as exogenous events in the network model. As a result of this nonresponsiveness treatment process, the final percentage of missing data is 15.7%, which was in the relatively safe range between 10% and 20% to reach stable estimations as specified in the RSiena Manual (Ripley et al., 2012).

LONGITUDINAL MODELING OF NETWORK DYNAMICS USING RSIENA

I adopted longitudinal network modeling to investigate the rules governing communication network evolution over time. Specifically, this involves stochastic actor-oriented models that analyze network dynamics (Snijders, 1996; Snijders, van de Bunt, & Steglich, 2010). The modeling process is based on the paradigm of statistical inference implemented via the RSiena package in the statistical software language R.[2] The dependent variable is the observable communication networks over the specified three time waves: before the earthquake, during emergency response period (up to one year after the

---

[2]In this study, I focus on the most basic longitudinal model specification with the objective function depicted as follows.

$$f_i(\beta, x) = \sum_{k=1}^{L} \beta_k S_{ik}(x)$$, the symbol i represents the "ego" or the focal actor in consideration. The weights β_k are statistical parameters indicating strength of effect $s_{ik}(x)$ (linear predictor). $f_i(\beta, x)$ represents the value of the objective function for actor   depending on the state x of the network, which is a state being perceived from the focal actor's point of view. Such state can be in terms of relationships and in actor covariates. The network effects are all included in the function $S_{ki}(x)$. The objective function thus determines the probabilities of change in the network, given that an actor has the opportunity to make a change. It can be depicted as the rules of network behavior as actors make their decisions to make or terminate a tie based on their overall evaluation of how they view the current state of the network and the effects of covariates.

disaster), and during the disaster recovery period (up to five years after the disaster). This research was conducted at the three-year mark after the earthquake and thus included retrospective data.

The network evolution dynamics are functions of four general categories of independent variables: (1) basic structural effects (outdegree, reciprocity, and transitivity-related effects); (2) degree-related structural effects capturing the trend of popularity of certain actors based on their current tie-building activities (indegree popularity, outdegree popularity, outdegree activity); (3) covariates capturing the effects of actor attributes and activity types (actor registration status, actor recovery activity type); and (4) interaction effects. I include actors' registration background as the explanatory actor variable to investigate whether the formality of actor institutional status has an effect on the formation and evolution of the communication network, controlling for reciprocity and transitivity. The constant covariate such as actor registration status is dichotomous with values of 1 and 2. The value 1 indicates a registered organization, and value 2 represents unregistered.

For recovery period, in particular, the types of activities that the actors engaged in are included as explanatory constant dyadic variables. In the original survey, the respondents were asked to name the type(s) of disaster recovery activity their groups or organizations had been engaged in after the earthquake into the recovery period. The variable *activity* is created as an aggregate constant dyadic covariate so as to examine its overall effect on communication network evolution. Each one of the activity types is also being singled out as constant covariates in order to see their respective effects on the dynamics of the communication network. These constant covariates are dichotomous with a value of 0 meaning not participating in the specified activity and a value of 1 meaning confirmed participation. The purpose here is to differentiate the various types of earthquake recovery-related activities and examine their explanatory power in the maintenance of communication networks over time.

I investigate two types of covariate effects: covariate ego and covariate alter. A positive covariate ego effect, such as a *registration ego* effect, means a tendency for actors with higher values (in this case, the nonregistered actors) to increase their outdegrees more rapidly. A positive *registration alter* effect implies a tendency for indegrees (i.e., nominations) of actors of higher values (the nonregistered actors) to increase more rapidly. A positive *housing ego* effect implies that those who participated in housing reconstruction to have a tendency to develop even more outreach activities. A positive *housing alter* effect means that there is a pattern for those participated in housing reconstruction to attract even more connections from others. Similar interpretation follows activity-related covariate ego and alter effects: elderly-disabled ego and elderly-disabled alter, women-children ego and women-children alter, environment ego and environment alter, psychology ego and psychology alter, as well as *livelihood ego and livelihood alter*.

I examine two types of interaction effects. One is the ego–alter interaction effect, which is designed to look at whether actors tend to have greater preference for others with similar traits or conducting similar type of activities. Variables that can be placed into this category include: registration ego × registration alter, housing ego × housing alter, elderly_disabled ego × elderly_disabled alter, women_children ego × women_children alter, environment ego × environment alter, psychology ego × psychology alter, livelihood ego × livelihood alter. The other is the interaction effect of registration similarity with reciprocity (Int. *Registration similarity × reciprocity*) and this is to look at the tendency for communication behavior to be reciprocal among organizations with similar registration status. Through the model development process, I particularly focus on the social selection (Steglich, Snijders, & Pearson, 2010) in the basic model. This process postulates that actors make their choice of ties based on attributes and the network embeddedness of the actor as well as those others in the network.

RESULTS
SOME DESCRIPTIVES

Table 10.1 shows some of the basic network descriptive measures for the three periods before and after the earthquake. I present the average indegree, reciprocity, and clustering measures here because comparing the changes in these basic network measures before and after the earthquake provides preliminary ideas on the degree of shifts in network building and sustainability.

From before the earthquake to the emergency response stage, the average indegree for information exchange increased from 2.7 to 9.8, representing a dramatic shift in the level of communication activity among civil society actors. Figs. 10.1 and 10.2 further demonstrate this change pattern. Note that not only were the isolated actors who appeared in the period before the earthquake drawn into the main connected network immediately after the disaster, but the network also became more compact and dense over time with increased reciprocity and clustering activities.

The tendency continued through the long-term recovery stage with a slight increase of average indegree to 11.1. Both reciprocity and clustering measures experienced further increase from emergency response to recovery stage. The communication network remained to be integrated as shown in Fig. 10.3. Observing the change patterns in standard deviation of indegree and outdegree measures, it can be concluded that after the earthquake, both the indegrees and the outdegrees of communication networks became increasingly variable with the latter higher than the former throughout the three time stages. This means that on the one hand, the agency motivations of civil society actors were being activated as prompted by the earthquake and sustained through the long-term recovery. On the other hand, the response of the activation may be differentiated.

DYNAMICS OF THE EMERGENCY RESPONSE COMMUNICATION NETWORK

With an intention to distinguish the patterns that govern the changes during the emergency response stage and the disaster recovery stage, the dynamics of the period from before the earthquake to immediately after event (t1 to t2) was estimated separately from the period between emergency response and long-term recovery (t2 to t3). Practically, knowing which structural effects were consistently making

Table 10.1 Communication Network Descriptive Measures

	Communication t1	Communication t2	Communication t3
Average indegree	2.7	9.8	11.1
Standard deviation in/out	2.9/5.2	6.8/13.5	6.9/14.0
Reciprocity	0.11	0.15	0.17
Clustering	0.17	0.24	0.28

"Communication t1" represents preearthquake period. "Communication t2" represents emergency response. "Communication t3" represents disaster recovery.

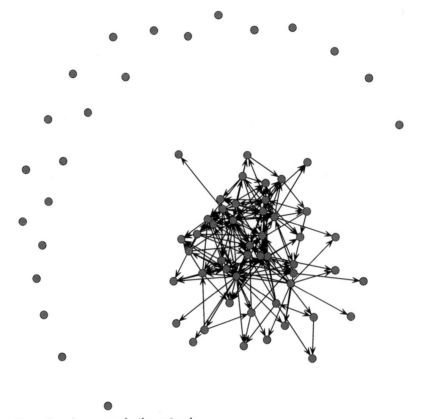

FIGURE 10.1 Preearthquake communication network.

Lu, J. (2013). The Wenchuan Earthquake Recovery: Civil Society, Institutions, and Planning. *Retrieved from ProQuest Digital Dissertations. (UMI No. 3609950).*

statistically significant contributions to network evolution for these two time periods also allows the study to initiate policy implications for both short- and long-term periods after a disaster, thus effectively incorporate the social aspect of constructing resilience by distinguishing characteristics of a change process. Table 10.2 shows the model estimation results for the emergency response period and the recovery period.

For the emergency response period, actors started out by contacting others that were also involved in the process to seek out further information immediately after the earthquake. The communication network at this period of time showed tendencies toward reciprocity (parameter value 0.41). Reciprocated responses were valued positively, and this reveals initial evidence of the emergence of a Chinese civil society triggered by the Wenchuan earthquake.

Aside from reciprocity, the emergency response communication network also showed tendencies toward various types of clustering, or network closure, effects. The most prominent one is the *transitive triplet* effect. This effect postulated that more intermediaries will add proportionally to the tendency to transitive closure. From Table 10.2, we can see that the transitive triplet effect turns out to be positively

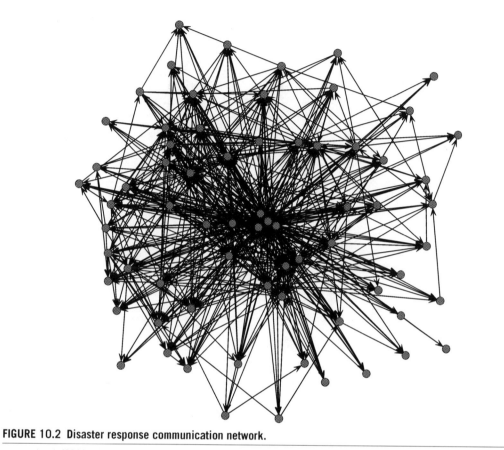

FIGURE 10.2 Disaster response communication network.

Lu, J. (2013). The Wenchuan Earthquake Recovery: Civil Society, Institutions, and Planning. *Retrieved from ProQuest Digital Dissertations. (UMI No. 3609950).*

significant (parameter value 0.11). This means that "friends of friends tend to be friends." The communication relationship triangle had a tendency to be closed. But it is not reinforced by *transitive ties* effect. The transitive triplet effect is often discussed together with the *three-cycle* effect, which can be interpreted as generalized reciprocity as the opposite of hierarchy. The findings show that the emergency response network did not have a tendency to develop three cycles that would work against the hierarchical ordering exemplified by transitive triplet effect.

In the context of emergency response after the earthquake, this pattern can be interpreted as a sense of eagerness for sociability among civil society actors to build up ties with others. Actors on the initiation end of a triadic relationship also tended to close up the transitive circle on their own action, rather than waiting for the actor on the other end to reach out.

Another network closure effect being tested was the *balance effect*, which measures the tendency for a network to have and create ties in such a way that other actors tend to make the same choices as the ego (actor i). This effect is significant but negative (parameter value −0.09). This means that there were few balanced triadic closures that would reveal a network tendency to have and create ties to other actors who

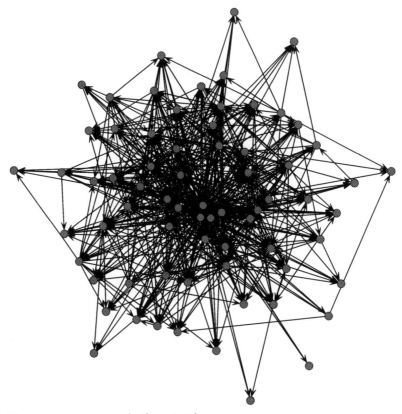

FIGURE 10.3 Disaster recovery communication network.

Lu, J. (2013). The Wenchuan Earthquake Recovery: Civil Society, Institutions, and Planning. *Retrieved from ProQuest Digital Dissertations. (UMI No. 3609950).*

made the same choices as ego. In other words, the role formation of Chinese civil society actors during the response period might be at a stage of diversification when considering the ways in which actors reach out to others. When looking at the number of outgoing choices and nonchoices that actors had in common, few actors can replace the role of the others or have structural equivalence with respect to out ties.

The *indegree popularity effect* is positive (parameter value 0.93), indicating actors with higher indegree were more attractive for other actors to send further incoming ties. In other words, high indegrees reinforce themselves, and there was a tendency for differentiated actor indegrees in the communication network. The *outdegree popularity* effect is negative (parameter value −0.17), indicating that those actors who nominated many others in the communication network were actually less popular when considered by others as potential information exchange partners. At this stage of time, even though the tendency for proactivity in exchanging information was prevalent in the network, such behavioral pattern had little effect on actors receiving further incoming ties.

The result also shows a negative effect in terms of the interaction between actors' reciprocity and them having the same registration status (parameter value −1.39). At first glance, this is rather

Table 10.2 Rules Governing Emergency Response and Recovery Periods Communication Network Evolution

Effect	Response (t1-t2; From Before Disaster to Emergency Response)		Recovery (t2-t3; From Emergency Response to Recovery)	
Within Network	**Parameter Estimate**	**(s.e.)**	**Parameter Estimate**	**(s.e.)**
Outdegree	−3.30**	(0.18)	−4.18**	(0.69)
Reciprocity	0.41†	(0.23)	0.91**	(0.17)
Transitive triplets	0.11**	(0.04)	0.08*	(0.04)
Three cycles	−0.12	(0.07)	0.01	(0.05)
Transitive ties	−0.89**	(0.21)	1.01**	(0.35)
Balance	−0.09**	(0.01)	−0.01	(0.01)
Indegree popularity (Sqrt)	0.93**	(0.07)	0.43**	(0.06)
Outdegree popularity (Sqrt)	−0.17*	(0.08)	−0.29*	(0.12)
Outdegree activity (Sqrt)	–	–	0.15	(0.13)
Int. Registration similarity × reciprocity	−1.39**	(0.40)	−0.48†	(0.27)
Registration alter			−0.06	(0.13)
Registration ego			0.41**	(0.12)
Registration ego × registration alter			0.12	(0.25)
Activity			−0.07	(0.11)
Housing alter			0.03	(0.12)
Housing ego			−0.28*	(0.13)
Housing ego × housing alter			0.84**	(0.31)
Elderly_disabled alter			0.00	(0.11)
Elderly_disabled ego			−0.22†	(0.11)
Elderly_disabled ego × Elderly_disabled alter			0.14	(0.22)
Women_children alter			0.06	(0.15)
Women_children ego			0.74**	(0.15)
Women_children ego × Women_children alter			0.18	(0.22)
Environment alter			−0.10	(0.11)
Environment ego			−0.50**	(0.15)
Environment ego × environment alter			0.31	(0.23)
Psychology alter			−0.16	(0.13)
Psychology ego			0.40**	(0.12)
Psychology ego × psychology alter			0.33	(0.23)
Livelihood alter			0.17	(0.12)
Livelihood ego			−0.02	(0.14)
Livelihood ego × livelihood alter			0.32	(0.22)

†$p<.10$, *$p<.05$, **$p<.01$ (two-sided).
"s.e." stands for standard error. Coefficients from standard SIENA (4.0) analysis implemented in the R statistical system of directed network matrix. All statistics converged with t-statistic <0.1. The use of dashed lines in certain cells indicates that the variable was dropped due to insignificance during the model selection process. For period t1-t2, the blank cells registration ego, registration alter, and interaction of registration ego and registration alter indicate that these effects were dropped due to the significant instability of model estimation by including them. All the activity-related variables were estimated only during the recovery stage, thus their corresponding cells for the emergency response period are blank.

counterintuitive because, normally, we expect reciprocity based on homophily in actors' traits. In other words, actors with similar characteristics would tend to reciprocate more with each other. For this study context, the negative parameter can be interpreted as follows. Tie reciprocation might be easier to form if two actors had similar registration status, but doing so won't bring much benefit or "satisfaction" to the actors. When a tie was reciprocated for two actors with different registration status, the other actor would develop more appreciation and also a sense of accomplishment in terms of reaching out for those that were of different traits.

Overall, the civil society communication network immediately after the Wenchuan earthquake showed the following evolutionary characteristics. There was a relatively strong tendency for reciprocity as demonstrated by the significant reciprocity parameter. Evidence for transitive closure was found as seen in the significant effects of transitive triplets. The positive sign of this parameter shows that there was a tendency for civil society actors to develop their proactivity immediately after the crisis event. The examination of network closure effects also reveals that few structural equivalences tended to develop with respect to outgoing ties, as demonstrated by negative balance effect. The tendencies toward closure were not completely egalitarian and showed some evidence for local hierarchical formation, as seen in significant positive indegree popularity effect and negative outdegree popularity effect. In addition, the emergency period did not show evidence of tendency to reciprocation being segregated by registration status.

DYNAMICS OF THE DISASTER RECOVERY COMMUNICATION NETWORK

The parameter values for the disaster recovery network model are shown in the third column of Table 10.2. When compared to its emergency response stage counterpart, the recovery period network had a stronger tendency for reciprocation (parameter value 0.91). This provides evidence that the initiation for communication ties immediately after the earthquake was not just impulsive acting on the civil society part. The persistence of such proactivity is a sign of commitment and sustainability into the longer term. The recovery network also had stronger tendencies for transitive closure, demonstrated by positively significant transitive triplets (parameter value 0.08) as well as transitive ties effects (parameter value 1.01). The balance effect turned out to be insignificant during the recovery period. This shows that the communication network might start to stabilize over time. At the same time, the tendency for local hierarchicalization continued to be evidenced by a positive indegree popularity effect (parameter value 0.43) with negative outdegree popularity effect (parameter value −0.29). This means that over the long term, active civil society actors with higher outdegrees were still less likely to be chosen as information exchange partners, demonstrating some kind of status effect in both the emergency response and recovery network dynamics.

The tendency to reciprocate among actors with diverse registration status sustained during the recovery stage. In the Chinese context, the nonregistered actors also tended to be those that were most grass roots and domestically originated after the Wenchuan earthquake. These actors had relatively fewer resources and smaller networks compared to the formally established registered actors (Lu, 2013). Thus, the maintenance of a strong tendency to reciprocate among actors with differentiated registration status during recovery means that civil society actors began to develop a general nurturing environment to help the growth of nonregistered social groups. Despite the fact that China is known for its harsh institutional environment for newly formed social groups to establish at the time when this research was conducted, particularly after the government established a three-year time limit for earthquake recovery, civil society actors showed sustainability for cross-registration status ties. As a result, the communication network was less likely to be segregated by registration status and there was a sense of reaching toward a common goal of disaster recovery.

The registration ego effect (parameter value 0.41) is found to be positively significant for the recovery model. Since higher values of registration status represented nonregistered actors, the result infers that informal social groups tended to engage in building and initiating more communication and information-sharing relationships. This is a distinct feature that pertains only to the disaster recovery period. The result demonstrates that over the long term, the nonregistered grassroots actors started to become the driving force in the development of communication networks. These actors, despite their informal institutional status, became the backdrop for information outreach.

When testing for the significance of a variety of types of actor activities during the long-term recovery period, the following results can be concluded. Civil society actors who participated in the activities of housing recovery (parameter value −0.28), caring for the elders and disabled population (parameter value −0.22), and environmental protection (parameter value −0.50) tended to communicate less with others in the network. On the other hand, the outreach tendency of the recovery network seems to be driven by those who participated in the social work areas such as caring for women and children (parameter value 0.74) and psychological counseling (parameter value 0.40). Furthermore, actors who engaged in housing recovery activities also tended to have greater communication preference for other actors who also were practicing in the same field. This homophily effect (parameter value 0.84) only existed for the area of housing and not with any other types of activities. This indicates that housing recovery may be a sector that requires more sharing of information among actors long term after the earthquake when compared to other types of activities.

EMERGENCE OF STRUCTURAL RESILIENCE

The dynamics of civil society—investigated through the process by which Chinese nongovernmental organizations responded to a catastrophic disaster—demonstrated Chinese civil society's active role in building sustainable interorganizational communication relationships. Despite the registration barriers at the time of this study, locally grown citizen-led informal groups played a significant part in establishing and maintaining communication ties among network actors. As soon as the joining and inclusion processes took place after the earthquake, the communication network was activated toward a high level of structural cohesiveness. This is demonstrated by the immediate structural integration during the emergency response stage.

The dynamics of the recovery process proceeded with signs of action persistence from actors with informal institutional status, thus demonstrating the source of commitment long after the disaster event. Patterns of interaction such as reciprocity and transitivity persisted throughout the recovery stage. The actors that established a high level of indegree communication ties during emergency response continued to occupy their favored position in attracting more communication ties from others. The tendency for heterogeneity in actor institutional statuses is a strong predictor in whether a communication tie was being reciprocated or not.

Civil society actors appeared to prefer to choose information exchange partners that were of different registration status from themselves. This willingness to engage in diversity building over the long run supported people's ability to withstand catastrophic change in the Chinese disaster recovery context. Thus, although the institutional environment in China did create obstacles for nongovernmental organizations to gain formal status, it did not inhibit the development of communicative interaction among organizations. In the Chinese case, the structural dynamics of civil society continued to evolve while the state continued to play a role in both the institutional and the network environments.

Facing a catastrophic disaster, Chinese civil society experienced a dramatic increase in outreach activity concentration. One of the most distinguished network development features was the strengthening of the resilient structure during the recovery stage, which essentially involved three factors. One is the sustainability of the dyadic and triadic structures in explaining the persistence of communication network cohesion. Not only did the tendency for tie reciprocation continue to exert its impact on recovery network evolution but this type of reciprocation was accompanied by actors seeking institutional status diversity. This type of recovery trend facilitated the proactivity of informal civil society organizations and helped them to be further integrated into the long-term communication network. Second, there was a strong behavioral tendency for actors to exchange information with those already having high indegrees. This shows that in the long run actors sought information exchange partners that had already gained popularity among others. Third, the varying degrees of significance in how recovery activities affect communication network development demonstrate that the chosen types of activities do make a difference in building resilient social structures.

CONCLUSIONS AND POLICY IMPLICATIONS

Actions of relationship building and their patterns of maintenance in the long term might reasonably be conceived as supporting a general resilience. A dynamic perspective is not only important for tracing the emergence and growth process of social resilience but also the analysis procedures of this study demonstrated that it can also be evaluated through observable measures for possible policy implementations. The process starts with actions taken on the part of civil society actors as the source of power to reconstruct the social structural conditions despite the distresses brought by the happening of the disaster. Civil society actors became more engaged when transitioning to the long-term social recovery stage and started seeking a communication environment that can strengthen the bonds established short term after the disaster event. Because the change process is partially motivated by the civil society actors' drive to bring about action, the continued growth and strengthening of such a structure is dependent on the collective awareness of actors recognizing each other's roles and positions, and consciously acting to expand the overall capacity of the structure.

One significant policy implication of this study derives from the ability of the model to identify the trends of activity types through distinct periods after a disaster, particularly the recovery period for this research study. The tendency for Chinese civil society actors engaged in assistance to women and children and psychological counseling services to initiate more communication ties may send a signal to policy makers that these are emerging sectors for further social assistance. In this case, the public agencies may need to plan for more engaged actions to partner or collaborate with civil society organizations practicing these types of earthquake recovery assistance. More relaxed institutional constraint such as the ease for registration may need to be implemented so that these organizations can grow to survive financially in the longer term. The tendency for civil society actors in the housing recovery to seek out each other for information exchange can be interpreted as another signal of the greater need in facilitating the coordination among organizations in this sector.

Overall, future research needs to further engage in the structural and the dynamic theoretical formulations of network development in the context of an extreme event. Detailed specifications that can translate theories into practices require more longitudinal modeling studies to discover significant behavioral trends of civil society actors to enhance the capacity of the coordinated disaster response, recovery, mitigation, and preparedness. Expansive cross-cultural examinations may deem it fruitful to develop network theories in particular relevancy to disaster recovery policy making, planning, and risk management.

ORGANIZATIONAL SUPPORT NETWORKS AND RELATIONAL RESILIENCE AFTER THE 2010/11 EARTHQUAKES IN CANTERBURY, NEW ZEALAND

11

Joanne R. Stevenson[1,2], David Conradson[2]

[1]Resilient Organisations Ltd., Sheffield, New Zealand; [2]University of Canterbury, Christchurch, New Zealand

CHAPTER OUTLINE

INTRODUCTION

Organizations that are resilient—those able to absorb or mitigate the negative impacts of a crisis and to recover, adapt, and thrive in its aftermath—are more likely to reach their goals in the face of challenges and foster resilience in the places and communities they inhabit. Resilience requires that organizations maintain a range of capabilities, practices, and behaviors that can improve their coping capacity, learning, and innovation in the face of unpredictable events (Kendra & Wachtendorf, 2003; Linnenluecke & Griffiths, 2010). Organizations can extend their access to supportive resources, information, and other assistance through interorganizational and interpersonal networks.

Studies of interorganizational networks in postdisaster response and the early phases of recovery provide insights into the ways interorganizational networks support individuals and communities (Hutter, 2011; Raab & Kenis, 2009), but few offer empirical insights about how networks support the recovery of nonresponder private organizations and businesses (eg, Bowden, 2011; Doerfel, Chewning, & Lai, 2013; Doerfel, Lai, & Chewning, 2010; Graham, 2007).

The research presented in this chapter examines the kinds of networks and relationships that support organizational recovery. Drawing on in-depth case studies of 32 organizations affected by the 2010/2011 earthquakes in Canterbury, New Zealand, we explore the organizations' postdisaster networks, including the types of support mobilized and the kinds of relationships. We also identify network features that facilitated positive and adaptive organizational responses post disaster. We address three research questions related to these organizations' networks:

- What kinds of support did organizations mobilize post disaster to aid their recovery?
- What kinds of relationships comprised postdisaster support networks?
- In what ways do the networking behaviors of organizations experiencing positive postdisaster trajectories differ from those with degenerative trajectories?

The discussion begins by establishing the theoretical foundations of this research, which is situated at the convergence of work on resilience, organizational embeddedness, and support networks. Then we describe the sequence of earthquakes that struck Canterbury, New Zealand, in 2010 and 2011, and the case study research methodology. We then present the results, describing the nature of support organizations received and assessing the ways in which organizations with different postdisaster trajectories engaged with their support networks. We conclude with a discussion of the ways organizations could cultivate their networks in order to enhance their resilience.

THEORETICAL FRAMEWORK

Theories of organizational resilience have identified a number of traits and processes that enhance an organization's ability to survive and thrive in a crisis. These include a culture of transparency and trust (Weick, 1993), planning and risk awareness (Boin & Lagadec, 2000), the ability to access and mobilize resources (Burnard & Bhamra, 2011; Somers, 2009), strong and engaged leadership, and an engrained culture of organizational learning (Seville, Van Opstal, & Vargo, 2015). James Kendra and Tricia Wachtendorf (2003, p. 49) characterize organizational resilience as a type of craftsmanship that requires artistry, perception, and the ability to sense changes and continually adjust responses in a way that incorporates learning.

A growing number of studies and commentaries highlight the importance of collaborative action and the relevance of interorganizational and interpersonal networks for resilience (Doerfel et al., 2013; Johnson & Elliott, 2011; Johnson, Elliott, & Drake, 2013; Stevenson et al., 2014). These networks can be characterized as part of an organization's extended resource base (Horne

& Orr, 1998; Lee, Vargo, & Seville, 2013; Mallak, 1998). Organizations, especially small- and medium-sized enterprises, often do not have the capacity or inclination to create redundant systems, maintain high levels of inventory, and have staff trained for every disruptive eventuality. Instead, the additional resources, operational capacity, and information can be accessed through extraorganizational relationships.

Pioneering work in the 1990s linked relational network structures (i.e., the patterns of exchanges between people or organizations) to organizational advantage (Baker, 1990; Nahapiet & Ghoshal, 1998). These studies found that performance differences between firms often reflected differences in a firm's ability to create and exploit social capital (Nahapiet & Ghoshal, 1998). Subsequent research has shown that social capital (ie, the actual and potential resources accumulated and available through relationships) between organizations is able to ease transaction costs with suppliers (Baker, 1990; Uzzi, 1997), improve the acquisition of strategic resources (De Wever, Martens, & Vandenbempt, 2005), shape patterns of collective innovation (Ahuja, 2000), and help firms acquire new skills and knowledge (Inkpen & Tsang, 2005; Podolny & Page, 1998). Social networks also play an important role in buffering in the effects of stress on employees' mental well-being in organizations (Monge & Contractor, 2001).

How do networks enable organizations to meet their needs in the time and resource-constrained postdisaster environment? Researchers have observed organizations engaging in a range of exchanges that facilitate recovery and adaptation, such as information transfer, exchange of manpower, donations of material or financial support, or delegation of authority (Butts, Acton, & Marcum, 2012). Networks are used for circulating intelligence to organizations in rapidly changing emergency environments, such as after the 2001 World Trade Center terrorist attacks (Tierney & Trainor, 2004). Another important function of organizations' networks after a disaster is brokerage, the act of linking two otherwise unconnected actors (Doerfel et al., 2013; Spiro, Acton, & Butts, 2013).

Disaster-affected organizations are not passive recipients of support. Organizations have differing abilities to perceive emerging issues, reciprocate support, improvise, and connect with others. Achieving strategic change and social innovation requires agency within a network (Moore & Westley, 2011). Organizations must intentionally pursue network activities by, for example, relationship building, knowledge and resource brokering, and guiding and encouraging network members toward a desired outcome. Networks are both potential strategic resources and sources of unique challenges. Some organizations navigate this complex relational landscape better than others.

THE CANTERBURY EARTHQUAKE SEQUENCE

The Canterbury region of New Zealand has faced years of disruption and multiple cycles of response and recovery following a series of earthquakes that began in September 2010 (Fig. 11.1). Between September 2010 and December 2011, Canterbury experienced six significant earthquakes and thousands of aftershocks.

The largest earthquakes in September 2010 (7.1 Moment Magnitude) and February 2011 (6.3 Moment Magnitude) had very different impacts. The September 2010 earthquake caused significant shaking and serious, but localized, liquefaction damage. Due in part to its timing and location, however, there were few serious injuries and no fatalities attributed to this earthquake. The February

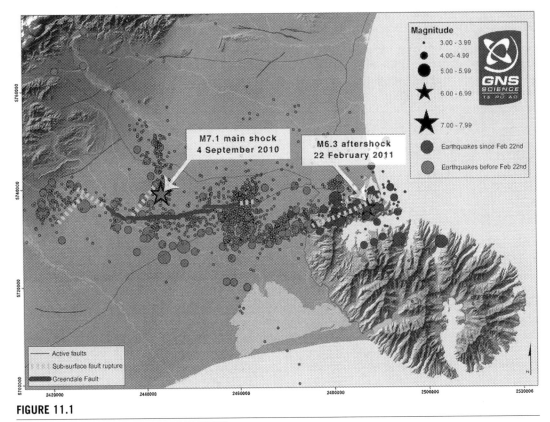

FIGURE 11.1

Canterbury earthquake sequence 2010/2011.

GeoNet (2011). Canterbury quakes. Retrieved from http://info.geonet.org.nz/display/home/Canterbury+Quakes.

2011 earthquakes occurred on a fault approximately 10 km east of the city of Christchurch's central business district (CBD). This earthquake led to 185 fatalities, thousands of injuries, and widespread damage to infrastructure and buildings caused by seismic shaking, liquefaction, rock fall, and landslides (CERA, 2012; NZ Police, 2012).

Access to property was a major challenge for businesses after the earthquakes. After the February 2011 earthquake, the city erected a cordon around the Christchurch central business district, and substantial parts remained in place for nearly 2.5 years. The Canterbury Earthquake Recovery Authority implemented a zoning system for all residential land in Christchurch. Additionally, the Canterbury Earthquake Recovery Authority required post-earthquake engineering evaluations that eliminated access to buildings, with sometimes only a day's notice, as late as 2012 and 2013. Over 7850 residential properties were deemed unsuitable for continued residential occupation, and approximately 10,000 required rebuilding or significant repairs (Statistics New Zealand, 2013), causing significant localized depopulation, yet development and population growth in other parts of the city. Badly damaged suburban areas of Christchurch lost up to 63% of their population between 2011 and 2013 (Statistics New Zealand, 2013).

METHODOLOGY
CASE STUDY SELECTION

We collected data at several points throughout 2010–13 in order to track organizations' recovery trajectories, and to assess the range of strategies and support they employed to navigate the postdisaster environment. The case study organizations were selected from three heavily impacted town centers in the Canterbury region: Christchurch, Kaiapoi, and Lyttelton.[1] The 32 organizations were drawn from culture, recreation, and social services ($n=4$); hospitality ($n=5$); retail ($n=12$); professional and personal services ($n=5$); manufacturing and wholesale ($n=3$), and the information, communications, and technology sector ($n=3$). The case studies were initially identified from a larger sample of 366 Canterbury organizations surveyed in 2010.[2] Each of the selected organizations had to: (1) exist prior to the Canterbury earthquakes (as opposed to those that emerged in the response and recovery phase), (2) have a physical site or base in the study area prior to the earthquakes, and (3) be willing to participate in a longitudinal study.

DOCUMENTING THE CASE STUDIES' JOURNEYS

Representatives from the 32 case study organizations filled out questionnaires at four points in time: 2010 (following the September earthquake), 2011 (following the February earthquake), 2012, and 2013. Site visits and participants' observations at each organization's premises—whether closed or operational—provided additional context and the ability to observe how organizations were engaging with their environments and customers. Semistructured interviews were undertaken with one or more representative(s) from each case study organization, and these interviews lasted between 1.5 and 4 h. Using a combination of financial data and qualitative assessment,[3] we were able to describe each organization's postdisaster trajectory in broad terms. *Trajectory* in this chapter refers to the organizations' health over time.

NETWORK DATA COLLECTION

During the interviews, the researcher and research participant recorded detailed information about the organization's postdisaster support networks using participant-aided sociograms (Fig. 11.2). As part of the data collection for these sociograms, participants were asked to identify the people and organizations that helped them, in any way, to run their business and adjust to changes following the Canterbury earthquakes. Using a large paper version of the participant-aided sociograms with the case study organization (ego) at the center of the network, we recorded: (1) the names of the people or organizations that helped the case study

---

[1]Both the September 2010 and February 2011 earthquakes heavily impacted the Christchurch Central Business District. Kaiapoi, approximately 12 miles (20 km) north of Christchurch, was most heavily damaged in the September earthquake, with severe liquefaction and lateral spreading. Lyttelton, a port town about seven miles (11 km) east of Christchurch, was most heavily damaged in the February 2011 earthquake, principally as a result of shaking and rock fall.

[2]For more on this larger study, see Kachali et al. (2012), Stevenson, Seville, & Vargo (2012), and Whitman et al. (2014).

[3]We included information gathered during the 2013 survey, where organizations reported either their revenue or "the financial resources available to support your mission for each financial year" each year for a five-year period (2008–2012) on a scale from very poor to excellent. Organization leaders also indicated whether they felt their organization was significantly better off, slightly better off, the same, slightly worse off, or significantly worse off than before the earthquakes, along with an explanation.

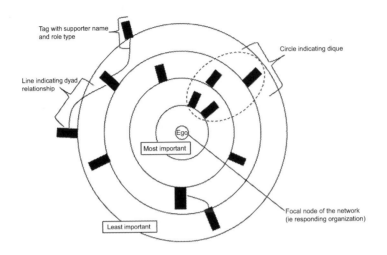

FIGURE 11.2

Schematic diagram of participant-aided sociogram (network data collection tool).

organizations (ie, its supporters) written on moveable tags, (2) the role of the supporter relative to the participant (eg, family member, supplier, government agency), (3) the perceived importance of that support for the organization (indicated by closeness to the center of the sociogram), (4) how long the participant or their organization knew the supporter, (5) where the supporter was based, (6) the type of support provided, and (7) approximately when support was initially received. This questioning can be repetitive and tiring for participants. To reduce this burden, questions 3–7 were only recorded for up to 15 supporters in each network (chosen using a standardized selection process). Thus, for the 457 supporters collectively listed by participants, analyses of questions 3–7 were based on a subset of 372 supporters.

The participant then placed each of the tags on the paper sociogram, which consisted of the participant organization at the center of four concentric circles. The circles represent the importance of the supporter to the organization (Fig. 11.2). The participant then indicated which of the supporters knew one another by drawing lines between pairs of supporters that knew each other (i.e., dyad relationships) and circles around groups of supporters in which each supporter knew every other supporter in the circle (i.e., cliques).

After gathering the network data on paper in the field, we coded the data into matrices for analysis in SPSS 19 and in UCINET. Additionally, we recorded, transcribed, and coded qualitative sociogram responses (eg, descriptions of support types) in QSR NVivo 9. The responses to surveys and interviews also aided interpretation of the network results.

FINDINGS
TYPES OF SUPPORT ORGANIZATIONS ACCESSED TO AID RECOVERY

The first research question concerned the kinds of support that organizations mobilized. Each organization in our study accessed some support from extraorganizational actors following the earthquakes. Not only were organizations faced with finding new premises or accessing cordoned buildings to recover critical data but many were trying to deal with disrupted suppliers, shifting markets, and stressed and exhausted employees.

Table 11.1 Types of Support Organizations Accessed Following the Earthquakes

Type of Support	Description	Proportion of All Received Support (%)
Services, assistance, and labor provided to the organization	Retail shop assistance, construction, doing taxes, assisting with relocation, cleaning	37
Support given to aid emotional coping of people within the organization	"Checking in," sympathy, supportive emails or calls	31
Monetary/financial support given to the organization or staff	Loans, donations, grants, purchasing from participant organization, insurance payouts, employee wage subsidies, bank overdrafts, debt forgiveness, credit, discounts	30
Nonmonetary material resources	Accommodations, storage space, vehicles, equipment, furnishings, office supplies, food/drinks, staff housing	25
Advice or information offered to the organization	Business mentoring; legal/financial advice; where to apply for aid	23
Social influence exerted by the tie	Brokerage, ie, bridging (socially "horizontal") or linking (socially "vertical") the organization to a wider range of people/organizations. Also included advocacy for the organization.	10

Across the organizations, participants reported receiving a range of support types, and these were coded into six mutually exclusive categories (Table 11.1).[4]

"Assistance and services" was the most common type of support (37% of supporters offered assistance in some form to case study organizations), while "brokering and advocacy" was by far the least common (10%).

Despite its low reported frequency, some organizations used "brokering and advocacy" to form new ties after the earthquake, including locating new suppliers at short notice. This category also included benefiting from another person or organization's political capital, such as gaining access to influential people or securing connections to critical resources.

THE RELATIONSHIPS THAT MAKE UP SUPPORT NETWORKS

The second research question asked about the kinds of relationships that characterized these support networks. Participants from the 32 case study organizations listed a total of 457 supporters in their networks. We examined network composition in terms of several variables, including geographic distribution, relationship duration, role description, and the nature and timing of the support offered.

Over 70% of the supporters came from within Canterbury (Table 11.2), meaning that they were also facing earthquake disruption. The duration of relationships with supporters was skewed toward long-term relationships. Organizations, organization owners, or employees had a relationship with about 47% of supporters for more than 10 years. In total, over 80% of relationships with supporters had been formed at least a year before the earthquakes.

---

[4]To be reported on the sociogram, "support," in the participant's opinion, had to have aided their organization's response and recovery at some point between September 2010 and mid-2012.

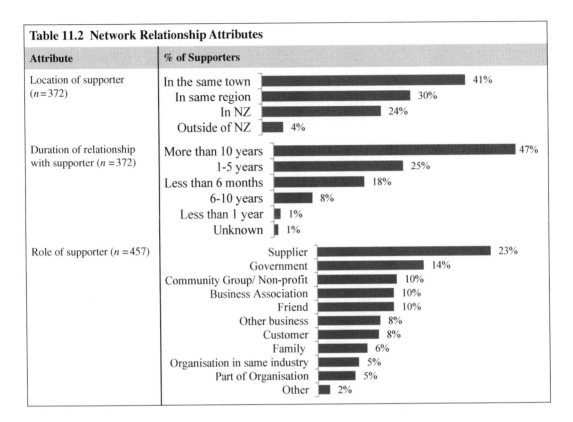

Table 11.2 Network Relationship Attributes

Attribute	% of Supporters	
Location of supporter (n = 372)	In the same town	41%
	In same region	30%
	In NZ	24%
	Outside of NZ	4%
Duration of relationship with supporter (n = 372)	More than 10 years	47%
	1-5 years	25%
	Less than 6 months	18%
	6-10 years	8%
	Less than 1 year	1%
	Unknown	1%
Role of supporter (n = 457)	Supplier	23%
	Government	14%
	Community Group/ Non-profit	10%
	Business Association	10%
	Friend	10%
	Other business	8%
	Customer	8%
	Family	6%
	Organisation in same industry	5%
	Part of Organisation	5%
	Other	2%

About 17% of organizations had not interacted with their supporters before the earthquakes. Often these supporters were newly formed agencies, such as the Canterbury Earthquake Recovery Authority or Recover Canterbury, which were able to provide unique earthquake specific assistance such as access to the cordon, grants, and mentorship. Some of the new supporters were contacted in order to address needs that emerged as a result of the earthquakes, such as new landlords or potential partners for temporary space sharing.

In the sociogram exercise, we also gathered information about the supporter's role relative to the organization (e.g., supplier, customer, friend of an employee). Suppliers of goods and services were by far the most common source of support reported by organizations following the earthquakes (Table 11.2). Organizations also received support through many formal and informal relationships, including with government agencies, community groups, friends, and family members.

Nearly every kind of support was significantly associated with a particular kind of supporter.[5] Unsurprisingly, emotional support was most likely to come from family members, information and advice from business associations and other businesses, and physical resources from suppliers. Brokerage and advocacy most frequently came from business associations and government agencies. In part this was because these types of supporters likely had the greatest access to broad and potentially powerful networks to which organizations in the sample were not directly connected.

---

[5]Each network member could provide between one and six types of support. Cross-tabulation and chi-squared analyses were run for each support type separately.

POSTEARTHQUAKE NETWORKING BEHAVIORS

The final research question looked more closely at the networking *behaviors* of organizations. We were interested to know whether organizations that recovered well interacted with their support networks differently from organizations that performed poorly.

ASSESSING ORGANIZATIONAL PERFORMANCE

By charting an organization's health over time, we found that its trajectory could be placed in one of three trajectory categories (Fig. 11.3A–C):

- *Developmental change* – positive adaptations, leading to reduced vulnerability or a sustained improvement in revenue following the earthquakes. This was also categorized as a "resilient" response.
- *Restoration* – organizations experienced no sustained increase or decrease in revenue or performance beyond that suggested by predisaster trends.
- *Degenerative change* – negative changes that increased an organization's vulnerability and which were accompanied by a sustained decrease in revenue.

Somewhat surprisingly, organization size, age, predisaster financial health, insurance status, and industry sector were not significantly associated with positive outcomes after the earthquakes. Similarly, the degree of direct impact and number of days an organization was closed did not significantly predict organizational trajectories. The difference between the three organizational trajectory groups was only explicable when we examined how the organizations subsequently engaged with their social and economic contexts.

Organizations experiencing developmental change ($n=8$) made significant operational and location changes throughout the study period (September 2010; April 2013). Decision makers in these organizations tended to be opportunity seeking (as opposed to restoration or survival seeking) and leveraged their networks to optimize outcomes. Organizations experiencing restoration ($n=6$) also adapted considerably following the earthquakes, however, they tended to be less proactive and were focused on restoring core operations. Finally, organizations experiencing degenerative change ($n=18$) demonstrated a lesser ability to change in response to postdisaster challenges. These organizations were less likely to form beneficial partnerships after the earthquakes and often struggled to understand how they fitted into changing local environments. Organizations that did well exhibited more of the beneficial networking behaviors described in the next section.

Sensors in the Community

The ability to make proactive decisions, innovate, and adapt following the earthquakes differentiated organizations that experienced developmental change from those that experienced degenerative change. In this regard, network connections served as environmental "sensors," which delivered a range of information to the organization. Local sensor networks were composed of embedded interorganizational relationships (ie, relationships reinforced by trust, friendship, and/or reciprocity); affiliations with local business associations, customers, and owners; and staff members' personal networks. The cases in Box 11.1 illustrate how an organization's local networks were able to deliver vital information about the local environment, even when the organization was not actively seeking that information.

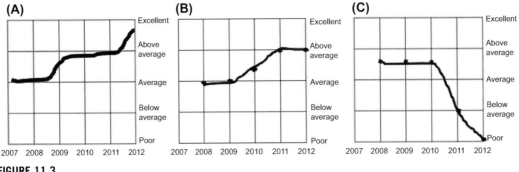

FIGURE 11.3

Representative drawings made by case study organization participants of their financial performance before and after the 2010/2011 earthquakes. Examples of (A) developmental change, (B) restoration, and (C) degenerative change.

Trust and Decision Evaluation

Following the earthquakes, trust both facilitated and encouraged the exchange of resources and information between organizations. Trust in relationships is enhanced by transparent communication, identification with a common cause or place, and a history of reliability, commitment, and consistency (Chow & Chan, 2008; Kramer & Tyler, 1996). Where trust exists in relationships the probability of opportunism is lower, and, as a result, trust increases the willingness of actors to engage in informal social exchange (De Wever et al., 2005; Nahapiet & Ghoshal, 1998).

For small business owners in particular, being part of discussion networks characterized by strong, trusting ties enabled them to evaluate potential adaptations, share ideas, and discuss problems as these arose.[6] Decision makers were more likely to heed advice if it came from a source they knew and trusted before the earthquakes. As the cases in Box 11.2 illustrate, the divergent outcomes for *The Attic* and *Elegance* were shaped by both access to external resources (specifically insurance) and the ability to critically evaluate decisions.

An organization's trust in its support networks was especially important when exchanging information and advice, outsourcing work, and receiving certain kinds of financial support. Decision makers often corroborated or discussed information and advice they received from official sources with network members whom they trusted.

Network Density and Understanding Connectivity

Network density ranged from no supporters in the network knowing each other to a maximum of 60% (0.6) of the supporters knowing each other. Analyses showed very low-density networks were associated with organizations that were not particularly effective at utilizing external sources of support. Four businesses in this study reported having completely unconnected networks (i.e., density of 0.0). Three of these organizations experienced degenerative change and one experienced restoration.

---

[6]Micro- (fewer than five full-time employees) and small-case study organisations (fewer than 20 full time employees) were more likely than medium and large organizations to seek external advice about business and recovery decisions following the earthquake.

BOX 11.1 CASE COMPARISON: FAILURE AND A SUCCESS IN PERCEPTION[1]

Amherst Retail: Developmental Change

Amherst Retail experienced a high degree of damage and disruption as a direct result of the earthquakes, but was able to implement adaptations that helped them avoid degenerative change in the earthquake aftermath. Amherst Retail had a greater capacity to perceive the need for change in part because of its internal and external "sensors," which included a high proportion of local staff aware of issues concerning local residents, staff attending postearthquake business information and networking meetings, and staff participating in the local promotions campaigns.

Amherst Retail's local networks allowed them to perceive potential disruptions and opportunities for growth and to act early. As a result of this awareness of the local area's current and future situation, the organization was able to fill a gap in the retail market by adding popular products that nonoperational local retailers had previously offered. Additionally, even though Amherst Retail did not lose its building, it opted to relocate to a new site to avoid probable future issues with the earthquake resistance of its prior building.

Kaiapoi Corner Store: Degenerative Change

After losing a significant competitor in the September 2010 earthquake, Kaiapoi Corner Store experienced a substantial increase in business in the final quarter of 2010 and through much of 2011. Yet in 2013 the store experienced falling revenue and reported being "worse off" than before the earthquakes. This downturn reflected a significant loss of local customers from nearby residential red zone areas. The owners had not perceived the potential impact of this change on the organization and had not taken adaptive action. The owners' lack of awareness regarding when and how the red zoning decision would affect their business was caused, in part, by their lack of local networks in the area. Kaiapoi Corner Store was not a member of local business associations or groups. The owner's personal friendship and kin networks were located outside of Kaiapoi, and the owners did not participate in local events, attend information nights, or otherwise engage with the community.

[1]To provide anonymity, these organizational names are pseudonyms.

BOX 11.2 CASE COMPARISON: EVALUATION AND ADAPTATION

Both *Elegance* and *The Attic* lost their retail stores in the Christchurch central business district as a result of the February 2011 earthquake, and both chose to relocate to the North Island in the aftermath. This decision nearly bankrupted *Elegance*, while *The Attic* successfully reestablished itself in Wellington and was able to return to preearthquake levels of income and growth. There were two main differences that contributed to these divergent outcomes. The first was that *The Attic* had business interruption insurance, while *Elegance's* business interruption insurance had lapsed between the September 2010 earthquake and February 2011 earthquake. The businesses also differed in the way they discussed the relocation and evaluated their options.

The Attic was owned by business partners with a close trusting relationship, and this facilitated collaborative evaluation and decision-making. The owners also discussed matters with a close-knit group of Christchurch business owners who they knew prior to the earthquakes, seeking advice and support as they considered their options for the future. *Elegance*, on the other hand, was a sole proprietorship and the owner sought advice about relocating from a mentor provided by Recover Canterbury. Having never met the mentor prior to the earthquake, the owner was less inclined to trust his advice not to proceed with the move to Auckland. The owner of *Elegance* ultimately made a series of choices without adequate evaluation, and these had severe and lasting negative impacts on both the organization's finances and the owner's personal well-being.

An unconnected network is perfectly "efficient" in technical terms,[7] which means that it assumes that supporters who do not interact with one another tend to provide unique resources and information to the focal organization. Completely unconnected networks, however, required a greater amount of input from the organization requiring support. In this research, unconnected networks tended to reflect the participant's lack of awareness about who was connected to whom, and how the organization might utilize its networks to reach others to whom it was not connected. These organizations' leaders were less effective at managing and mobilizing support from their networks, and they struggled to supplement their internal capacity adequately with external support. Unconnected networks had several features in common: a tendency to rely on formal or market ties, a lack of engagement with business associations or other groups that had important bridging and bonding roles, and relatively low levels of reciprocity between the focal organization and their contacts.

Managing Reciprocity

Reciprocity is the informal credit system in social networks, where an actor that provides support creates an expectation of future support, which the recipient feels obliged to repay at some point (Coleman, 1988). Reciprocating assistance can increase trust in networks, and the expectation of reciprocity can speed interactions by forgoing negotiations over reimbursement (Uzzi, 1997). Reciprocity based on obligation, however, may actually tax organizational resources and constrain access to resources and information (Ahuja, 2000; Bloch, Genicot, & Ray, 2007; Doerfel et al., 2013).

In this study, organizations were more likely to reciprocate support to organizations with which they had close relationships, to friends and family, and to other organizations in the same industry. Support was more likely to flow one way, without reciprocity, from actors with which the organizations had more formal relationships (e.g., government agencies, suppliers, and business associations). On average, organizations that experienced developmental change reciprocated, in some way, the support of nearly 50% of their network members. The distribution of the amount of reciprocated support provided by organizations with degenerative trajectories was bimodal. The bottom quartile of organizations in the degenerative category reciprocated support to approximately 7% of their supporters on average, and the upper quartile reciprocated support to just over 60% on average (Fig. 11.4). Organizations experiencing degenerative change that had very high levels of reciprocity may have been demonstrating "amplified reciprocity," in which network members act against their own best interests out of obligation (Doerfel et al., 2013; Gargiulo & Benassi, 2000, p. 185).

Participants with unusually high levels of reciprocity were typically from organizations that tended to accept high levels of support in areas not directly relevant to organizational response and recovery activities. Most of this support came from local actors with whom organizational leaders had close personal relationships. They, therefore, felt obliged to receive the support even if it was not helpful—and to offer support in return.

Conversely, organizations with very low levels of reciprocity received a relatively high proportion of their support through formal and market relationships (i.e., suppliers, government agencies, and business associations). While, close (bonded) relationships were more likely to involve obligations of reciprocation, "arms-length" or more formal ties (e.g., government agencies and business associations)

---

[7]The calculation for network efficiency norms the effective size of the organization's network by its actual size. Effective size is calculated as the number of network members that an organization has, minus the average number of ties that each network member has to other network members (Hanneman & Riddle, 2005).

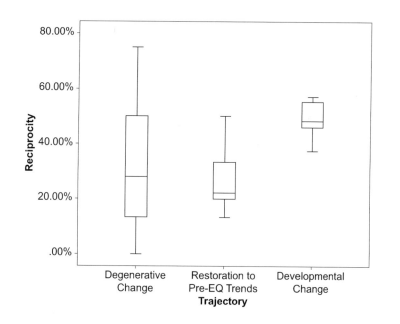

FIGURE 11.4

Percent of support reciprocated by trajectory group.

were often less flexible and tended to be considered less important by organizations. Organizations, therefore, benefited from having a mix of formal and informal ties in their support networks.

Devolved Responsibilities

Some organizations very effectively used connections within their networks to manage tasks and reduce their burden of communication. Two network structures—a particularly well-connected supporter (which we refer to as a "coordinator") and network cliques (ie, a subsection of the network where all of the supporters know each other)—were particularly useful for some organizations. Support coordinators reduced the frequency with which the organization needed to interact directly with others in their network. The coordinators tended to have a high degree of connectivity in the network prior to the earthquake, but also coordinated with new network members.

Support coordinators took a range of forms within organizations' networks, but always served to reduce network complexity for the focal organization. The coordinator in retailer *Tailor Made*'s network, for example, was an active industry association that gathered and distributed information to the associated organizations affected by the earthquake. The coordinator in this network also reported the status of suppliers and provided information about support offered by industry members and the government. The support coordinator in *Executive Sweets'* café network played a much different role. This family member of the owner coordinated contact with agencies offering support (e.g., the Internal Revenue Department and Recover Canterbury) and communicated with mutual friends who wanted to assist this disrupted business. In both of these cases, the role of the coordinator was to give the focal organization a single point of contact to connect with support and information in a much broader support network.

Other organizations reduced the complexity of their network by collectively coordinating with cliques in their network. A clique is a subset of a network in which the members are all relatively closely tied, such as the clique in Kaiapoi Society's support network. This clique was a formally affiliated group of organizations to which *Kaiapoi Society* also belonged.

Kaiapoi Society tended to communicate in a group setting (group emails, conference calls, and meetings) with these supporters, again reducing the need for redundant communications and ensuring that information was appropriately distributed. Network features such as coordinators and cliques reduced the communication burden on already strained organizational resources. Coordinators and cliques created additional capacity for organizations while allowing them to retain access to resources and information from a large number of supporters.

DISCUSSION AND CONCLUSIONS

Much of the support discussed in this study has also been highlighted as important in postdisaster interorganizational response networks, including brokerage (Spiro et al., 2013), access to information about emerging issues and trends (Newman & Dale, 2005; Tierney & Trainor, 2004), and emergent coordination of communications (Drabek & McEntire, 2002; Petrescu-Prahova & Butts, 2005).

We found that support tended to come from local supporters and from those with whom organizations had long-term relationships. These findings support observations that geographic proximity facilitates network formation and postdisaster activation (Brouwer, 2004; Zakour & Gillespie, 1998). It also confirms that organizations tend to draw on preexisting relationships where they have trust and established communication pathways (Doerfel et al., 2013; Kapucu, 2006a; Tierney, 2013).

Network composition influenced the type of support organizations received, as well as the norms that governed interaction (eg, obligations of reciprocity). Close and informal relationships with friends and other parts of the intracorporate network were the quickest to provide support, provided more kinds of support, and were more adaptable to emerging postearthquake needs. Other relationships delivered specialized support. For example, organizations in the same industry as the case study organizations were often in the best position to provide specialist supplies, labor support, advice, and locations from which to operate.

RELATIONAL RESILIENCE

Our examination of support networks in the aftermath of the Canterbury earthquakes demonstrated that some of an organization's resilience potential is held outside of the organization, in its social network. We refer to this concept as *relational resilience*. For example, proactively engaging with suppliers and customers throughout the recovery process was helpful for maintaining, renegotiating, and mobilizing these network connections. Similarly, organizations that actively associated with affiliation groups (e.g., business associations) prior to the earthquake were able to access resources and information from a wider base of support following the earthquake.

An aspect that has received very little coverage in the organizational literature is emotional support. Nearly one-third of the case study organizations' supporters provided some form of emotional support for organization members. This corresponds with findings indicating that managing staff well-being and dealing with chronic stress were among the greatest ongoing challenges for organizations following the earthquakes (Kachali et al., 2012). Organizations that experienced developmental change mobilized more emotional support from their networks than organizations experiencing restorative or degenerative change.

FUTURE WORK AND IMPLICATIONS

Going forward there is great potential in better understanding the relational nature of organizations' postdisaster outcomes. There is evidence that social networks can have a greater influence on behavior than do many organizational characteristics (e.g., age or industry sector) (Rayner & Malone, 1998). Thus, the efficacy of interventions geared toward enhancing resilience may increase if they were to be implemented at the network level. Another important concept for future investigation is whether efforts to improve place-based community resilience and organizational resilience should be pursued jointly. As a collective resource, organizational resilience can have positive externalities for disaster-affected people and places, while simultaneously creating strong and agile organizations.

WELL-BEING AND PARTICIPATION IN NEW SOCIAL NETWORKS FOLLOWING A DAY CARE FIRE IN HERMOSILLO, MEXICO

12

Maria L. Rangel[1], Eric C. Jones[2], Arthur D. Murphy[3]

[1]*University of Texas School of Public Health, Houston, TX, United States;* [2]*University of Texas Health Science Center at Houston, El Paso, TX, United States;* [3]*University of North Carolina-Greensboro, Greensboro, NC, United States*

CHAPTER OUTLINE

INTRODUCTION

Nearly 4% of the world's population and about 8% of the United States population are estimated to have suffered from posttraumatic stress disorder (PTSD) in the previous year,[1] and about 60% of men and 51% of women in the United States have been exposed to one or more traumatic event at least once in their lives

---

[1]Posttraumatic stress disorder for public health/epidemiological purposes is assessed/diagnosed by lay interviewers with a clinical interview protocol.

(Kessler et al., 2005; Kessler & Ustun, 2008). These traumatic events can be disasters, physical attacks, active-duty combat, child or sexual abuse, or extreme life events such as car accidents, house fires, or medical emergencies. Studies and interventions concerning recent disasters, mass shootings, and terror attacks have shone a light on the issue of the mental and emotional repercussions of exposure to traumatic events.

Traumatic events increase the risk of a population developing symptoms of posttraumatic stress and depression (Kessler et al., 1995). However, the risk of developing PTSD or its symptoms varies depending on the experienced trauma. For example, one study found that life-threatening illness of a child or unexpected, sudden death of a friend or family member increases the risk by 10% and 14%, respectively (PTSD Alliance, n.d.).

The ABC day care fire occurred in Hermosillo, Sonora, Mexico, on June 5, 2009. About 142 children were in the building at the time of the incident.[2] The fire caused the death of 49 children, and over 40 children were hospitalized with burns. The remainder of the children were treated and released from local hospitals. The official cause of the fire was attributed to a malfunction of the air conditioning unit in a government-owned warehouse next to and attached to the day care, according to a commissioned report from a United States team. Another forensic examination by a different US team suggested intentional burning of paper in that state-owned warehouse next door, speculated to be records destruction for the upcoming election and change of leadership—not an uncommon practice in Mexico's history. There were also a number of irregularities found in both the day care and the warehouse, such as a locked emergency exit, a false ceiling, limited day care staff at the time of the incident, and not being sufficiently well equipped with fire extinguishers, water sprinklers, or smoke alarms. After the fire, many of the affected parents and caretakers blamed the government since the day care was a government subsidized—but privately operated—service. The event was seen by many Mexicans, and especially people in Hermosillo, as another example of the corruption and injustice in Mexico.

The purpose of the ABC study was to document the incidence of grief, depression, and posttraumatic stress among parents and caretakers whose children were injured or killed in the ABC day care fire. First in this chapter, we set up a framework for understanding how groups respond to tragedy via protest and social movements at a broader level plus how social networks influence individual well-being at a more microlevel. The methodology then covers the network and well-being questionnaire we used. Next, the results show the differences between parents of deceased children and parents of injured children, and we explore the role of social networks and political participation by these two broad kinds of parents/caretakers after the ABC day care fire. Our discussion then engages the ways in which participation in social movements and networks might be protective by allowing an outlet and social support and/or serves as a risk factor.

THEORETICAL FRAMEWORK
GROUPS SEEKING SOCIAL JUSTICE IN EXTREME SETTINGS

Social science disaster scholars have known for years the importance of community support and social networks in helping individuals and communities cope with and recover from traumatic events or disasters (Cook & Bickman, 1990; Faas & Jones, Chapter 2). Some extreme events result in human-caused

---

[2]An exact count is not available because the fire occurred during siesta and some parents had come early to take their children home for the afternoon. All records were lost in the fire.

fatalities and crime that compels the pursuit of social justice. However, there is limited research on the relationship between how individuals and communities come together to seek social justice following a traumatic event, and even less research considers the psychological effect of participating in these disaster-related social justice movements. The psychological consequences due to a traumatic event may include depression, posttraumatic stress disorder, anxiety, or other psychological conditions. Our interest is in the degree to which, and how, people are helped by participating in public memorialization and justice seeking versus hindered by the ongoing reminders of the loss plus the possible stress of political struggle. The next paragraph provides some examples of how communities come together to seek social justice to set the stage for our question about how this participation relates to well-being.

When human fatalities or crimes are involved in traumatic events, individuals affected by the event often seek social justice for themselves and their loved ones. In this traumatic recovery period, individuals seek social support from their community members and, in some cases, unite as a group for social justice (Ferguson & Gupta, 2002). Various traumatic events have led to the creation of social groups seeking justice for their loved ones. During the 1970s, mothers would meet regularly in the Plaza de Mayo seeking a resolution for the disappearance of their loved ones at the hands of Argentine dictatorships (Humphrey & Valverde, 2007; Kordon, Edelman, & Lagos, 1988). Similarly, the 1984 Bhopal chemical disaster in India produced groups that mourned loved ones and sought accountability (Fortun, 2001; Lapierre & Moro, 2002; Mukherjee, 2010). Also, Yael Danieli (2009) studied the search for reparative justice among Holocaust survivors and compared it to other cases of genocide/warfare with justice seeking by survivors. Despite the potential for corrosive communities—conflict ridden due to impacts of a disaster—evolving out of the search for justice in the form of reparations (e.g., Picou, Marshall, & Gill, 2005), Danieli noted the importance of justice seeking as a collective process as long as there is sharing of grief, mourning, and memory among those impacted.

Social justice-seeking groups can provide a means of social support during the grieving period after a traumatic event. Social support has been closely related to mental health. Particularly, social support has been shown to be a protective factor against the negative mental health consequences in the face of a traumatic event (Cohen & Wills, 1985; Kessler & McLeod, 1984; Thoits, 2011). Some of these negative mental health consequences include symptoms associated with depression and posttraumatic stress. Two meta-analyses have shown that lack of social support was an important risk factor for posttraumatic stress reactions following a traumatic event (Brewin, Andrews, & Valentine, 2000; Ozer, Best, Lipsey, & Weiss, 2003). Similarly, posttraumatic stress risk, symptom severity, and recovery have been shown to be related to social support (Charuvastra & Cloitre, 2008). The relationship between social support and mental health can also vary depending on the definition of social support. For instance, received social support is the actual provision of tangible, emotional, or informational support, whereas perceived social support denotes the perceptions that emotional, informational, or tangible support would be available if the individual required it. Perceived social support has been shown to be associated with better physical and mental health. However, there are inconclusive results for received support (Thoits, 2011).

Groups may also aid in eliminating social support barriers, which may include individuals' perceptions that other people were tired of hearing about the trauma, that others had their own problems to face, that other people would consider the survivors too preoccupied with the traumatic event or that friendships might be overburdened, and the feeling that other people were not able to understand their suffering since they had not experienced it themselves (Thoresen, Jensen, Wentzel-Larsen, & Dyb, 2014). Siri Thoresen et al. (2014) concluded that social support barriers were highly related to posttraumatic stress reactions and psychological distress in a sample of survivors of the terrorist attack on Utoya Island, Norway.

However, the mechanisms involved are still poorly understood. There are limited studies that look at temporal relationships between various aspects of social support and mental health (Kaniasty & Norris, 2008; Robinaugh et al., 2011). More longitudinal studies are needed to fully understand the mechanisms involved in this relationship, especially when social injustice is added in the relationship.

Research suggests that smaller social networks, fewer close relationships, and poor social support are linked to depressive symptoms (Barnett & Gotlib, 1998). A pathway in which the participation in social networks can affect psychological well-being is through social influence. Social ties can also increase with participation in groups for social justice. As a member of a social group, the individual obtains normative guidance about health-relevant behaviors such as physical activity or smoking. These health-relevant behaviors, in turn, influence mental health. Also, the participation in social groups can also produce a positive psychological state by improving the individual's sense of belonging, security, and self-worth (Cohen, Underwood, & Gottlieb, 2000). These positive psychological states may result in improved mental health. Lastly, the participation of these groups enhances the accessibility to community resources and support that can protect against distress. While the association between social ties and mental health is well documented, more research is needed looking at the mechanisms of various aspects of social ties that can improve the maintenance of psychological well-being (Kawachi & Berkman, 2001).

Some traumatic events involve a violent crime and social injustice. Crime victimization has been associated with the participation of a political group for social justice (Bateson, 2012). Victims who personally experienced wartime violence and trauma in Burundi exhibited increased political participation such as rates of voting, community leadership, and civic engagement (Voors et al., 2012). Studies of groups have also found heightened social capital, altruism, and political participation in war-affected communities (Gilligan, Pasquale, & Samii, 2011; Kage, 2011). Crime victims also talk more about politics to their social network, attend marches, political meetings, and community meetings. The participation in political groups by crime victims or grieving individuals seeking social justice can mitigate the emotional consequences of victimization. While these individuals may sustain psychological harm, they are able to channel their agony into activism and contribute to their healing process (Humphrey & Valverde, 2007; Macmillan, 2001). For example, a study of homicide survivors in Ciudad Juarez found that political organizing following a traumatic event was a source of social support for the victims (Bejarano, 2002). This social support can be reinforced when an entire community has suffered a similar crime, thus creating solidaristic ties between the affected or grieving individuals.

The effects of victimization and suffering may yield a phenomenon referred to as altruism born of suffering (Staub & Vollhardt, 2008). Contrary to the negative consequences associated with traumatic events, affected individuals may become more caring and helpful. The participation in political groups is one way in which individuals can express their altruism and foster healing in the wake of trauma. Posttraumatic growth, or being able to incorporate adversity and heal in a way that promotes personal growth, has been associated with reduced posttraumatic stress symptoms and less aggressive behavior (Staub & Vollhardt, 2008; Tedeschi & Calhoun, 2004). High levels of altruism and prosocial behavior have also been documented after natural disasters. The emergence of an "altruism community" after natural disasters creates solidary, fellowship, and acts as a protective factor against the negative psychological effects such as caused by Hurricane Hugo in the American southeast (Kaniasty & Norris, 1995).

William Axinn, Dirgha Ghimire, Nathalie Williams, and Kate Scott (2015) also noted the importance of nonfamily social organizations in providing social support, a key component of social capital to enhance mental health. They explained that these informal groups have the potential to build human capital as social relationships, increase access to goods and services, and enhance support networks. Along with schools, markets, and health services, the study found that social support groups can substantially reduce the odds of depression, PTSD, intermittent explosive disorder and anxiety disorders. Trauma had strong associations with all of these disorders except for anxiety. More research is needed to fully understand the community integration of these nonfamily social organizations to promote mental health, which may often go unobserved (Axinn et al., 2015).

Various aspects of political groups may be influential in shaping individual psychological well-being. For example, some political groups have strong religious components. For the past two decades, there has been a growing recognition of the protective health effects of religious involvement on both physical and psychological well-being (Aranda, 2008). For example, religious involvement was directly associated with better health outcomes such as lower levels of psychological distress, fewer depressive symptoms, and anxiety. A study that examined the relationship between religious involvement and depression in a low-income clinical sample of 230 older United States citizens and immigrant Latinos found that higher levels of religious attendance were associated with lower risk of depressive illness after adjusting for physical functioning, stress exposure, and social support (Aranda, 2008). This protective role may be due to the "public" or communal nature that parallels the social nature of Latinos and help-seeking/giving behaviors they expect during times of distress. This may be an indication that social justice groups with a strong religious component may have protective effects on psychological well-being following a traumatic event.

POSTDISASTER MENTAL HEALTH AND SOCIAL NETWORKS

PTSD is an anxiety disorder that develops in response to physical injury or severe mental or emotional distress, such as military combat, violent assault, natural disaster, or trauma that the person feared as life threatening for themselves or others. The symptoms of posttraumatic stress are characterized in three separate categories: reexperiencing or intrusion, avoidance, and hyperarousal. The symptoms of reexperiencing or intrusion include having flashbacks, bad dreams, or frightening thoughts about the traumatic event. A person with PTSD can also experience avoidance symptoms such as staying away from things that are reminders of the traumatic experience, losing interest in things they used to enjoy, and feeling emotionally numb, strong guilt, depression, or worry. Also, a person who went through a trauma can experience hyperarousal symptoms, which include feeling tense or "on edge," having difficulty sleeping, being jumpy, and having angry outbursts. Gendered differences occur in the incidence of postdisaster mental health, with females usually about twice as susceptible as men to posttraumatic stress, and experiencing more symptoms than do men (Tolin & Foa, 2006). For a disaster like a day care fire, the loss or injury of a child and not knowing what happened to their children in the hours after the fire constitute the trauma for the parents and caretakers. As women bear greater responsibility in this community for raising children, it is worth having an idea of the relative gendered mental health burden in this case as we begin to examine what people do publically in response to their pain and frustration.

In posttraumatic grief, supportive interactions have been shown to be a protective factor against the health consequences of life stress (Cohen et al., 2000). Research has long shown that social support is a moderator of life stress. For example, social support was shown to reduce the amount of medication required, accelerate recovery, and facilitate medication compliance for grieving individuals (Cobb, 1976). Additionally, lack of social support has been associated with increased grief and loneliness among older people who recently lost a loved one (Utz, Swenson, Caserta, Lund, & deVries, 2014).

A. J. Faas and colleagues (2014) found gendered differences in receiving and giving various forms of social support in postdisaster and disaster-induced resettlement settings, and Eric Jones and colleagues found gendered differences in the association of social network factors with mental health, even when incidence and degree of posttraumatic stress and depression were not that different between males and females (Jones et al., 2014; Tobin et al., Chapter 16).

The forms of social support typically measured in community psychology, while not social network analysis per se, typically include:

- interindividual support (e.g., Norris, Baker, Murphy, & Kaniasty, 2005)
  - the support that people perceive they have from others in general or from various types of relationships like family, friends, spouses
  - whether or not people received certain kinds of support—like tangible, emotional, informational—from specific kinds of people
  - the general degree of one's embeddedness in social groups and associations
- participation in larger movements or groups (e.g., Humphrey & Valverde, 2007)
- personal feeling about a community like attachment to place, or sense of community (Long & Perkins, 2007)

Currently, there is limited evidence on whether the participation in an interest or political group will have a protective effect on physical and mental health outcomes such as posttraumatic stress. Our study therefore addresses each of these major facets of social support typically pursued in community psychology, but this chapter focuses on participation in larger movements or groups and its relationship to postdisaster mental health. We use both network analysis and standard survey questions to evaluate the role of participation in groups with postdisaster mental health. The collection of social network data is important to identify the social support system of grieving individuals and understand its influence on their physical and mental health. Social network data can also help identify how the participation in interest or political groups in protest of the traumatic event can increase social support and influence health outcomes such as symptoms associated with posttraumatic stress and depression.

RESEARCH QUESTIONS

In this study, we looked at a broad research question regarding how participation in political, public, and justice-seeking activities in the year prior to being interviewed (roughly between the first and second anniversary of the fire) was associated with the physical and mental health of the parents and caretakers of children killed or injured in the incident. Specifically, we looked at the role of participation in: (1) marches; (2) talking to the media; (3) being involved in one of four formal groups that formed in the two years since the fire; and (4) being connected to one of the four informal subnetworks of parents and caretakers based on interpersonal relationships in the past year. We hypothesized that parents and caretakers who participated in these formal groups would exhibit higher posttraumatic stress and depression symptoms. One explanation posits their well-being as cause of participation—people with more need are more likely to participate due to the anger, pain, or frustration they feel. A different explanation posits participation as cause of well-being, specifically that being involved in such activities may reduce the likelihood of personally dealing with the trauma, rather people would be projecting the issue instead of inwardly engaging the grief.

METHODOLOGY

As previously mentioned, this study used data collected from the ABC day care fire study, which occurred at two points: 8–11 months after the fire and 20–23 months after the fire. Most of the data presented in this chapter come from the second wave.

Out of the 204 parents and caretakers that had complete interviews for the second wave, 59% were parents and caretakers of children that were affected or injured, and 41% were parents and caretakers of children that died during the incident (Table 12.1). This included 27 mothers, 20 fathers, and 37 caretakers of deceased children and 52 mothers, 32 fathers, and 36 caretakers of injured/affected children.

The data included information on the wellness of parents and caretakers of injured and deceased children, including symptoms of posttraumatic stress and depression, perceived social support (family, friend, or spouse), and perceived social constraints (e.g., people not wanting to listen to them). Also, social network information was collected—we asked parents and caretakers to name up to seven people that they knew who had children enrolled at the ABC day care when the fire occurred. These lists overlapped to generate a network of the parents and caretakers of children. No person who was not a parent or caretaker was interviewed or included in the final network calculations.

Information on participation in marches and meetings plus participation in political groups created after the traumatic incident was collected at 20 months post-fire, but unfortunately not at 8 months post-fire due to our perception that political participation would be more of a touchy issue requiring the establishment of trust that might develop through our repeated interactions with the participants. Interviews typically lasted 70–95 min and mainly took place inside or just outside the homes of the respondents. Sometimes arrangements were made to meet somewhere else. Seven interviewers with Masters degrees participated in each of the two waves, with six of them helping us out both times. All of the field team was from Hermosillo, except the research coordinator who was from southern Mexico and had worked with us on other disaster research projects.

Variables for justice-seeking involvement include active participation in marches and public meetings about the incident, active participation in group and private meetings about the incident, participation in speeches to crowds about the incident, participation in talking to the press about the incident, participation in a group related to the incident, and participation in first-anniversary activities related to the incident.

We calculated frequencies and correlations using the statistical software SSPS 23. We conducted cross-tabulation and Chi square tests when variables were categorical. In the course of analysis, we recoded the number of marches and public meetings attended after the day care fire and the frequency of attending group meetings related to the day care fire to ordinal variables: highly active participation versus low participation. We recoded participants with more than 10 times of participation in marches and public meetings as highly active participation in marches and public meetings held to bring justice

Table 12.1 Interviewees by Child Status After the Day Care Fire

Child Status	Mother	Father	Caretakers (Mostly Female Relatives)	% of Total
Affected, injured	52	32	36	59
Died	27	20	37	41

for the children. Those attending less than 10 marches or meetings were coded as low levels of participation. In terms of participation in group meetings—separate from marches and public meetings—we coded participants with five or more times attending group meetings as highly active members of the group meetings, and less than five meetings indicated low activity.

VisuaLyzer 2 (Medical Decision Logic, 2014) served as the tool to analyze or describe the network data. For this study, the Girvan–Newman clustering algorithm generated mathematically distinct clusters of interactions (Girvan & Newman, 2002) between parents and caretakers that were part of the postfire network.

RESULTS

First, we present results that gradually paint a picture of the mental health of mothers, fathers, and caretakers of deceased and injured children. We then turn to the degree to which people participated in public memorialization and justice seeking, the relationship of this participation to posttraumatic stress and depression, and finally how participation in the new networks that formed relate to mental health.

PRIOR EXPOSURE TO TRAUMATIC EVENTS OTHER THAN THE ABC DAY CARE FIRE

In order to have a sense of how much of people's well-being might be due to prior trauma, we collected data on events other than the ABC day care fire. Out of 204 parents and caretakers interviewed in both wave 1 and wave 2, all provided information of exposure to traumatic events. A total of 33% of the participants indicated that in their lifetime, they had experienced the sudden unexpected non-ABC fire-related death of someone close to them in the past year and a half. No difference was noted between parents and caretakers of injured children (16% of total) versus deceased children (17% of total). Out of 204 of participants, 3% reported that they had experienced a disaster other than the ABC fire, such as a flood, hurricane, tornado, or earthquake in which they were injured or their property was damaged.

TOTAL NUMBER OF PTSD SYMPTOMS AND DEPRESSION

The total number or score of posttraumatic stress symptoms and depression of parents and caretakers of injured versus deceased children was calculated using the statistical software SPSS. The results at nearly two years post-fire found a higher mean of the numbers of posttraumatic stress symptoms in parents/caretakers of deceased (mean 9.6) children versus parents/caretakers of injured (mean of 7.9) (Table 12.2). Out of the 204 parents/caretakers, 27% reported more than or equal to 12 symptoms of PTSD. However, parents/caretakers of injured children (23%) reported more than or equal to 15 symptoms of posttraumatic stress compared to 32% of parents/caretakers of deceased children.

In regards to depression scores, the parents/caretakers of deceased children were more likely to have higher scores of depression symptoms compared to parents/caretakers of injured children, a mean of 8.9 versus 6.3 (Table 12.3).

When controlling for role (mother, father, or caretaker) and child outcome (dead or injured), the number of posttraumatic stress symptoms was predicted by participation in the first anniversary of the event ($B = 1.409$, SE $= 0.665$, $p < .035$; $R^2 = 0.14$), but not by active participation in marches and public

Table 12.2 Number of Posttraumatic Stress Symptoms for Parents and Caretakers by Child Outcome

Number of PTSD Symptoms	Parents/Caretakers of Injured N (%)	Parents/Caretakers of Deceased N (%)
0–5	34 (33%)	13 (16%)
6–11	53 (44%)	44 (52%)
12–17	28 (23%)	27 (32%)
Total	120 (100%)	84 (100%)

Table 12.3 Depression Score for Parents and Caretakers by Child Outcome

Depression Score	Parents/Caretakers of Injured N (%)	Parents/Caretakers of Deceased N (%)
0	32 (27%)	7 (8%)
1–9	54 (45%)	42 (50%)
10–19	27 (23%)	31 (37%)
20–27[a]	7 (6%)	4 (45%)
Total	120 (100%)	84 (100%)

[a]Patient Health Questionnaire-9 Depression scale goes to 27, but the highest score was 25.

meetings, talking to press about the incident, participation in group membership, active participation in group meetings, or speaking to a crowd about the incident. Furthermore, the participation in the first anniversary of the incident was also a predictor of depression score when controlling for role and child outcome ($B = 2.32$, SE = 1.064, $p < 0.03$; $R^2 = 0.137$). No other variables related with political participation were significant predictors of the depression score.

PARTICIPATION IN GROUPS FOR MEMORIALIZATION AND JUSTICE SEEKING

We used two measures of participation in the network of parents as predictors of stress and depression. First, we looked at whether people said they were part of one of the four formal/named groups that had formed by 23 months after the fire. Second, we created clusters within the overall network (Fig. 12.1).

Tables 12.4 and 12.5 illustrate the number of posttraumatic stress and depression symptoms by membership in one of the four formal or named groups. There is overlap of stress/depression with child outcome, such that each of the groups largely has parents and caretakers of children with one outcome (e.g., group of mostly parents of injured children). For anonymity purposes, we do not share which of these groups is associated with which child outcomes. For the same reason, we give these groups pseudonyms as 1, 2, 3, and 4, and we do not provide the numbers of individuals who said they were a member of a named group.

For posttraumatic stress levels by group membership, the members of Groups 2 and 3 had more posttraumatic stress symptoms on average. Group 4 and people who said they were not a part of any of these named groups have the lowest mean levels of posttraumatic stress symptoms. For the depression

Table 12.4 Posttraumatic Stress by Named Group Membership

Named Group	PTSD Symptoms Overall Mean	Standard Deviation	PTSD Symptoms Male Mean	PTSD Symptoms Female Mean
1	9.3	4.4	5.6	10.6
2	10.1	3.2	8.3	11.4
3	10.5	4.3	11.8	9.7
4	9.0	2.1	7.0	9.5
Nonmembers (n = 125)	7.8	4.3	6.4	8.4
All (n = 204)	8.6	4.2	7.1	9.2

Table 12.5 Depression by Named Group Membership

Named Group	Depression Score Mean	Standard Deviation	Depression Score Male Mean	Depression Score Female Mean
1	9.3	8.6	3.4	11.3
2	10.6	6.2	7.4	12.8
3	7.9	5.9	7.8	8.0
4	5	4.2	1.0	6.0
Nonmembers (n = 125)	6.2	6.4	3.8	7.1
All (n = 204)	7.4	6.9	4.8	8.6

by group membership, members of Group 2 again had higher depression mean score compared to the rest of the groups. A score of greater than 10 indicates moderate to severe depression. Again, Group 4 and nonmembers had the lowest mean scores for depression. However, an examination of the gendered mental health burden and the distribution of that burden across the named groups suggests that men and women are having different experiences in some of these groups.

In some groups, there is no difference between men and women in average scores, while other groups show great differences, with men's posttraumatic stress scores even being higher than women's stress scores in Group 3. Men have a much wider range of variation (5.6–11.8) in posttraumatic stress symptoms across the groups than do women (8.4–11.4), and their average is lower with women experiencing 30% more symptoms on average. This difference between males and females is similar to results from Graham Tobin and colleagues (Chapter 16), who found posttraumatic stress differences of around 15–30% between men and women in the several sites they studied in Mexico and Ecuador. For depression of parents and caretakers of ABC children, both genders have a similar range on variation—1–7.8 for males, and 6–12.8 for females—but their scores are wildly different, with women experiencing 80% greater symptomatology.

PARTICIPATING IN THE NEW NETWORKS

Fig. 12.1 illustrates the ABC network of parents and caretakers of injured and deceased children. Five distinct clusters appear, as generated by the Girvan–Newman clustering algorithm in VisuaLyzer 2 (Medical Decision Logic, 2014). These are four clusters visible in the connected network, plus one set comprised of

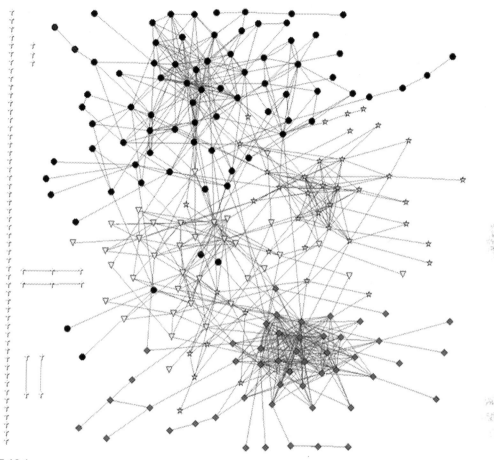

FIGURE 12.1

ABC Network at Time 2 (20–23 months after fire). Four clusters (two large, two small) are easily identified by the color-shape combination for each individual in the network.

isolates—individuals not tied to other individuals in the network. The four clusters are black circles, light gray stars, very light gray upside-down triangles, gray squares, and the isolates are white crosses. It is clear that there are two larger clusters at the ends, and two smaller clusters in the middle. We do not present which clusters are associated with which outcomes in the graph to help maintain anonymity.

Next, in Table 12.6, we present the levels of stress and depression by gender for each of the four clusters and the fifth cluster of isolates seen in Fig. 12.1. There is a high level of overlap between the four named groups and the four clusters that emerged over the two years following the fire (Pearson's $R = 0.55$; $p < 0.001$). While this moderately high association indicates overlap between clusters and named groups—which accompanies the relationship between these groups/clusters and child outcome for participating parents—we limit how much detail we present on these overlaps in order to better achieve anonymity for the named groups and their members.

Table 12.6 Average Number of Posttraumatic Stress Symptoms by Network Cluster

Cluster	PTSD Symptoms Overall Mean	PTSD Symptoms Male Mean	PTSD Symptoms Female Mean
A	8.8	6.6	9.4
B	9.2	7.2	10.8
C	8.5	7.0	9.0
D	11.7	12.3	11.5
E	6.9	5.6	7.5
All	8.6	7.1	9.2

Table 12.7 Average Depression Score by Network Cluster

Cluster	Depression Score Overall Mean	Depression Score Male Mean	Depression Score Female Mean
A	8.1	3.8	9.4
B	9.2	6.8	11.1
C	6.1	2.5	7.2
D	10.8	11.3	10.6
E	5.0	2.3	6.4
All	7.4	4.8	8.6

The respondents with the lowest levels of posttraumatic stress symptoms are those (Cluster E) who did not name anyone else from the day care when asked to list up to seven people. In Table 12.7, depression symptoms vary similarly to posttraumatic stress overall, but males in Clusters A, C, and E have much lower depression scores compared to the other two clusters, and not so much difference between clusters for posttraumatic stress. Cluster D has high rates of depression symptoms and posttraumatic stress for men, while both Clusters B and D present high rates of both kinds of symptoms for women. Cluster C presents somewhat lower levels of depression than do the other three clusters connected to the network (remember, Cluster E is not connected to the others in the network).

What these findings in Tables 12.6 and 12.7 suggest is that depression and posttraumatic stress are behaving similarly within genders across the clusters but not between genders. Men and women are not experiencing their informal clusters exactly the same. Tables 12.4 and 12.5 indicate the same thing—that although named groups are not fully isomorphic with informal network clusters, there are similar patterns within gender for depression and posttraumatic stress across the named groups, but that any given group is not producing the same relative impact for males and females. For example, Group 3 has the highest posttraumatic stress for males in Table 12.4, while Group 2 has highest for females; Group 1 for males is the lowest, and Group 1 for females is the second highest.

DISCUSSION

Study findings suggest that there is an important association between mental health and participation in a network of justice-seeking/memorialization in the ABC Day Care fire case. While public activities—except participation in first anniversary events—did not appear to predict stress or depression after controlling for child outcome and parent gender, there was a relationship between mental health and (1) being part of certain named groups and (2) being part of the identifiable clusters that emerge from the network.

Potentially one of two possible dynamics is playing out in the data we collected, though it is unclear which. First, it is possible that network involvement may be what it takes to keep the most affected parents from experiencing even worse levels of stress or depression, such that participation becomes a protective factor for posttraumatic stress by increasing the individual's social support and positively impacting their network. Parents described to us how they felt alone and helpless after the fire. They began congregating at the Plaza Zubeldia and there encountered not only other parents but also locals who had heard of the fire and had come out in their shock. The parents told us that citizens helped them form groups to organize for protest and find support with each other. Victims' individual (non-social) responses ranged from wanting to be left alone to go on with their lives, to demands for medical and financial support from the government, to incarceration of the responsible parties. This is similar to what others have found when people respond collectively to an extreme event imbued with human negligence (Erikson, 1976; Oliver-Smith, 2002) or malice (Wagner, 2008).

Second, it is also distinctly possible that participation in these activities keeps reminding people of their loss and keeps people from dealing psychologically with the trauma. However, we found no significant relationship between number of posttraumatic stress symptoms and being a member (or not) of any named group that developed after the ABC fire. Our colleagues who served as therapists and counselors for affected families have told us the same thing based on their general observations. However, depression scores are lower among nonmembers of named groups and among the members of one of the groups. Again, the question is whether social selection (choosing activities because of one's status) or social causation (activities influencing one's status) or both are at play.

Finally, social network data provided an illustration of the distribution of posttraumatic stress and depression symptoms in the network of parents and caretakers of injured and deceased children. It depicts the social networks created after the ABC day care fire. The network is clearly highly connected, although a number of people do not consider themselves members and/or cannot name any other parent for caretaker from the ABC day care. The connectivity of the network may be the result of the creation of groups to protest against the government, to bring justice, and to memorialize the injured and deceased children. The participation in the networks associated with these activities may influence posttraumatic stress and depression. Each group may provide different resources or support to the parents and caretakers of injured or deceased children. For example, a group of parents with injured children may seek more health services for its members. They also might provide different kinds of constraints, approaches, and reminders of the pain. The study findings provide graphical evidence on how social networks play a role in the grieving process, perhaps notably when networks are created for memorialization and social justice.

FUTURE IMPLICATIONS

The study results can be used in the development of interventions for other grieving populations such as those affected during military combat, mass shootings, natural disasters, and harsh immigration journeys. These are vulnerable populations at risk for posttraumatic stress and depression. Therefore, understanding the importance of social network analysis would be vital in the development of interventions for this population and other at-risk populations.

Also, social networks of grieving individuals can be expanded by the use of indirect communications such as social media, online grieving support groups, and texting. Each of the identifiable groups has developed a Facebook page where they announce events such as marches and meetings. Many parents now post remembrances of their child on his or her birthday. A recent Facebook development is a flood of posting on the fifth day of each month reminding the world of how many months have passed "*sin justicia.*" These indirect forms of communications can reach a mass audience in providing vital social support and group membership to alleviate the symptoms of posttraumatic stress and depression. Future research can focus on whether this online social support and group membership through social media can be a protective factor against symptoms of posttraumatic stress and depression for grieving individuals or whether it maintains levels of stress or other problems.

The results can also contribute to the new care protocol for posttraumatic stress from the World Health Organization (WHO), which recommends that helping affected people should include identifying and strengthening positive coping methods and social supports. The use of social network analysis can help improve interventions to reduce posttraumatic stress and depression in grieving individuals after a traumatic event. Social support-based interventions can be used to reduce the use of antianxiety medications for posttraumatic stress and depression. For example, the World Health Organization discourages the use of benzodiazepines, an antianxiety drug to reduce acute traumatic stress symptoms and sleep problems after a traumatic event.

Finally, with the recent events of police brutalities, mass shootings, and terror attacks in the United States and around the world, it is important to bring awareness of the physical and mental repercussions of the exposure to traumatic events. These current events involved politics and the creation of interest groups similar to the ABC day care fire. Also, characteristics such as religion or violent protesting by the members of newly formed groups have been identified as drivers affecting posttraumatic stress and depression symptoms, though violent protests did not occur in the ABC case. Therefore, it is important that in the future more research looks at the impact of participation in interest and political groups created after a traumatic event on the physical and mental health of the affected individuals.

ACKNOWLEDGMENTS

We would like to thank study collaborators Dr. Fran Norris, the principal investigator of this project (Prolonged Grief and Functioning of Parents after the Hermosillo Day Care Fire; NIH-NIMH R21MH090703; AD Murphy-CoPI) and Dr. Kathy Sherrieb of the National Center for Posttraumatic Stress Disorder; our collaborators Migdelina López Reyes, Diana Luque Agraz and Cristina Taddei Bringas at the Center for Investigation of Nutrition and Development (CIAD) in Hermosillo; the field coordinator Fabiola Juárez Guevara; data entry helpers Brittany Burke and Joaquín Barragán Rosas; interviewers María de los Ángeles Cabral Porchas, Manuel Acuña Zavala, Jesús Robles, Miguel Angel Torres Avila, Dolores García, Delisahé Velarde Hernández, and Daniela Hernández Ramos; the psychologists Sonia Valentina Javalera Celaya and Verónica Lizeth Martínez Álvarez who supported interviewees in need and helped keep the team centered; Guadalupe A. de Majul and her therapist colleagues at Tech Palewi in Mexico City for their help in interpreting some of the results; and the Centro de Investigaciones en Alimentacion y Dessarrollo (CIAD-Hermosillo) for hosting this project and providing logistical support.

NETWORKS IN HAZARD MITIGATION AND ADAPTATION

NETWORKS AND HAZARD ADAPTATION AMONG WEST AFRICAN PASTORALISTS*

Mark Moritz

The Ohio State University, Columbus, OH, United States

CHAPTER OUTLINE

INTRODUCTION

Livestock transfers are a total social fact (Mauss, 1925)—rich in meaning, complexity, and morality, integrating all aspects of African pastoral societies. However, recent research on these transfers has focused narrowly on the economic question of whether or not livestock transfers serve as effective social risk management strategies (e.g., Aktipis, Cronk, & Aguiar, 2011; Bollig, 2006; Johnson, 1999;

---

*This chapter has been edited and reused with permission from Springer Publishers. A version less focused specifically on networks was originally published in *Human Ecology*. For a discussion of the complexity of risk, quantitative and qualitative risk assessments in the three communities, the complexity of property rights over livestock, and the focus on livestock transfers as ideal types, I refer readers to the original article: Moritz, M. (2013). Livestock transfers, risk management, and human careers in a West African pastoral system. *Human Ecology, 41*(2), 205–219.

Social Network Analysis of Disaster Response, Recovery, and Adaptation. http://dx.doi.org/10.1016/B978-0-12-805196-2.00013-3

McPeak, 2006). Pastoralists risk losing their livelihood overnight due to drought, disease, and other disasters. They employ different strategies to minimize these risks, including mobility, diversification, herd maximization, and social strategies (Halstead & O'Shea, 1989). Social strategies, in particular livestock transfers, have been considered critical because they provide not only a safety net during disasters but also contribute to the resilience of pastoral societies by allowing pastoralists to rebuild their herds after disasters. It is for these reasons that systems of livestock transfers have been described as a pastoral moral economy (Bollig, 1998; Potkanski, 1999).

In the last 20 years, a number of studies have systematically examined the cultural ecology of livestock transfers—in particular, whether and how these transfers might serve as risk management strategies (e.g., Aktipis et al., 2011; Bollig, 2006; Johnson, 1999; McPeak, 2006). The studies have shown that the role of livestock transfers in risk management is more complex than the theory suggests and that transfers are often not functioning as effective risk management strategies. The question is why do these studies fail to support the theory that livestock transfers serve as effective social risk management strategies? I think that there are a few reasons. Foremost, the theory and measurements are too narrowly focused on dyadic material transfers—rather than larger networks—and on the utilitarian concept of risk management rather than risk management as only one concern in the set of costs and benefits. As such, the studies do not consider the social and cultural contexts and dimensions of decisions that individuals make when they engage in these transfers. In this chapter, I draw on Walter Goldschmidt's (1969, 1990) conceptual framework of the *human career* and on his ethnographic work with the Sebei to examine whether and how livestock transfers serve as risk management strategies among FulBe pastoralists in the Far North Region of Cameroon. Goldschmidt's anthropological approach is particularly useful because it focuses on decision-making that individuals make within a context of multiple sets of overlapping social responsibilities, which, I argue, is key to understanding livestock transfers in pastoral societies.

THE HUMAN CAREER OF AN AFRICAN HERDSMAN

Social risk management strategies carry special importance in pastoral systems because they contribute to the social integration of pastoralists as well as to the survival of their societies by allowing pastoralists to rebuild their herds after disaster. As a result the livestock transfers and the family herds they produce are a reflection of the social life of pastoralists. The idea that herds can be read as social biographies is well described in Goldschmidt's book *Kambuya's Cattle: The Legacy of an African Herdsman* (1969). It is based on a unique data set comprising descriptions of a two-week event with extended discussions about the settlement of Kambuya's estate and an extensive social and genealogical history of his herd, reflecting his human career. *Kambuya's Cattle* is a good illustration of Goldschmidt's theoretical approach. The conceptual framework focuses on the individual within a cultural community and the pursuit of their human career, i.e., the lifetime pursuit of satisfactions, both physical and social, which is captured well in his concept of "affect hunger" (1990, p. 3). In other words, it focuses on what drives and motivates individuals, how they make strategic choices within a social and cultural context, and how their aspirations for the future guide their decision-making.

In *Kambuya's Cattle*, Goldschmidt describes how Sebei individuals strategically pursue their human careers, including the maintenance of meaningful social relations within their community. One of the ways in which the Sebei and other pastoralists create networks of meaningful social relations is through livestock transfers. One of the most important livestock transfers among the Sebei is called *namanya*, in which a man who needs to ceremonially slaughter an animal takes a bullock from another man in exchange for a heifer. The immediate reason for the exchange is that it is not good to kill one's own bullock (Goldschmidt, 1972, p. 189). But the exchange also binds the men for a number of years as the original heifer is only returned when she produces a heifer, which then replaces the original bullock. This and other livestock transfers serve multiple functions in addition to giving Sebei the intrinsic pleasure of having cattle. They build up herds, provide milk, spread risk, resolve labor shortages, create public investment in the herd, and they create networks of social obligations, which can be used in many social contexts, including getting free beer. Risk management is but one of many reasons why Sebei engage in livestock transfers.

LIVESTOCK TRANSFERS AND RISK MANAGEMENT IN BROADER SOCIAL NETWORKS

Pastoral societies have been considered the epitome of risk management. The risks to pastoral systems derive from pastoralists' reliance on livestock in marginal environments. For example, in dryland ecosystems rainfall variability is at the root of the problem as it directly determines pasture productivity and droughts can easily wipe out pastoral households. But pastoralists can also lose their livelihood overnight to the hazards of raids and diseases, and they employ various strategies to minimize the likelihood of these threats and increase the reliability of their production strategies.

The hypothesis that livestock transfers serve as social risk management strategies is an appealing one because of the concreteness of the livestock transfers, which allows for quantitative measurements of social support networks. However, the hypothesis is also extremely difficult to test because it requires a longitudinal data set about a sensitive and often secretive topic—livestock ownership and transfers. Ideally, the longitudinal data set also includes a drought period in which the social support networks come into action. Recently, a number of comparative studies of risk management in pastoral systems have systematically examined whether and how livestock transfers serve as risk management strategies, including Michael Bollig's comparison of the Himba and the Pokot (2006) and the Global Livestock-Collaborative Research Support Program on Pastoral Risk Management (McPeak, Little, & Doss, 2011).

These studies show that the role of livestock transfers in risk management is more complex than the theory suggests. First, the broader cultural meanings of these transfers (e.g., support, trust, community) do not accurately reflect the behavioral reality (e.g., lack of support for the poor). For example, Bollig's (1998) ethnographic study of the Pokot before and during a drought showed that the concept of *tilyai* guides emotions and behavior of livestock transfers, and involves affection, trust, and compassion, and also that this is not directly reflected in the practice of livestock transfers. In general, the studies show that transfers are not functioning as effective risk management strategies. Bollig (1998) found a very weak correlation between exchange networks and aid given during a disastrous year among 37 Pokot households. Similarly, John McPeak's (2006) econometric analysis from a study of livestock transfers among the Gabra showed that wealthier households were able to have larger support networks than poorer households and that livestock transfers are not effective risk management strategies for the poor because they are excluded from these networks (see also Little, McPeak, Barrett, & Kristjanson, 2008, p. 598).

Such findings dovetail with those of other studies of pastoral systems, which have shown that, in general, the poor will be poor and the wealthy will be wealthy simply because of the dynamics of herd growth (Bradburd, 1982; Fratkin & Roth, 1990), which over time leads to distinct low herd size and high herd size stable equilibria (Lybbert, Barrett, Desta, & Coppock, 2005). Thus, there may be two reasons why livestock transfers do not work for poorer households: either they are excluded and/or the number of livestock transfers may simply be too small to have any significant effect on herd longevity. Computer simulations support this explanation. Recently, C. Athena Aktipis and colleagues (2011) developed an agent-based model of Maasai livestock transfers to examine whether these *osotua* transfers were an effective way of risk pooling. The computer simulations showed that the system of transfers is an effective way of risk pooling, but that the median herd duration is only 18 years. Thus livestock transfers may be effective in the short term but are not sufficient for long-term survival.

It is also possible that livestock transfers may not be necessary for the maintenance of support networks. Scholars have argued that livestock transfers create social capital that can be used to demand aid during disasters and provide economic capital to reconstitute herds after disasters (Bollig, 1998; Little et al., 2008, p. 598; McPeak, 2006). However, transfers are but one of many ways in which pastoralists develop social support networks. For example, Brooke Johnson's (1999) ethnographic study found that Turkana social networks consist of relationships that are maintained through joint activities, food sharing, mutual affection, and only occasionally livestock transfer. Livestock transfers may be instrumental in solidifying some of these social relations, but they do not necessarily create strong interpersonal bonds. In the end, it is the quality of the social relationship and the mutual affection that is critical, not the transfer of livestock, which happens less frequently than other social activities like social visits and food sharing (Johnson, 1999).

These studies intriguingly suggest that livestock transfers are neither sufficient nor necessary for support networks in pastoral societies. Nevertheless, research continues to focus on these social strategies of risk management—in part because they are institutionalized and expressed materially and thus relatively easily studied compared to other qualitative and quantitative aspects of social relations, like visits, food sharing, and the intangibles of friendship. However, there are disadvantages to this narrow focus on institutional forms of livestock transfers, especially when they are described as ideal types (Moritz, 2013).

METHODS

The data for this article were collected in a comparative, ethnographic study of three pastoral communities in the Far North Region of Cameroon in 2000–2001. The communities represent different pastoral systems: peri-urban, agropastoral, and mobile pastoralist (see Table 13.1). The Far North has a semiarid climate with a single rainy season. Rainfall is characterized by high spatiotemporal variability with frequent drought. During the eight-month dry season from October to May cattle lose considerable weight and become more susceptible to diseases. The primary goal of pastoralists is to overcome the difficulties of the dry season.

I conducted multiple household surveys throughout the year to collect demographic, agricultural, and consumption data on individual households. I also conducted herd surveys in which data were collected on herd management, production costs, as well as ownership and transfers of animals. The data on FulBe family herds also allowed me to examine whether livestock transfers function as effective risk management strategies. In 2010, I visited the communities and recorded which family herds still existed in order to assess long-term herd viability.

The survey began by identifying all the cows that had calved by name and recording their age, offspring, and what happened to their offspring, i.e., whether the offspring was still in the herd, dead, or sold. Then, I recorded all the animals whose mothers were not or no longer in the herd: bulls, bullocks, and heifers that either had been bought or whose mother had died or been sold. In subsequent

Table 13.1 Characteristics of the Three Communities, 2000–2001

		Wuro Badaberniwol (WB)	Wuro Ladde (WL)	Wuro EggoBe (WE)
Community	Pastoral system	Peri-urban pastoralists	Agropastoralists	Mobile pastoralists
	Distance from Maroua Center	10 km	50 km	75 km
	Administrative subdivision	Maroua	Bogo	Moulvoudaye, Guirvidig
	Population density	100–149 km²	25–49 km²	1–24 km²
	Number of households	26	16	8
	Number of people	226	109	71
	Ethnic groups	FulBe, RiimayBe, Kanuri	FulBe	FulBe
	Number of pastoral households	6	16	8
Households	Household size	17.0 (±8.1)	6.8 (±3.0)	8.9 (±4.0)
	ACE per household	12.7 (±6.4)	5.1 (±2.3)	6.6. (±3.1)
	Number of cattle	642	306	789
	Average size village herd	28.0 (±5.6), n=6	9.3 (±5.2), n=16	75.4 (±44.6), n=6
	Average size bush herd	77.1 (±69), n=5	17.6 (±9.7), n=8	NA
	Average size entrusted herd	NA	NA	46.8 (±17.2), n=3
	TLU per ACE	8.1 (±7.1)	4.0 (±4.4)	16.4 (±16.2)

The household data in the peri-urban community concern only the six pastoral households. Standard deviations are given in parentheses. ACE, adult consumer equivalents; TLU, tropical livestock unit. 1 TLU = 1 camel; 0.8 cattle; 0.1 small stock.

interviews, I recorded the changes in the herd over the year 2000–2001 starting with the beginning of the rains in 2000, for example, births, deaths, sales, loans, strays, thefts, inheritance, entrustments, and purchases. The data on structure and changes of the family herds were collected in several interviews throughout the year. I also conducted a survey to document property rights of animals that were presently owned and/or managed by each household member, including those animals that were in other corrals.

I used descriptive statistics, chi-square, multiple regressions, and binary logistic regressions to analyze livestock ownership and transfers within and between communities and whether and how livestock transfers contribute to short-term survival of households and long-term viability of herds. In designing the study, I opted for documenting detailed property rights and transfers within individual herds, which meant that the sample size is relatively small (28 herds with 1442 cattle in total), and this has implications for statistical analyses.

RESULTS

In order to examine whether and how livestock transfers serve as risk management strategies in a larger set of social networks among FulBe pastoralists in the Far North of Cameroon, I will discuss the variation in livestock transfer practices within and between the three communities, including the relationship between transfers and risk management. Finally, I will discuss what motivates pastoralists to engage in livestock transfers. I focus here on two types of livestock exchanges—*nannganaaye* loans and herding contracts—which are common among FulBe pastoralists across West Africa. They have generally been narrowly described as ideal types emphasizing the dichotomy of the morality of the loans and the amorality of the loans, respectively (Scott & Gormley, 1980; White, 1990). However, my study of livestock transfers in three FulBe communities shows that there is considerable variation and some overlap in these transfers (see also Boutrais, 2008; Moritz, 2013). First, I will discuss the risks that pastoralists faced in the three communities.

RISK ASSESSMENT

The study was conducted in a drought year (2000–2001) and thus provided a good opportunity to study the risks that pastoralists faced in the Far North of Cameroon and how livestock transfers and the networks they create may be used to support households that face disasters. A quantitative risk assessment showed that the drought did not result in a reduction in herd size. In fact, pastoralists in the peri-urban and mobile communities saw their herd size increase during the drought year, whereas average herd size neither increased nor decreased in the agropastoral community (Moritz, 2013). A qualitative assessment of the risks that pastoralists faced in the three communities supports the previous quantitative descriptions. First, no household lost their livelihood overnight; the disasters were slow and gradual in the making. The poorer households knew well in advance that they would be hungry next rainy season when their savings would be finished. The same was true for livestock losses; the risks of losing animals to diseases and raids were real but no one lost their entire herd overnight. Second, the immediate reason for the hunger was the failed rains, but the ultimate cause was poverty. Poorer households simply did not have the capital to cope with the failed harvest, whereas wealthier households had no problems dealing with the failed harvest and the rising cereal prices. Third, the greatest tragedies in the communities were personal and had nothing to do with the drought. Finally, while the main concern for the peri-urban and agropastoral poor was to feed the household members, the greatest concern for herders in the bush was to their lives, not their livelihoods. They ran the risk of being killed by cattle thieves or armed bandits.

THE COMPLEXITY OF NANNGANAAYE TRANSFERS

The *nannganaaye* is a term used to describe many different practices of livestock transfers, which have changed over time. In my study of FulBe family herds and livestock transfers, I did not come across the prototypical WoDaaBe *nannganaaye* delayed exchange in which a cow is only returned to the original owner after it has calved three times. Instead, I documented other forms of the *nannganaaye* as well as many other types of livestock transfers, some of which overlapped with the *nannganaaye*. The transfers did not represent mutually exclusive categories. For example, the *nannganaaye* did not always involve a change in ownership as no calves were given to the recipient of the loan. Instead, the *nannganaaye* served as a source of milk for the receiving household, similar to a short loan of a lactating cow (*diilaaye*).

However, among mobile pastoralists, the current practice of the *nannganaaye* was somewhat similar to the WoDaaBe prototype in that one offspring was given to the receiver of the loan, which was a recent borrowing from the Daneeji, a group of FulBe named after their white cattle, with whom the mobile pastoralists shared transhumance routes in the 1970s, but who have since left for Chad and the Central African Republic. The older practice of the *nannganaaye* among mobile pastoralists was quite different; it was all about milk, not about giving animals. Moreover, *nannganaaye* animals and their offspring would remain indefinitely in the receiver's herd, which made it critical that the recipients were skilled herders so that the number of animals increased. Informants argued that in the past, *nannganaaye* was all about herding (*ngaynaaka*), while now it was all about love (*enDam*), trust (*amaana*), and aid (*ballal*), i.e., about social relations between people rather than the economics of herd growth.

The different forms of *nannganaaye* were not associated with different communities, i.e., among mobile pastoralists, the *nannganaaye* is also used to describe short-term loans of lactating cows like the *diilaaye*. Thus, one cannot study livestock transfers by simply counting the different types of transfers because the terms cover multiple forms that overlap, e.g., a short-term loan of a lactating cow is sometimes referred to as *diilaaye* and sometimes as *nannganaaye*. Moreover, one of the most generous transfers, a simple gift without any strings attached, is hypocognized in the sense that there is no term for this type of transfer. Informants would say, "I just gave the animal" (*mi hokki meere non*), and the animal was simply referred to as "the animal given" (*hokkaange*). Again, simply counting ideal types of transfers, rather than describing practices of transfers, would overlook this transfer.

HERDING CONTRACTS IN A SOCIAL CONTEXT

A ubiquitous livestock transfer was the entrustment of animals and herds by absentee owners. However, this category of long-term transfers comprised multiple types of transfers, ranging from herding contracts in which a hired herder received a monthly wage and usufruct rights over the animals in the herd (Moritz, Ritchey, & Kari, 2011), to the entrustment of a few animals without compensation save for usufruct rights (*goofalye*). In fact, in some cases, transfers that were called *goofalye* were very similar to the *nannganaaye* exchange and sometimes even referred to as such. One could make a distinction between *goofalye* and *nannganaaye* in terms of the benefits for the different parties, i.e., whether the transfer benefits the receiver (food aid) or the donor (labor). However, in many cases, the transfer of animals served both purposes, which makes it difficult to use this as a criterion to make a distinction between the two.

Herding contracts are arrangements between an owner and a herder who cares for the herd and is compensated with a wage and/or livestock. Although there is considerable variation in herding contracts, one can distinguish two main types of contracts in West Africa: hired herding and entrustment. Hired herding is a labor contract in which an owner pays a herder a monthly wage and provides him with herding equipment (e.g., shoes, clothes, stick). Entrustment is a leasing contract in which an owner entrusts animals to a herder who has usufruct rights over milk but is not paid a wage, although there may be other forms of compensation, including cash (Moritz et al., 2011).

The herding contracts were not simply labor contracts; they are better described as patron–client relationships in which the owner has responsibility for the herder and his family. In fact, owners are called *jaagordo* or patron in Fulfulde. Herders and absentee owners had longstanding relations. Although many contracts ended within one year, the majority of the hired herders we interviewed had been working for the same owner for over 5 years and a few had been working for the same owner for over 20 years (Moritz et al., 2011). The aforementioned herder involved in the car accident was bedridden for more than 6 months and during that time the absentee owner's family cared for him, while his patrilineal kin managed the absentee-owned herd. Thus, herding contracts also provided social support for poor pastoralists, in addition to *nannganaaye* transfers from fellow pastoralists.

SOCIAL INFLUENCES ON VARIATION IN TRANSFERS BETWEEN COMMUNITIES

The multiple and overlapping forms of transfers result in great complexity, which makes it difficult to measure and compare transfers within and between communities. I have therefore organized the types of transfers in four broad categories in which I used the descriptions of the practices rather than the names of the transfers to classify the transfers: (1) **reciprocal loans** in which there is an expectation of reciprocity, e.g., the *nannganaaye*; (2) **gifts** in which there is no expectation of reciprocity, e.g., from parents to children (*sukkilaaye*) or between friends (*hokkaange*); (3) **short-term loans** to cover a specific need of the receiving household, e.g., the loan of a milk cow (*diilaaye, nannganaaye*), carrying bull (*garwaari*), or breeding bull (*kalhaldi*); and (4) **long-term entrustments** in which the receiving party gains usufruct rights over cattle in return for labor, e.g., direct entrustment (*goofalye, nannganaaye, halfiinge*) and herding contracts (for detailed descriptions of the different types see, Moritz, 2012). The first three types of exchanges generally occur between friends and family within communities (i.e., strong ties), whereas the last type generally involves patrons that are living outside the community (i.e., weak ties).

There is considerable variation between communities in the percentage of cattle transferred in the different categories (see Table 13.2). Although it seems that peri-urban pastoralists are more involved in livestock transfers than agropastoralists and mobile pastoralists (81%, 73%, and 60%, respectively), this is not the case if one excludes long-term entrustments, most of which are herding contracts (2%, 17%, and 32%, respectively for the three communities).

In the peri-urban community only half of the households were engaged in livestock transfers—not counting the herds under contract that are entrusted to mobile pastoralists—and these transfers only involved 2% of all the animals, while in the mobile community more than 20% of the animals were involved in a *nannganaaye* transfer. The mobile community was the only one with short-term loans of milk cows, breeding bulls, and pack oxen (5% of the animals). The agropastoral community fell somewhere between the other two communities in terms of the number of livestock transfers. Whereas in the peri-urban community all long-term transfers were given and in the mobile community all long-term transfers were received, agropastoralists both gave and received long-term entrustments (28% and 27% of the animals, respectively), and few of them were herding contracts with wages.

Table 13.2 Number of Livestock Transfers by Category in Three Communities, 2000–2001			
	Peri-Urban Pastoralists (*n*=6)	**Agro Pastoralists** (*n*=16)	**Mobile Pastoralists** (*n*=6)	**Total Per Transfer Category**
1. Reciprocal loans	5 (1%)	45 (13%)	120 (24%)	170 (13%)
2. Gifts	5 (1%)	19 (6%)	27 (5%)	51 (4%)
3. Short-term loans	0 (0%)	0 (0%)	27 (5%)	27 (2%)
4. Long-term entrustment	410 (98%)	275 (81%)	328 (65%)	1013 (80%)
Total per community	420 (33%)	377 (27%)	546 (40%)	1261 (100%)

The table shows the number of cattle that were either given or received by households in the three communities per category (e.g., reciprocal loans). Transfers per category differed significantly across communities (whole table Chi-square = 170.76, df = 6, p < .0001).

THE EFFECTIVENESS OF TRANSFERS AS RISK MANAGEMENT

To examine whether there were any patterns in livestock transfers received by poor versus wealthy households (see Table 13.3), I combined the data for all three communities, as the number of households in each community was small, and used multiple regressions and binary logistic regressions (using community as a dummy variable in which communities with agriculture were combined). The results show that the number of all transfers received can be explained to some extent by herd size and community (Moritz, 2013). In other words, households with smaller herds were more likely to receive livestock transfers (all categories), and this also holds when loans and gifts (categories 1–3) and long-term entrustment (category 4) are examined separately.

I also examined whether pastoralists who have herds or animals under contract (category 4) from absentee owners were excluded from loans and gifts within the community (categories 1–3) and found that this is not the case (Moritz, 2013). In other words, pastoralists with entrusted animals were not excluded from loans as has been described for the WoDaaBe in Niger (White, 1990). On the contrary, three of the four households that received the most loans and gifts in the mobile community had also herds under contract (see Table 13.3). All this indicates that poorer households received more transfers than wealthy households in the mobile community and that support came from within the community as well as from outside absentee owners.

However, the question is whether the livestock transfers are effective in terms of risk management, which in the case of pastoral systems has been defined in terms of short-term survival (i.e., food aid) and long-term sustainability (i.e., rebuilding herds) (Bollig, 1998). I measured short-term survival in terms of whether a household has enough livestock to provide for its members, i.e., whether it was self-sufficient in terms of herd size. This measure does not apply to agropastoral and peri-urban households because they also rely on agricultural production for their subsistence needs. Additionally, this measure does not account for the extent to which people rely on remittances or other extra-pastoral contributions by family members. Fratkin and Roth (1990) estimated that 4.5 tropical livestock units (TLUs) per person would provide an individual with sufficient calories in pure pastoral systems, TLUs in which one unit = 1 camel; 0.8 cattle; 0.1 small stock (Dahl & Hjort, 1976). I used adult consumer equivalents (ACEs) to adjust for age of household members. Using these criteria, three of the mobile households would not have been self-sufficient were it not for the livestock transfers (see Table 13.3).

However, these three poorest households were no longer part of the mobile community in 2010 (10 years after the original study was conducted), while the other three households that had more than 4.5 TLUs/ACEs were doing well in 2010. I used a binary logistic regression to examine whether herd size or number of transfers received could explain the long-term viability of family herds. The results show that households that received the most transfers were less likely to be around 10 years later in 2010, meaning that the household had left pastoral society and settled in a village or that the household no longer existed because its members had joined other households (Moritz, 2013). Thus, livestock transfers, including herding contracts, were effective in providing subsistence for households in the short term, but the transfers did not aid in rebuilding their herds beyond the threshold of herd viability. The main reason is that the livestock transfers, including the *nannganaaye*, provided the receiving households with usufruct rights over the animals, but not necessarily with the right of disposal, and so households did not see their herd size increase beyond the viability threshold.

Table 13.3 Transfers Received, Herd Size, and Consumers per Household in Three Communities, 2000–2001

Households	Herd Size Without Transfers	Loans and Gifts (Cat. 1–3)	Long-term Entrustment (Cat. 4)	Herd Size With Transfers	Ratio of Transfers to Herd Size	ACE	TLU/ACE Without Transfers	TLU/ACE With Transfers
WL6	0	10	0	10	1.0	4	0	2
WL8	0	0	13	13	1.0	5	0	2
WL5+	1	0	18	19	0.8	3	0	6
WL7	1	4	0	5	0.9	5	0	1
WL9	1	4	8	13	0.0	6	0	2
WL3	3	0	0	3	0.6	3	1	1
WL13	4	3	3	10	0.6	7	0	1
WL4	6	4	6	16	0.0	11	0	1
WL16+	6	0	0	6	0.6	2	2	2
WL2	11	5	10	26	0.5	6	1	3
WL10	17	0	15	32	0.3	7	2	4
WL14	22	0	9	31	0.2	5	4	5
WL15	23	0	6	29	0.3	2	12	15
WL11	24	0	8	32	0.2	8	2	3
WL12	41	1	9	51	0.0	4	8	10
WL1	81	0	0	81	0.0	4	15	15
WB6	26	0	0	26	0.1	8	3	3
WB1	49	0	4	53	0.0	19	2	2
WB3	66	0	0	66	0.0	4	14	14
WB5	77	0	0	77	0.0	12	5	5
WB2	79	0	2	81	0.0	21	3	3
WB4	214	0	2	216	0.7	12	14	14
WE2+	18	6	34	58	0.7	4	3	11
WE6+	26	18	54	98	0.7	7	3	11
WE1+	37	25	53	115	0.4	12	3	8
WE5	42	17	16	75	0.1	5	7	13
WE4	75	4	0	79	0.1	7	8	9
WE3	105	16	0	121	1.0	5	17	19

WL, agropastoral community; WB, peri-urban community; WE, mobile pastoralists. The plus signs (+) indicate those herds that no longer existed 10 years after the original study.

PASTORALISTS' SOCIAL ENGAGEMENT IN LIVESTOCK TRANSFERS

There are numerous reasons why pastoralists transferred livestock. One of the purposes of the transfers across communities was to support poor households with food aid or to support them in other ways—for example, by transferring ownership rights over entrusted animals so that the recipient household could sell the animal to buy sorghum or pay hospital bills. But livestock were also transferred because of practical considerations on the part of the lender (e.g., weaning of calves, lack of corral space). Others transferred animals to a good herder whose skills (*ngaynaaka*) or luck (*risku*) would increase their wealth. This was the main reason why mobile pastoralists engaged in the *nannganaaye* before they adopted the Daneeji cultural model of the *nannganaaye*, which involves a reciprocal exchange and the gift of offspring (*sukkaaye*).

In the agropastoral village, people were also transferring livestock to fellow villagers without animals of their own to enable them to participate in the social life of the village. Animal husbandry was one of the major topics of conversation and entailed a number of common everyday activities in villages (e.g., herding, watering). People without animals were unable to fully participate in the social life of the village because they did not share common concerns and could not participate in socially meaningful everyday activities.

Finally and foremost, pastoralists gave *nannganaaye* to reinforce and deepen friendships (*enDindirgo*). *EnDam* (love, affection) was a key component of the *nannganaaye*, also when it was given as food aid. Some poor households did not receive *nannganaaye* because the people were not liked much in the community. The functions and motivations overlapped synchronically and diachronically, e.g., one pastoralist gave surplus animals to an impoverished leader with whom he was befriended. Another transferred an animal for practical reasons, which later became food aid for the recipient household. To add to the confusion, all these livestock exchanges were often lumped under the same term of *nannganaaye*.

The motivations for engaging in livestock transfers also conflicted; for example, while helping the poor was an important reason for livestock transfers such as the *nannganaaye*, it conflicted with pastoralists' strategic goal of increasing their own herd size. In discussions, pastoralists argued it was better to give *nannganaaye* to wealthy friends because the poor do not take good care of the animals (e.g., they do not leave enough milk for calves or do not give supplementary feed). Moreover, it would be shameful to take the animals back because the poor need the animals. Wealthy pastoralists on the other hand would take good care of your animals and might even reciprocate the loan and give you offspring (*sukkaaye*). There was thus a tension between pastoralists' social responsibilities and the strategic goals of increasing one's own herd. Exchanging livestock with wealthy friends allowed pastoralists to develop their social network and pursue their strategic goals, but that was not the case when transferring livestock to poor pastoralists because they were less reliable as exchange partners than wealthier pastoralists.

Exchange relations were not without problems. I recorded many stories of exchange relations that had soured and ended in traditional courts, either because transferred animals had been sold without permission from the lender or because it was unclear who had which rights over what animal. Ultimately, exchange relations were all about trust since the stakes were high. The monetary value of cattle has remained high in the last decade; the market value of a cow was approximately 150,000 FCFA (or $300) in 2000–2001. This was one of the reasons why many recipients of *sukkaaye* animals (offspring of the *nannganaaye* cows) immediately sold the animals at local markets in order to avoid future conflicts over property rights—for example, with the lender's heirs. When animals from livestock transfers were sold, they did not aid in rebuilding family herds, unless other animals were bought with the revenues, which happened in some cases. In most cases, however, the money was spent on sorghum and other necessary expenses. This may be one of the reasons why livestock transfers contributed to short-term survival but not long-term sustainability.

DISCUSSION

In my comparative study of livestock transfers in three FulBe communities, I found that pastoralists did not lose their livelihood overnight and that poorer households were most at risk of falling below subsistence level (Moritz, 2003). I also found that livestock transfers, including herding contracts, provided short-term support for these households, but did not contribute to long-term viability of herds. I am aware that the sample size is small, but the patterns are robust, statistically significant (Moritz, 2013), and supported by qualitative analysis.

The question is why livestock transfers provide short-term support but do not allow pastoralists to rebuild their herds? I offer a few interrelated explanations, which I will discuss in some detail next. First, risk management cannot be reduced to livestock transfers, which are neither necessary nor sufficient for support networks. Second, the nature and number of livestock transfers are not enough to overcome herd growth dynamics. Third, individuals transferring animals are making strategic decisions that are aimed not only at supporting the poor but also at increasing their own herd size and advancing their own human careers.

STUDYING SUPPORT NETWORKS

This and other studies have found that the material transactions of livestock transfers are neither necessary nor sufficient for social support networks or risk management (e.g., Bollig, 1998; Johnson, 1999). My original study did not examine the relationship between support networks and risk management. It examined the relationship between livestock transfers and risk management. I argue now in this chapter that this is problematic because transfer networks are not the same as support networks. It is more critical that individuals have family and friends that love and trust them and want to support them. These sentiments are developed over time through myriad activities, including visits in which people develop personal affinities. Livestock transfers may help to solidify those social relations, but again, in themselves, those material transfers are never enough for developing support networks. In other words, networks of livestock transfers cannot be equated with support networks, which are much broader and more encompassing.

An alternative approach to the study of social risk management strategies may be the study of how the quality and quantity of individuals' social networks matter for risk management. Livestock transfers are then considered but one of the ways pastoralists invest in support networks, but not synonymous with the network. One would expect that larger networks are better but that the quality of the relationships and the resources of its members are critical, too. For example, the quality of the relationship—described in terms of love, trust, and friendship—was critical for support networks, as FulBe pastoralists did not transfer animals to people whom they did not love or trust.

However, there is the methodological problem of collecting social network data from pastoralists, which I personally found even more challenging than collecting data about livestock ownership and transfers. Moreover, social networks—just as livestock transfers—are not just about managing risks but also about other facets of social life (e.g., status, politics, and friendship). A social network approach could run into similar limitations as the current focus on livestock transfers. Johnson's (1999) study of Turkana social networks describes how a social network approach could address some of these challenges, for example, by documenting everyday social practices of sharing the elders' tree, social visits, food sharing, and herding partnerships as well as structural relationships like kinship relations, including affinal relations, and, of course, the various livestock transfers. However, such a comprehensive approach may not be feasible. It is telling that no comprehensive study of African pastoralists' social networks (e.g., not just livestock transfers) has been conducted since the South Turkana Ecosystem Project (Johnson, 1999).

SOCIAL NETWORKS AND THE TRANSFERS OF RIGHTS

The literature on the livestock transfers among FulBe pastoralists has focused on ideal type of the *nannganaaye* exchange, which is often contrasted with herding contracts between absentee owners and hired herders (White, 1990). In this literature, the *nannganaaye* involves the gift of three offspring to the receiving household, which allows it to rebuild the family herd. In contrast, herding contracts only provide wages and usufruct rights over the milk of the animals, but no access to capital, i.e., reproductive animals. Moreover, Cynthia White (1990) has argued that poor WoDaaBe are excluded from reciprocal loans when they engage in herding contracts because other WoDaaBe are concerned about care for their *nannganaaye* animals.

This was not the case in my study. There was considerable overlap in form and function of these transfers, and they are often referred to by the same term. Households with herding contracts were also not excluded from exchange networks; poorer households were supported through herding contracts and *nannganaaye* transfers. Moreover, herding contracts were not market exchanges but patron–client relationships in which absentee owners had social obligations toward their herders. However, most livestock transfers involved the transfer of usufruct rights but not the transfer of the right of disposal, which would give recipients full ownership over the animal. This is true for herding contracts in which hired herders were paid wages rather than animals (Moritz et al., 2011) as well as for the *nannganaaye*, which did not always involve the gift of offspring. Although the ideal type of the *nannganaaye* in the literature involves the transfers of the right of disposal over three offspring (Dupire, 1962), this does not describe transfer practices among FulBe pastoralists in the Far North of Cameroon.

DYNAMICS OF HERD GROWTH

The pastoral moral economy is often contrasted with the market economy in which market dynamics lead to greater economic inequality. However, herd dynamics may have the same effect in pastoral societies. Previous studies on pastoral wealth have shown the dynamics of herd growth are due to the advantage of pastoralists with larger herds and that those same dynamics work against pastoralists whose herd size is below a certain threshold (Bradburd, 1982; Fratkin & Roth, 1990; Lybbert et al., 2005). Moreover, a recent comparative study shows that intergenerational transfers of wealth also contribute to persistent inequality (Borgerhoff Mulder et al., 2010). If the dynamics of herd growth are such that the wealthy remain wealthy and the poor remain poor (Fratkin & Roth, 1990; Sieff, 1999), then livestock transfers only make a difference if they push herd size over this viability threshold. However, crossing this viability threshold in the long term does not seem to be common among poorer FulBe pastoralists in this study. The evidence also suggests that livestock transfers further consolidate the position of the wealthy, at least among the Pokot and the Himba (Bollig, 2006). Finally, livestock transfers may be simply too few in number to have any significant effect on herd growth dynamics (Dyson-Hudson & McCabe, 1985; Sieff, 1999, p. 8), although Danny de Vries, Leslie, and McCabe (2006) argue that transfers make a difference in herd demography among Turkana pastoralists.

Livestock transfers and other forms of assistance may prevent pastoral households from going below subsistence level, but they do not allow the poor to rebuild their herds beyond the viability threshold, which means that they are slowly removed from the pastoral system. This does not necessarily mean that herd dynamics threaten the viability of pastoral systems. The system may be resilient, even if not all the elements that make up the system are resilient, i.e., poorer households. Paying attention to the dynamics of economic inequality endogenous to pastoral systems is critical to understand why, when, and for whom social risk management strategies are effective.

STRATEGIC DECISIONS IN THE PURSUIT OF A HUMAN CAREER

The theoretical narrative of risk management has primarily focused on needs of the poorer households and the resilience of the pastoral system as a whole. But what about the wealthier households that are transferring animals? What are their reasons for engaging in livestock transfer? The assumption has always been that disasters strike all households and that wealthier households also need a support network. However, the studies on herd dynamics show that if herd size is above a certain threshold, households are buffered against most risks (Bradburd, 1982; Fratkin & Roth, 1990). In addition, livestock transfers may not be compatible with the risk management strategy of herd maximization (Fratkin & Roth, 1990). Wealthier pastoralists in the Far North argue that they are better off engaging in livestock transfers with other wealthier pastoralists in order to ensure growth of their own herd and the pursuit of their human career, rather than loaning livestock to poorer pastoralists.

Wealthy pastoralists assisted poorer households with livestock transfers and other forms of aid. However, they were also concerned with the growth of their own herd. Most transfers to poorer households involved only usufruct rights and thus provided only short-term support. Goldschmidt's conceptual framework of the human career also helps to understand why livestock transfers are not effective social risk management strategies in the long term, as it focuses on the motivation of individual pastoralists and explains why and how they engage in livestock transfers. Goldschmidt (1990, p. 2) writes in the *Human Career* that individuals are "motivated to a sense of self which means the attainment of social worth—prestige—in the context of community values." In addition to maintaining relationships of love and trust, FulBe pastoralists want to be someone; they want to be valued and respected by friends and others in their community, and this means foremost to amass herds. However, leadership-based status did not play much of a role at all in the livestock transfers. It was really all about wealth. FulBe pastoralists are concerned with supporting poor households in need, but not to the point that it reduces their own herd growth and their own human career.

CONCLUSION

Livestock transfers are often portrayed as total social facts; however, somehow this often gets lost in the studies that have found that transfers are not effective as risk management strategies in the long term for poorer households. Anthropological understandings of livestock transfers have been hampered by the narrow focus on the utilitarian function of risk management and the methodological focus on material transfers, rather than a holistic approach of individuals' social relations within cultural communities. It is no surprise that the studies found that the role of livestock transfers in risk management is limited, because the transfers were never primarily or solely about managing risks—they were about people making strategic choices about their human career within their cultural community. To understand support networks and risk management in pastoral societies is not sufficient to study livestock transfers. Instead, it is critical to use a holistic approach to study pastoralists' networks broadly and how these are used as hazard adaptation.

CYCLONES ALTER RISK SHARING AGAINST ILLNESS THROUGH NETWORKS AND GROUPS: EVIDENCE FROM FIJI

14

Yoshito Takasaki

University of Tokyo, Tokyo, Japan

CHAPTER OUTLINE

INTRODUCTION

Whereas disaster aid for relief and reconstruction plays a central role in disaster victims' recovery (e.g., Amin & Goldstein, 2008), disaster aid is essentially a temporary public transfer targeted toward disaster-specific damages, such as those on housing, livelihoods, productive assets, infrastructure, and so forth. During post-disaster periods, survivors continue to rely on private safety nets to deal with adverse shocks not directly related to the disaster. In particular, in developing countries, informal risk-sharing institutions play a central role in how people cope with health shocks, because neither health insurance nor public safety nets are available among the poor (Strauss & Thomas, 1998). In this chapter, I focus on ex post risk sharing in response to realized shocks as a risk coping strategy; my data precludes me from examining ex ante risk sharing against future potential shocks. Numerous studies in the field of economics have shown that informal risk sharing against illness (especially an idiosyncratic one, but not a covariate one like epidemics) is available in developing areas, although such risk sharing is far from complete (e.g., Dercon & Krishnan, 2000; Gertler & Gruber, 2002; Kochar, 1995).

This chapter explores adaptations to natural disasters among poor populations by examining how natural disasters alter their informal risk sharing against illness. Although adverse effects of natural disasters on various dimensions of well-being, such as consumption, child nutrition, and public health, have received much attention from researchers (e.g., Noji, 1997), systematic studies that address the link between natural disasters and informal risk sharing against subsequent non-disaster shocks are extremely scarce. The analysis is based on original household survey data collected in 43 Fijian villages more than 2 years after a tropical cyclone, when the region was still in the reconstruction phase. My companion paper explores the same question based on a panel subsample of households in seven Fijian villages (Takasaki, 2011c).

Informal risk sharing is built on social networks (Barrett, 2005; Jackson, 2008, 2010). Households exchange private transfers in the forms of cash, in-kind, and labor to help those who are experiencing adverse shock within a network. The Fijian survey data capture not only private transfers exchanged among households within a network (*network-based transfers*), but also transfers exchanged directly with groups to which the household belongs (*group-based transfers*), such as ritual gifts for kin groups, village communal work, and church donations. Although economists have extensively studied network-based transfers (see, for example, Cox & Fafchamps, 2008 for review), group-based transfers have received very limited attention in developing countries; in developed countries, in contrast, transfers to community institutions in general (e.g., charitable giving) have been well studied (see, for example, Schokkaert, 2006 for review). In Fiji, group-based transfers constitute a much greater amount than network-based transfers, because of significant household contributions to groups for the provision of local public goods (Takasaki, 2011b). Partha Deb, Cagla Okten and Una Osili (2010) conducted a similar comparison using Indonesian Family Life Surveys, though risk sharing is not their focus. Group-based transfers can contain a significant risk-sharing component, such that group members with adverse shock contribute less than do others (Takasaki, 2011c).

Although economists often highlight the village as a risk-sharing pool because of its information and enforcement advantages, recent works directly address the question of among whom people share risk. Some researchers focus on preformed groups other than the village, such as kin, caste, and ethnic groups (e.g., Grimard, 1997; Morduch, 2005), while others study the formation of risk-sharing groups and networks (e.g., De Weerdt & Dercon, 2006; Fafchamps & Gubert, 2007). In Fiji, not only group-based transfers are exchanged with different types of groups—kin, religious, and social groups as well as village—but also these groups form household networks within which network-based transfers are made.

In reciprocity-based risk-sharing arrangements with limited enforceability, current transfers depend on the past history of transfers (Ligon, Thomas, & Worrall, 2002). Andrew Foster and Mark Rosenzweig (2001, p. 390) demonstrate that "the existence of binding imperfect commitment constraints implies that households that have made net transfers in previous periods are less likely to provide subsequent transfers, given the current state of the world, than are households that have been net recipients of transfers." This is referred to as the norm of reciprocity in sociology (Gouldner, 1960). Eliana La Ferrara (2003) examines this pattern in credit transactions among kin members in Ghana. This "reciprocity effect" suggests that disaster victims may be "less" insured against illness during the reconstruction phase than nonvictims if victims reciprocate the help that they already received from nonvictims for their recovery. In such a case, the norm of reciprocity places an undue burden on those most in need (i.e., victims) as they should be on the receiving end of altruistic transfers rather than beholden to an expectation to use their own resources to repay some implicit debt they incur by receiving help (see Takasaki, 2011c for more discussions about the reciprocity effect in the reconstruction phase).

Whether the reciprocity effect is significant in network- and group-based transfers is an empirical question. A comparison of network- and group-based transfer responses to illness between disaster victims and nonvictims reveals that although group-based transfers show patterns consistent with the reciprocity effect, network-based transfers show the opposite. Specifically, although victims are "less" insured against illness among kin and social group members than nonvictims, victims are "more" insured against illness through kin and religious networks than nonvictims. As such, the general thesis of this study is that social networks enable disaster victims to partly compensate for their weakened private safety nets against subsequent non-disaster shocks in Fiji.

The rest of the chapter is organized as follows. The next section describes the study area, the cyclone, and health shock. Section "Household Private Transfers" provides a description of household private transfers. Section "Empirical Analysis" examines how household private transfers respond to health shock. The last section provides interpretations of the results plus conclusions.

DATA, DISASTER, AND HEALTH
STUDY AREA AND DATA

In June-September 2005, I conducted a household survey among 906 households in 43 native Fijian villages in Cakaudrove Province in the northern region of the country.[1] In each of 16 districts in the province, villages were intentionally chosen to cover distinct environmental and economic conditions, but not cyclone damage, because of a lack of comprehensive village-level damage data. In each village, households were stratified by *tokatoka* (smallest kin-group unit, as discussed later) and a combination of individual leadership status (e.g., kin leader) and major asset holdings (e.g., shops) (all tokatoka were sampled); in each stratum, households were randomly sampled (50% of the population in each stratum, on average). The analysis sample is 879 households for which there is complete information of household-level cyclone damage and household private transfers, as well as basic household characteristics. At the time of interviews in 2005, the mean annual total income earned by sample households was about F$11,000, or about F$2000 per capita (F$1 = US$.60); farming and fishing, respectively, account for 64% and 11% of income earned. Households employ traditional farming practices, using no mechanized equipment or animal traction, to produce taro, cassava, coconut, and kava plants and engage in artisanal fishing, using lines and hooks, simple spear guns, or rudimentary nets.

---

[1]Fiji is divided almost evenly between native Fijians and Indo-Fijians. My study focuses on native Fijians. Cakaudrove Province, consisting of part of Vanua Levu, all of Taveuni, and other small islands, lags significantly behind the main island, Viti Levu, where the state capital and most tourism businesses are situated.

DISASTER AND AID

On January 13, 2003, Cyclone Ami swept over the northern and eastern regions of the country, including the whole province of Cakaudrove (Ami was the only cyclone in the northern region from 1991 through 2005). All 43 villages in the sample were damaged. The cyclone caused no casualties, and permanent migration was virtually nonexistent in its aftermath. According to respondents' subjective assessments, over 60% of residents' dwellings—a main house and/or free-standing units, such as the kitchen, shower, and toilet (not all households have such units, as these facilities are often located inside the main house)—were damaged (Table 14.1).[2] Among households that experienced housing damage, almost 40% became refugees who stayed in others' residences in the same village (Takasaki, 2011d). About two-thirds of those refugees lived with households in the same kin group (*matagaqli*, defined later); that is, kin networks served as a major risk-sharing pool. In the remainder of the chapter, I call those households with housing damage *victims* and others *nonvictims*.

Table 14.1 Household Means and Standard Deviations (SD) of Shocks and Characteristics		
	Mean	**SD**
Shocks:		
Housing damaged (0/1)	0.62	(0.49)
Housing aid receipt (0/1)	0.25	(0.43)
Illness of any household member (0/1)	0.27	(0.44)
Illness of household head (0/1)	0.13	(0.34)
Illness of any adults other than household head (0/1)	0.14	(0.35)
Illness of any children (0/1)	0.05	(0.22)
Illness of any household member of the household whose head is not ill (0/1)	0.14	(0.35)
Characteristics:		
Annual earned income (F$)	11,010	(20357)
Number of working-age adults (ages 15–65)	3.12	(1.79)
Number of children (ages 0–14)	2.04	(1.85)
Number of elderly (ages 66 or above)	0.33	(0.60)
Age of household head	51.4	(14.5)
Female head (0/1)	0.09	(0.29)
Adult secondary education (0/1)	0.81	(0.39)
Land (acre)	2.83	(4.85)
Fishing capital (F$)	338	(2285)
No. observations	879	

Standard deviations are in parentheses.

---

[2]Retrospective errors in the damage incidence are minimal because the survey employed the same assessment scheme as relief officers' and the damage status of each house was common knowledge among villagers.

The Red Cross, other nongovernmental organizations, and governments provisioned relief, and almost all households received emergency food aid (Takasaki, 2011d). Housing reconstruction programs followed. At the time of interviews in 2005, one-quarter of households had received construction materials, and 12% had received these materials more than a year before the interview (mainly in 2004). Although there was at least one disaster victim in each village, a few villages had no recipients of housing aid. Although almost all aid recipients were victims (i.e., virtually no leakage of resources), about 40% of victims were recipients (i.e., large undercoverage) (aid was strongly targeted on the severity of the damage, Takasaki, 2011a). Nonvictims helped victims' housing rehabilitation, especially by providing labor time (Takasaki, 2011d). Among households with a completely destroyed main house (19% of households in the sample), not only more than half of aid recipients, but also 20% of nonrecipients, had built a new house by the time of interviews in 2005 (information about repaired houses is lacking) (Takasaki, 2012). Households also contributed significant labor time to rehabilitate the buildings of the groups to which they belong, such as village facilities (e.g., community halls), churches, and schools (Takasaki, 2011c).

ILLNESS

Public health problems were not a major issue after the cyclone in the sample villages, and respondents reported very limited injuries and illnesses caused directly by the cyclone. Respondents were asked about each household member's health conditions over the past year. Among the households, 27% had at least one sick member, 13% had a sick household head, 14% had a sick adult member other than the head, and 5% had a sick child (Table 14.1). Patterns of illness are almost the same between victims and nonvictims, suggesting that housing damage was unlikely to be a major reason for health problems occurring more than a year and a half after the cyclone. This does not necessarily mean that housing damage did not cause health problems right after the cyclone. In the panel subsample of households in seven Fijian villages, illness was more common among victims than nonvictims in 2003 (with a 0.13 correlation) (Takasaki, 2011c).

HOUSEHOLD PRIVATE TRANSFERS
GROUPS

This subsection provides the description of three types of groups that play major roles in Fijians' lives, along with villages: kin, religious, and social groups (Takasaki, 2011b). The hierarchical structure of kin groups underlies the sampling design discussed previously and frames the disaggregate analysis of network-based transfers discussed later. First, each native Fijian belongs to a lineage of the *vanua-yavusa-mataqali-tokatoka* hierarchy. Vanua consists of several yavusa; yavusa consists of several mataqali[3]; and mataqali consists of several tokatoka (Ravuvu, 1983). Roughly matching an old district in the administrative unit, vanua ranges over several villages; a village consists of one or a few yavusa, which includes several lower-order units, mataqali, and then tokatoka. The 43 villages in the sample cover 20 vanua, 53 yavusa, 146 mataqali, and 234 tokatoka (which formed strata in the sampling); an average village consists of 1.2 yavusa, 3.4 mataqali, and 5.6 tokatoka (Takasaki, 2011b, Table 1).[4]

---

[3]Land is communally owned by mataqali. About 83% of the country's total land is communal.
[4]Marriage across different kin groups in the village or different villages is common, and this chapter focuses on the kin groups to which households currently belong.

Many ritual activities, such as funerals and weddings, are organized by mataqali and yavusa; member households make contributions or pay de facto taxes to groups for their provisions of local public goods for rituals—in the form of cash, in-kind contributions (e.g., food, kava, handicrafts), and labor (e.g., cooking, setup, transport).[5]

Second, Christianity underlies Fijian society, and church donations are quite significant, as described later. A religious group formed for each church, which often also covers more than one nearby village, is available in all villages in the sample—3.3 church groups per village, on average—and virtually all households are members.

Third, social groups include women's, school, and youth groups in all villages (market-oriented groups, such as cooperatives, are almost nonexistent). Note that in this chapter social group is defined in this specific manner. Although membership is fixed for kin and religious groups (unless the household converted to a different religion), participation in social groups is based on individual decisions made prior to the cyclone among the eligible (determined by gender, child schooling, and age); although most eligible individuals belong to school groups, such is not the case for women's and youth groups. A total of 86% of households belong to at least one social group.

TRANSFER DATA

Respondents were asked not only about each major transfer that their household had received from and given to other households (network-based transfers), but also about the transfers their household contributed to and received directly from each kin, religious, and social group to which they belonged, as well as the village (group-based transfers), in the past year, i.e., from 17 months through 32 months after the cyclone.

Three contextual facets are worth mentioning. First, distinct from smaller island states, such as Tonga, where overseas remittances are common (e.g., Bertram, 1986), overseas remittances are almost nonexistent in this sample. Second, transfer measures capture not only cash and in-kind transfers but also labor time. In the following analysis, labor time is monetized based on men's daily wage in each village, the mean of which is about F$16. Third, although the transfers that the household offers to groups include all the resources it contributes, those it receives from the group capture only partial benefits, because it excludes those of local public goods that the group provides, such as social activities and village upkeep. Measuring such benefits is very difficult because they often include unobservable, noneconomic benefits and can be realized over a long time. In contrast, the network-based transfer data are balanced in coverage between receipt and giving.

NETWORK-BASED VERSUS GROUP-BASED TRANSFERS

Table 14.2 reports the proportions of participation in and mean amounts of household private transfers in the past year. Virtually all households had exchanged transfers with other households. The sample means of annual gross network-based transfers received and given are F$607 and F$425, respectively; thus, average households are net receivers of network-based transfers (net transfers received = gross transfers received − gross transfers given). Participation in group-based transfers is also almost uniform.

---

[5]The dominant symbol of Fijian culture is kava (a beverage infused from the root of a pepper plant, *Piper methysticum*), and kava rituals frequently involve exchanges of ceremonial goods, such as food, mats, and bark cloth (Turner, 1987).

Table 14.2 Household Annual Private Transfers

	No. Obs.	Participation (0/1)			Mean Amounts (F$)		
		Transfers Received or Given	Transfers Received	Transfers Given	Net Transfers Received	Gross Transfers Received	Gross Transfers Given
Aggregated Transfers:							
Both	879	0.99	0.95	0.99	−1486 (2313)	901 (925)	2201 (2383)
Network-based	879	0.97	0.94	0.89	183 (563)	607 (688)	425 (493)
Group-based	879	0.98	0.41	0.98	−1483 (2294)	294 (577)	1777 (2217)
Disaggregated Network-Based Transfers:							
Same village	878	0.93	0.90	0.87	7 (280)	377 (428)	369 (437)
Other village or city	879	0.37	0.37	0.17	171 (469)	227 (499)	55 (192)
Same tokatoka	860	0.85	0.82	0.77	118 (468)	420 (575)	305 (411)
Other tokatoka	860	0.52	0.50	0.44	69 (281)	193 (364)	125 (252)
Same religious group	876	0.82	0.79	0.75	95 (453)	416 (534)	321 (416)
Other religious group	876	0.29	0.28	0.23	83 (352)	185 (523)	102 (330)
Disaggregated Group-Based Transfers:							
Kin groups	879	0.81	0.34	0.80	−282 (844)	167 (389)	449 (761)
Village	879	0.89	0.23	0.87	−309 (527)	68 (193)	377 (483)
Religious groups	879	0.95	0.16	0.95	−427 (776)	40 (135)	467 (784)
Social groups	879	0.73	0.07	0.73	−465 (972)	19 (107)	484 (975)

Standard deviations are in parentheses.

Reflecting the imbalance in the group-based transfer data's coverage, transfers given to groups are much more common and greater than those received from groups; 41% of households received group-based transfers, and the mean amount received is about one-sixth of the mean amount given. Even though network-based transfers received are over two times group-based transfers received, group-based transfers given are over four times network-based transfers given. When network-based transfers and group-based transfers are combined, gross transfers given are about 2.4 times greater than gross transfers received, mostly because of the imbalance in the group-based transfer data's coverage.

DISAGGREGATED TRANSFERS

I consider group-based transfers exchanged with village, kin groups, religious groups, and social groups separately. To measure kin- and social-group-based transfers, I combine the four kin groups (vanua, yavusa, mataqali, and tokatoka) and the three social groups (women's, school, and youth groups), respectively (see Takasaki, 2011b for disaggregated transfers for each kin and social group).

Although most households contributed to village, kin groups, religious groups, and social groups (among the 86% of households who are members of at least one social group), receipt of group-based transfers is not common and varies across these groups (Table 14.2). Although gross transfers given to kin groups, religious groups, and social groups are similar to each other, those to the village are somewhat smaller, and gross transfers received from kin groups are much larger than those from other groups. On average, households are net donors of group-based transfers across these groups.

Respondents provided characteristics of each household with which they made transfers. Major transfer networks are in-village, kin, and religious ones. Transfers exchanged with other households in the village, in the same tokatoka (with the closest kin relationship), and in the same religious group, are more common and much greater than those outside the village, in other tokatoka, and in other church groups, respectively (because of a very small number of missing observations, the total number of observations for most of the disaggregated transfers are a little smaller than that of the whole analysis sample, 879). Thus, individual households' transfer network formation is directly related to kin and religious groups, as well as the village.

The difference between network-based transfers received and given comes mostly from across-village transfers. Although most households participate in within-village transfers, and the amounts received and given are almost balanced, across-village transfers received are more common and larger than those given. In contrast, there is no such considerable difference between network-based transfers received and given when they are disaggregated by kinship or religion.

EMPIRICAL ANALYSIS

This section examines how household private transfers respond to health shock. I first discuss the econometric specification and then the estimation results. I consider different definitions of health shocks, compare network- and group-based transfers, and analyze disaggregated transfers.

ECONOMETRIC SPECIFICATION

I estimate the following transfer equation:

$$y_i = \alpha h_i + \beta s_i + \eta X_i + V + e_i \tag{14.1}$$

where y_i is household i's net private transfer received; h_i is a dichotomous variable for illness (defined shortly); s_i is a dichotomous variable for housing damage; X_i is a vector of household characteristics that affect transfers (defined shortly); V is village fixed effects that capture village-level covariate shocks and aid supply, as well as village heterogeneity; and e_i is an error term. Eq. (14.1) identifies the effects of illness and housing damage if neither of them is correlated with unobserved household heterogeneity that affects household private transfers, conditional on observable household characteristics within villages. If private transfers are ex post risk-sharing arrangements against illness in the aftermath of the disaster, households with illness receive more in transfers on the net, i.e., $\alpha > 0$. To examine how risk-sharing arrangements are differentiated by housing damage, I estimate Eq. (14.1) for the nonvictim sample ($s_i = 0$) and the victim sample ($s_i = 1$) separately. This subsample analysis, which is equivalent to adding interaction terms of s_i with h_i, X_i, and V to Eq. (14.1), is the least restrictive way to compare victims and nonvictims.

For X_{it} I consider demographic size (working-age adults at ages 15–65, children, and elderly), age of household head, female head, and education attainment (secondary schooling by any adult) (their descriptive statistics are reported in Table 14.1). For a robustness check, I also control for agricultural land and fishing capital measured at the time of interviews (information about them held a year before is lacking), finding very similar results. I do not control for income, which is endogenous, because the household may adjust its earning efforts in anticipation of private transfers and any unobserved factors that determine income, such as skills, may also influence household transfer decisions; still, household permanent income is captured by demographics, education, and assets (if added as controls). Forms of disaster aid received by individual households are not included as explanatory variables because they are endogenously determined as part of private risk sharing within villages (Dercon & Krishnan, 2005; Takasaki, 2014). Adding the receipt of disaster aid as a control—either aid received more than a year before the interview (i.e., before transfers were exchanged) or aid received at the time of the interview, including aid received within a year before the interview (i.e., including aid received after some transfers)—does not alter the main results reported in the following discussion.

WHOSE ILLNESS?

Transfer responses to health shocks may vary depending on the magnitude of the shock. To capture that, I consider illness of different household members. First of all, I consider illness of any household member, ignoring who is ill. The ordinary least squares estimates of effects of illness (dichotomous) on net total transfers received—network-based and group-based transfers combined—are reported in Table 14.3, which reports robust standard errors (estimation results for controls are reported in Table 14.A1). Although the sizes of the estimated coefficients are considerable, especially among nonvictims, none of them are statistically significant at conventional levels in the whole sample, the nonvictim sample, or the victim sample.

Next, I compare illness of a household head and that of other household members. The dichotomous variable for the illness of others takes one if any household member (either adults other than the head

Table 14.3 Effects of Illness on Net Total Household Transfers Received - Whose Illness

	All ($n=879$)	Housing Undamaged ($n=334$)	Housing Damaged ($n=545$)
	(1)	(2)	(3)
Illness of any household member (0/1)	200	375	137
	(167)	(309)	(209)
Illness of household head (0/1)	268	591**	113
	(192)	(272)	(270)
Illness of any household member of the household whose head is not ill (0/1)	76	71	113
	(218)	(425)	(258)

*Dependent variables are net total household transfers received (F$). Robust standard errors are in parentheses. Controls not shown here are number of working-age adults (ages 15–65), number of children (ages 0–14), number of elderly (ages 66 or above), age of household head, female head (0/1), adult secondary education (0/1), and village dummies. Housing damage (0/1) is also controlled for in column (1). *10% significance, **5% significance, ***1% significance.*

or children) is ill among households whose head is not ill (this variable takes 1 for 14% of households in the sample, Table 14.1). Among nonvictims, net transfers received by those with a sick household head are greater by about F$590 compared to the level among those without a sick household head (F$-1390) (i.e., 42% increase), and this difference is statistically different from 0 at a 5% significance level. In contrast, the estimated effect for victims is small and not statistically significant. The estimated effects of illness of other household members are also small with no statistical significance across the samples.

NETWORK-BASED VERSUS GROUP-BASED TRANSFERS AGAINST ILLNESS OF A HOUSEHOLD HEAD

I compare net transfers received, gross transfers received, and gross transfers given for network-based transfers and group-based transfers separately, using Eq. (14.1) with corresponding dependent variables. The ordinary least squares estimates of illness of a household head are reported in columns one, two, and three of Table 14.4 (column four is discussed later). The estimated effect of illness on gross group-based transfers received captures the combined effect on the receipt of transfers and the amount of

Table 14.4 Effects of Illness of Household Head and Housing Damage - Network-Based and Group-Based Transfers Received and Given

	Illness of Household Head (0/1)			Housing Damage (0/1)
	All ($n=879$)	Housing Undamaged ($n=334$)	Housing Damaged ($n=545$)	All ($n=879$)
	(1)	(2)	(3)	(4)
Network-Based Transfers (F$):				
Net transfers received	70	21	98	27
	(69)	(146)	(74)	(40)
Gross transfers received	74	65	83	51
	(77)	(134)	(93)	(47)
Gross transfers given	4	44	−15	24
	(44)	(64)	(58)	(37)
Group-Based Transfers (F$):				
Net transfers received	144	496	−45	−242*
	(199)	(310)	(276)	(147)
Gross transfers received	−25	−47	−33	−123***
	(43)	(98)	(49)	(44)
Gross transfers given	−169	−543*	11	119
	(197)	(317)	(273)	(144)

Robust standard errors are in parentheses. Controls not shown here are number of working-age adults (ages 15–65), number of children (ages 0–14), number of elderly (ages 66 or above), age of household head, female head (0/1), adult secondary education (0/1), and village dummies. *10% significance, **5% significance, ***1% significance.

transfers received among those who received transfers (41% in the whole sample). The only statistically significant result is for group-based transfers given among nonvictims, and the estimated effect of illness is about F$-540 (30% reduction from the level among those without a sick household head); accordingly, the estimated effect of illness on net group-based transfers received among nonvictims is also large (almost F$500), though not statistically significant at conventional levels. All other estimated effects of illness on group-based transfers, as well as network-based transfers, are small (less than F$100 in magnitude among both nonvictims and victims) with no statistical significance across the samples. These results suggest that the strong effect of a household head's illness on the net total transfers received among nonvictims found previously comes mostly from a large decrease in contributions to groups.

DISAGGREGATED TRANSFERS AGAINST ILLNESS OF A HOUSEHOLD HEAD

The ordinary least squares estimates of illness of a household head on disaggregated net transfers received are reported in columns one, two, and three of Table 14.5 (column four is discussed later). Group-based transfers are disaggregated into the village, kin groups, religious groups, and social groups. The only statistically significant result is for kin-group transfers among nonvictims, and the estimated effect of illness is about F$285; the estimated effect for social groups among nonvictims is also considerable (about F$125), though not statistically significant at conventional levels. All other estimated effects of illness on group-based transfers are small with no statistical significance across the samples. When gross group-based transfers received and given are analyzed separately, qualitatively the same patterns appear only for gross transfers given (with the opposite sign) (results not shown). Hence, the significant reduction in contributions to groups among nonvictims with a sick household head found previously appear mostly in kin groups and social groups.[6]

Network-based transfers are disaggregated into two in the following three ways: (1) in-village versus out-village, (2) same tokatoka (with the closest kin relationship) versus other tokatoka, and (3) same religious group versus other religious group. Recall that in-village, kin (tokatoka), and religious networks are the main transfer networks. The estimated effects of illness are considerable (in the range of F$110–140) and statistically significant for out-village, kin, and religious networks among victims, but not nonvictims; all other estimated effects of illness on network-based transfers are small, with no statistical significance, across the samples. When gross network-based transfers received and given are analyzed separately, qualitatively the same patterns appear only for gross transfers received from households in the same tokatoka and the same church group and gross transfers given to households outside the village (with the opposite signs) (results not shown). Thus, victims with a sick household head receive more in transfers through kin and religious networks and give less in transfers to households outside the village.

ILLNESS OF A NON HEAD OF HOUSEHOLD

I repeat the analyses for illness of any household member other than the head among households whose head is not ill (results not shown). On one hand, the results of group-based transfers are similar to those

---

[6]When transfers exchanged with social groups only among members (86% of households in the sample) are considered, the results are very similar to those for the whole sample presented here.

Table 14.5 Effects of Illness of Household Head and Housing Damage on Net Transfers Received - Disaggregated Networks and Groups

	Illness of Household Head (0/1)			Housing Damage (0/1)
	All (n=879)	Housing Undamaged (n=334)	Housing Damaged (n=545)	All (n=879)
	(1)	(2)	(3)	(4)
Network-Based Transfers (F$):				
Same village[a]	−13	−25	−14	10
	(25)	(46)	(31)	(21)
Other village or city	82	48	109*	13
	(63)	(135)	(65)	(33)
Same tokatoka[a]	95	48	119*	43
	(64)	(130)	(68)	(32)
Other tokatoka[a]	−38	−39	−33	−17
	(27)	(58)	(35)	(22)
Same religious group[a]	97	50	140**	51
	(61)	(123)	(63)	(33)
Other religious group[a]	−15	−22	−29	−20
	(39)	(86)	(37)	(26)
Group-Based Transfers (F$):				
Kin groups	110	285**	49	−104*
	(68)	(142)	(82)	(61)
Village	14	33	−9	−80
	(40)	(71)	(52)	(50)
Religious groups	23	54	12	−29
	(74)	(126)	(97)	(54)
Social groups	−3	125	−97	−29
	(90)	(104)	(134)	(56)

*Robust standard errors are in parentheses. Controls not shown here are number of working-age adults (ages 15–65), number of children (ages 0–14), number of elderly (ages 66 or above), age of household head, female head (0/1), adult secondary education (0/1), and village dummies. [a]No. of observations are 860–878 in columns (1) and (4), 327–333 in column (2), and 533–545 in column (3). *10% significance, **5% significance, ***1% significance.*

for illness of a household head. Illness of others significantly decreases transfers given to groups, especially kin groups. On the other hand, illness of others does not significantly affect transfers through kin and religious networks or out-village networks.

EFFECTS OF HOUSING DAMAGE

Tables 14.4 and 14.5 also report the estimated impacts of housing damage on household private transfers in the whole sample (column four). These estimates capture how disaster damage experienced about two years earlier influences household private transfers. If victims who had received help for their recovery from nonvictims reciprocated transfer exchanges, victims should receive a lower gross amount in transfers and give a greater gross amount in transfers than nonvictims (i.e., the reciprocity effect). This pattern strongly holds for group-based transfers—the estimated impacts of housing damage on gross transfers received and given, respectively, are negative and positive with a considerable magnitude (over F$100, though the latter is not statistically significant at conventional levels), and the estimated impact on net transfers received is about F$-240. This reciprocal pattern holds across groups, and it is strong for kin groups and village (the result for the latter is statistically weak). At the same time, such reciprocal patterns do not exist for network-based transfers; the estimated coefficients for net transfers received and gross transfers received are both positive and small with no statistical significance. These results suggest that although victims reciprocated transfer exchanges with nonvictims among group members, on average, they did not do so through household networks.

DISCUSSION AND CONCLUSIONS

These results reveal a sharp contrast between network-based and group-based transfers about two years after the cyclone. On one hand, nonvictims, but not victims, with sick household members—either a household head or others—contribute less to groups, especially kin and social groups, than those without. That is, illness experienced by those who receive help for disaster recovery is less insured than illness experienced by others who offer such help, which is consistent with the reciprocity effect (Ligon et al., 2002). At the same time, the significant role of group-based transfers in risk sharing suggests that limited attention to group-based transfers in developing countries in the literature is a significant lacuna. On the other hand, victims, but not nonvictims, with a sick household head receive more in transfers from other households through kin and religious networks and give less in transfers to households outside the village, though overall network-based transfers are largely irresponsive to illness regardless of disaster victimization.

These results suggest that victims compensate for the limited risk sharing against illness within groups through household networks. The reciprocity of within-network risk sharing is more flexible than such within-group reciprocity, probably because distinct from the fixed membership of groups, mostly within villages, social networks are endogenously formed within and across villages. This pattern is consistent with my early work on the role of social hierarchies in household private transfers based on the same Fijian data—network-based transfers adjust to hierarchy bias in group-based transfers (Takasaki, 2011b). Network-based transfers also adjust to bias in group-based transfers caused by disaster victimization through the reciprocity effect.

Victims' adjustment in within-network risk sharing, however, is incomplete in two important ways. First, such adjustment is possible only for large health shocks. It works for the illness of a household head but not of others. Second, network-based transfer response to the illness of a household head for victims is much smaller in magnitude than group-based transfer response to this shock for nonvictims (F$140 at most versus about F$500; Tables 14.4 and 14.5). A possible reason for the relatively limited adjustment in within-network risk sharing is that the formation of household networks is closely related to group memberships determined by fixed social relations, such as kin and religious ones.

To conclude, disaster victimization weakens Fijians' reciprocity-based private safety nets against subsequent non-disaster shocks among group members, especially among kin and social group members. At the same time, this is partly mitigated through household networks, especially kin and religious networks and across-village ones. As such, social networks that underlie informal risk-sharing institutions among Fijians play an important role in their adaptation to natural disasters. Whereas the extent to which the finding in rural Fiji is generalizable is an empirical question, a better understanding of such adaptation patterns in specific contexts is critically important to designing policies that lower the constraints in the informal institutions that could provide significant safety nets for the poor aftermath of natural disasters. In Fiji, strengthening network-based risk sharing, for example, by promoting network formation based on broad social relations such as market-oriented ones may be promising.

Table 14.A1 Full Estimation Results for Net Total Household Transfers Received

	All (n=879)	Housing Undamaged (n=334)	Housing Damaged (n=545)
	(1)	(2)	(3)
Illness of any household member (0/1)	200	375	137
	(167)	(309)	(209)
Housing damaged (0/1)	−234		
	(149)		
Number of working-age adults (ages 15–65)	−132***	−194***	−88
	(44)	(69)	(57)
Number of children (ages 0–14)	−35	−55	−24
	(39)	(67)	(53)
Number of elderly (ages 66 or above)	−316**	−175	−368*
	(156)	(285)	(201)
Age of household head	18***	12	24***
	(6)	(10)	(8)
Female head (0/1)	336	459	290
	(216)	(322)	(320)
Adult secondary education (0/1)	−144	173	−219
	(182)	(266)	(263)
R-squared	0.26	0.37	0.26

*Dependent variables are net total household transfers received (F$). Robust standard errors are in parentheses. Other controls not shown here are village dummies. *10% significance, **5% significance, ***1% significance.*

ACKNOWLEDGMENTS

I wish to thank my field team—Jonati Torocake, Viliame Manavure, Viliame Lomaloma, and 16 enumerators—for their advice, enthusiasm, and exceptional efforts on behalf of this project. Special thanks are owed to the Fijians of the region who so willingly participated in the survey. The Cakaudrove Provincial Office in Fiji offered valuable institutional support for this project. This research was made possible through support provided by the Sumitomo Foundation, the Japan Society for the Promotion of Science, and the Ministry of Education, Culture, Sports, Science and Technology in Japan. Any errors of interpretation are solely the author's responsibility.

CHAPTER

STAY OR RELOCATE: THE ROLES OF NETWORKS AFTER THE GREAT EAST JAPAN EARTHQUAKE*

15

Young-Jun Lee[1], Hiroaki Sugiura[2], Ingrida Gečienė[3]

[1]Hirosaki University, Hirosaki, Aomori, Japan; [2]Aichi University, Nagoya, Japan; [3]Lithuanian Social Research Centre, Institute of Social Innovations, Vilnius, Lithuania

CHAPTER OUTLINE

INTRODUCTION

The March 2011 Great East Japan earthquake took 15,883 lives in Japan and left 2671 people still reported missing (Reconstruction Agency, June 2013). In the three most-stricken prefectures of Iwate, Miyagi, and Fukushima, more than 131,000 persons have relocated to other prefectures within one year following the disaster. Nearly 10% of the population were displaced from coastal areas of these prefectures since the disaster. See Fig. 15.1 for the map of Japan.

The main concern of this chapter is population migration associated with the disaster and the role of networks therein, but our original concern was prior continuous population outflow from rural areas. Because the Great East Japan earthquake struck depopulating rural areas, population decrease was accelerated and became a matter of the highest urgency in rural areas. The Great Earthquake revealed the inconsistencies in economies between urban and rural areas and may widen interregional economic gaps.

Fig. 15.2 shows the net out-migration from Noda village, Iwate prefecture, as a percentage of the village population of 2014[1] (see Fig. 15.1 for the location of Noda village). As we introduce later,

*This chapter is a modified and extended version of Lee and Sugiura (2014) in association with Ingrida Gečienė.
[1]The population for Noda was 5285 in 1990, while it becomes 4494 in 2014.

Social Network Analysis of Disaster Response, Recovery, and Adaptation. http://dx.doi.org/10.1016/B978-0-12-805196-2.00015-7

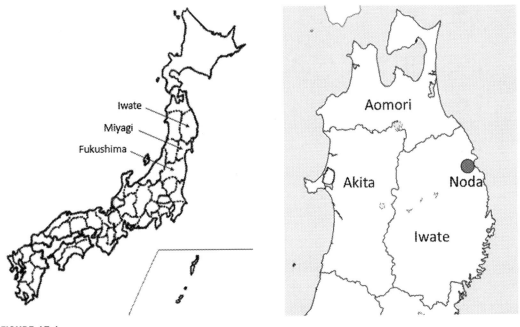

FIGURE 15.1

The maps of Japan and Noda village, Iwate prefecture.

Noda is a typical costal village in northern Japan. The population in Noda began to decrease in the 1970s and is consistently declining at a rate of approximately 0.9%. When the Great Earthquake hit the village in 2011, a net total of 2.3% of the population migrated out of the village. Although some migrants returned to the village within one or two years following the earthquake, the net population outflow has not stopped[2].

Population in rural Japan has been declining since the growth era of the 1960s. Substantial numbers have migrated to metropolitan areas to fill demands for labor in manufacturing. Not surprisingly, consumption, production, and birth rates in rural areas decline following out-migration of the younger population. Deprived of their economic force, many rural areas are becoming economically unsustainable. Making matters worse, persons younger than 29 years of age account for 55% of the net migration out of the three prefectures most stricken by the 2011 earthquake, according to *the Report on Internal Migration in Japan* in 2013 (Statistics Bureau). Population outflow is impeding authorities' efforts to restore sustainable areas through reconstruction.

Income differences that motivate people to move between regions still persist. Shioji (2001) examined whether the composition effect—i.e., migrants possess greater human capital than those

---

[2]According to *Population Census* (Statistics Bureau), the rural area of Japan has lost its population, irrespective of the earthquake. The rural part of Iwate prefecture that was stricken has decreased its population by 7.4% from 2010 (pre disaster) to 2015 (post disaster), while the rural part of Akita prefecture that was not stricken has decreased by 9.0% in the same period.

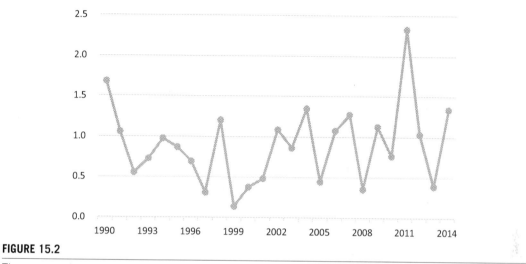

FIGURE 15.2

The net out-migration from Noda village, Iwate prefecture.

Iwate Prefecture Report on Internal Migration.

who remain—accounts for interregional income differences in Japan. He finds that when this effect exists, migration from low- to high-income regions does not contribute to a convergence—contrary to a simple theoretical model. He showed that this effect demonstrated slow convergence, but its magnitude was small.

With regard to regional characteristics of Japan's youth labor market, Ohta (2005) found that youth unemployment is higher in regions with few vacancies for the less educated, and workers commonly migrate elsewhere. He stressed the need for labor market policies to consider regional, economic, and social characteristics because interregional differences lead to differences in younger workers' attitudes towards work. Abe and Tamada (2007) investigated regional patterns in employment of less-educated Japanese men and find that the employment–population ratio declined from 1990 to 2007. Interregional wage growth displayed a distinctive pattern during this period: wage growth was high in low-wage areas during the 1990s, but high in high-wage areas during the 2000s.

Drawing on the authors' survey of Noda village in northern Japan's Iwate Prefecture, this study clarifies factors determining respondents' intention to relocate following the Great East Japan earthquake and highlights the role of social relationships and social networks. In this context, "intention to relocate" refers to the nature and extent of villagers' motives to remain in or leave the village after the disaster. Focusing on characteristics of questionnaire respondents and people in their personal networks, we examine which population groups intended to remain in or leave the village.[3]

---

[3]We can no longer conduct any follow-up survey, such as to what extent intention to relocate is correlated with actual relocation. Because the personal information protection law did not permit individual identifications from our survey, we cannot connect between expected intention and actual behavior.

THEORETICAL FRAMEWORK
MIGRATION DECISIONS UNDER NORMAL CIRCUMSTANCES

Much of the early work on migration was influenced by two theoretical approaches: (1) neoclassical theory with its basic assumption that migration is triggered primarily by economic comparisons of relative benefits and costs, including financial and psychological (De Haas, 2010) and, (2) modernization theory that contrasted sending and receiving areas and proposed push factors of out-migration and pull factors of in-migration (Brettell, 2008).

The basic model of neoclassical theory highlights that migration results from interregional wage differentials, distance between origin and destination, and labor market conditions such as the unemployment rate as factors determining migration. The seminal work of Sjaastad (1962) provides a theoretical framework for examining households' decisions to relocate from a geographical area, defining the issue as their desire to maximize their net economic return on human capital. The compensating wage hypothesis (Harris & Todaro, 1970) assumes that the expected income is equalized across regions by internal migration. High-wage areas attract more workers than the available jobs until the wage rate falls to the level of surrounding lower-wage areas and internal migration ceases.

Roback (1982) analyzed how wages and land rents are determined in an equilibrium that various quantities of local amenities are given. This formulation of household migration follows the compensating differentials model of Rosen (1974) in which interregional wage differentials can persist and reflect the value of location-specific amenities. Location-specific amenities include natural attributes valued by households such as temperature, rainfall, and clean air. Clark and Cosgrave (1991) found evidence for persistence of interregional wage differentials, supporting the human capital model of Sjaastad (1962) and the hedonic model of Roback (1982). Local amenities are likely to affect wages significantly and negatively.

However, this neoclassical approach has been subjected to criticism as it has been "viewed as mechanically reducing migration determinants, ignoring market imperfections, homogenizing migrants and migrant societies and being ahistorical and static" (Kurekova, 2011, p. 7). Also, it tends to view "migrants as atomistic, utility maximizing individuals, and tends to disregard other migration motives as well as migrants' belonging to social groups such as households, families and communities" (De Haas, 2010). Further, it appeared that the linear relationship in the wages-migration tandem does not explain the facts that neither do the poorest individuals migrate nor do the poorest countries send the most labor (Faist, 2000).

The other major theoretical approach explaining migration decision is modernization theory, which, according to Brettel, emerged from the rural-urban continuum model originally formulated by Redfield (1941), a model that opposed city and countryside and contrasted two distinct ways of life, one traditional and one modern (Brettell, 2008, p. 118). In comparison to the neoclassical approach, this theoretical framework proposes more factors for migration decision making—the so-called push and pull factors. Push factors include extraordinary natural phenomena, poor living standards, mechanized farming, unemployment, conflict, and war. Pull factors usually mirror push factors, and include higher incomes, better education, well-organized health care, urban facilities, and protection from conflicts or disasters. The modernization approach to migration was also criticized because of its inability to determine dominant factors (De Haas, 2010) and of the shortcomings of the equilibrium model of linear development with which modernization theory has been associated (Brettell, 2008).

Despite the critiques of both neoclassical and modernization theories, these approaches still prevail to frame discussions of why people migrate. The migration networks theoretical framework has responded to these criticisms to help explain how migration continues even when wage differentials or recruitment policies cease to exist due to changes in economic and public policies that serve to constrain or halt migration streams (Brettell, 2008; Kurekova, 2011).

Migration networks theoretical framework, or as Wilson (1994) has labeled "network mediated migration theory," reveals that networks can become self-perpetuating for migration because "each act of migration itself creates the social structure needed to sustain it" by reducing the costs for migrating family and friends (Wilson, 1994, p. 275). Therefore, migration networks must be perceived as facilitating rather than encapsulating, and as expanding and fluid rather than rigid and bounded (Brettell, 2008, p. 124).

In general, migrant networks are defined as interpersonal ties that connect family, relatives, friends, and community members in the places of origin and destination. However, many migrants have ties to various organizations—governmental, higher education, private employment, diaspora organizations, companies, clubs, religious groups, and other cultural institutions—that can help with the migration, finding work, or adjusting to their new life (Poros, 2011).

Migrant networks provide a variety of benefits, including kinds of connections and information needed to make migration possible, plus funds needed to cover the cost of migration (Dolfin & Genicot, 2010). Also, migrant networks help to find a job and housing, open up a business, and access health care and other services. Moreover, migrant networks "determine whether and to what extent immigrants integrate into their host countries while also maintaining a connection to their home countries" (Poros, 2011, p. 2).

Finally, network-mediated migration shifts the unit of migrant analysis more from individual to household. According to Brettell (2008, p. 125), households and social networks mediate relationships between the individual and the world system and this approach "not only brings the migrant-as-decision-maker back to focus, but also reintroduces the social and cultural variables that must be considered in conjunction with economic variables." This theory allows to analyze why some local communities become extensively involved in migration and others do not. Inequalities are also explained by this framework, as networks do not offer equal opportunities and resources to all of its members; "many studies have shown that transactions within migrant network ties can often include tensions, conflict, resistance, and capitulation as a result of wrongdoing or broken promises" (Poros, 2011, p. 3).

DECISIONS IN DISASTER-INDUCED DISPLACEMENT

Most migration theorizations deal with voluntary migration, but since the end of the Cold War, forced migration has gained some attention (Castles, 2003, p. 13). Forced migration often causes major social changes. It can destroy intangible local resources such as traditions and communities and thus become a factor in underdevelopment because it "reduces the capacity of communities and societies to achieve positive change" (Castles, 2003, p. 18).

While refugees and asylum seekers as cross-border forced migration have provoked considerable political debates in many countries, the United Nations Development Programme (2009) emphasizes that the overwhelming majority of movers relocate inside their own country. They have crossed no frontiers but may face special difficulties caused by conflict or natural disasters in a place away from home. Therefore we might miss a large impact on human development if we pay exclusive attention to international migration networks.

Some studies distinguish internal displacement and internal migration: internal displacement tends to emphasize push factors to leave and migration emphasizes pull factors at the intended destination, while in practice each is a mixture of both (Internal Displacement Monitoring Centre, 2015, p. 14). Displacement sits within a continuum including "(predominantly) forced displacement" and "(predominantly) voluntary migration," as well as "(voluntary or forced) planned relocation" (Kaelin, 2015).

Internal displacement can be caused by several reasons: natural disasters, drought or famine, war, and conflicts. In this chapter, we will pay more attention to natural disasters or hazards such as drought, earthquakes, tsunamis, hurricanes, and floods. In rapidly growing hazards, people are forced to flee their homes with little notice of the immediate threat. Gradual and long-lasting hazards allow people more time to take steps to avoid, mitigate, and adapt to the impacts on their homes, livelihoods, and communities (Internal Displacement Monitoring Centre, 2015, p. 14). These two different processes of evolving disasters can result in different decisions to migrate and return.

Many recent studies about hazard-caused internal displacement cover the case of Hurricane Katrina in the United States (Groen & Polivka, 2010; Landry, Bin, Hindsley, Whitehead, & Wilson, 2007; Paxson & Rouse, 2008). Our study will add new facets about internal displacement caused by multiple types of hazards in the case of the Fukushima Daiichi nuclear power plant reactor meltdown and radiation leak in the aftermath of the 2011 Tohoku earthquake and tsunami.

Christina Paxson and Cecilia Elena Rouse (2008) examined the decisions of residents displaced by Hurricane Katrina to return to their previous locations. They assumed that individuals' utility is a function of their incomes and of location-specific capital. Location-specific capital includes homes, communities, and networks of friends that are difficult to replace elsewhere.

Craig Landry et al. (2007) used classic variables and approaches discussed in the previous section to examine the evacuation-migration decisions of Hurricane Katrina survivors. Their results indicate that decisions to return after relocating are affected by age, education, household income, employment, marital status, and home ownership. They assess the effect of "connection to place" or "sense of place" by asking whether people were born where they lived before the hurricane. Contrary to their expectations, they found a negative correlation between birthplace and the decision to return and interpreted that persons with longstanding connections to a locale might have been more highly traumatized by the disaster.

There are some differences between evacuation and migration from the disaster. First, evacuation is basically a temporal phenomenon. Whether evacuees return to the home or not is of considerable interest to scholars in part because many evacuees have the intention to return home. In this respect, there are similarities between return migration after hurricanes and migration from the earthquake stricken areas. As Brian C. Thiede and David L. Brown (2013) suggest, although migration is distinct from evacuation, social networks influence disaster-related mobility decisions through cost-reducing functions. We think that the expected utility theory can be applied to both evacuation and migration.

SOCIAL NETWORKS AND DISASTER-INDUCED MIGRATION

In the light of the network-mediated migration theory assumptions, forced migration also needs to be "analysed as social process in which human agency and social networks play a major part" (Castles, 2003, p. 13). First, when hazards cause increased morbidity and mortality, survivors experience narrowing, disruption, or destruction of social and economic support networks. This can cause psychological trauma and the decision to emigrate. Loss of family and the prospect of rebuilding in a zone of devastation is a strong push (Hasegawa, 2013).

Second, evacuees frequently need to move a number of times during displacement time, especially in case of protracted displacement characterized as "a long-lasting situation in which progress towards durable solutions … is slow or stalled" (Internal Displacement Monitoring Centre, 2015, p. 48). For example, four years after the Tohoku earthquake in Japan, around 230,000 people are still displaced and unable to achieve durable solutions (Internal Displacement Monitoring Centre, 2015, p. 137). Many of these people had to move a number of times in the first six months after the disaster, which has also contributed to the splitting up of household members (Internal Displacement Monitoring Centre, 2015, p. 67). Therefore internal displacement commonly causes loss of social/community support networks (Brun, 2005; Esnar & Sapat, 2014).

Third, following their initial displacement, people's trajectories are influenced by the extent of their social networks. As was stated earlier, the social networks can offer assistance in migration. As Elizabeth Fussell (2006) explains in the case of Hurricane Katrina, alongside the inequality in education and income, residential segregation, and discrimination, social networks also shaped people's strategies to leave New Orleans or not.

Fourth, social networks play a key role in how internally displaced people are rebuilding their lives in a hazard's aftermath. According to Fussell, people with weaker or disrupted social networks tended to stay longer in provided shelters because they are not able to return to their communities or relocate (Brun, 2005; Fussell, 2006).

Fifth, existing structures, such as members of their family, village, community, or other social networks of internally displaced people, help maintain cultural continuity and resilience. As Brun points out, such cultural continuity and the continuance of everyday practices is often noted as a coping strategy and a starting point for life in a new place, especially where displacement is protracted. New "linkages are necessarily established in the place of refuge—both with internally displaced from other villages with and members of the host population—while those with their original home are simultaneously maintained" (Brun, 2005, p. 18). Therefore, migration, in turn, reproduces and extends personal networks.

Summing up, the scope and quality of social/community support networks plays a crucial role during hazards-caused displacement and aftermath. Despite migrant networks now being widely recognized as very influential in migration decisions, as Sarah Dolfin and Garance Genicot (2010) observe, there remain questions about how migrant networks function and what is "the actual mechanism by which networks affect the migration decisions of their members." Therefore, we aim to reveal some new roles of networks in migration decision-making.

SOCIAL NETWORKS IN JAPAN

Social networks and their operations in Japan have origins in centuries-old agricultural life centered on rice cultivation. To grow rice from planting to cropping, much of the labor was temporally required, but family members alone could not satisfy such a demand. Thus, relatives with blood relations plus local associates were precious sources of labor supply.

Ties for helping each other become the relationships where the rewards for labor are not paid by money but by another person's labor.[4] Such relations were not created in the short term, but rather these Japanese social networks have roots in the trust relationships that are built over the long term.

---

[4]Japanese call it *Otagai-sama*, which means cancelling each other, and value it.

Such social networks play an essential role in serving the requirements of labor and financial transactions, not only for agricultural activities but also for ceremonial occasions such as marriages and funerals. The long-term trust relationships also play an important role in daily information exchange and livelihood support—from food exchange to financial transactions. Such relations seem to have built stronger social networks in rural communities than they have built in urban areas.

Social networks built through these processes play an important role in a time of disaster, such as a huge earthquake. In the Great Hanshin-Awaji earthquake of 1995, among survivors who were buried alive or confined to building, 34.9% were rescued by themselves, 31.9% rescued by family, and 28.1% rescued by friends or neighbors (The Japan Association for Fire Science and Engineering, 1996). This fact suggests local relationships and broader social networks play vital roles in the stricken areas.[5]

Social networks in Japan play an important role as social infrastructure not only in everyday life but also in case of emergency, as is mentioned earlier. Daniel Aldrich and Yasuyuki Sawada (2015) suggest that the level of social capital affected mortality rates in the tsunami of the 2011 Great East Japan earthquake.[6] Social networks affect individual feelings of relief, satisfaction, and happiness and can produce positive effects for individual health. Networks are also essential in economic activities such as finding employment opportunities and supporting business.

A MODEL OF LOCAL-SPECIFIC CAPITAL IN DISASTER-INDUCED MIGRATION

Our analysis is based on the economic model of Paxson and Rouse (2008). The reason we adopt their model is that it introduces local-specific capital to assess the effect of a natural disaster. Our research focuses not only on economic aspects of local-specific capital but particularly covers the damage to social networks. From this viewpoint, Paxson and Rouse's (2008) framework is appropriate to estimate losses in nonfinancial assets.

Individuals' utility is assumed to be a function of their income (Y) and their stock of location-specific capital (C). Location-specific capital is defined as home ownership, human networks, and location-specific jobs such as agriculture and fisheries—i.e., attributes not easily obtained elsewhere.[7]

An individual who lives in a village receives income Y^S and has a location-specific capital C. If the individual decides to leave, he/she would receive income Y^O and have a location-specific capital of zero. However, the individual would decide to stay if the utility of staying in the village exceeds that of relocating. That is,

$$U\left(Y^S, (1-\lambda)\, C\right) \geq U\left(Y^O,\, 0\right),$$

where λ denotes the fraction of location-specific capital that is destroyed.

---

[5]Spatial proximity does not simply mean social connection. The social networks that had already been built before the disaster play vital roles after the disaster.

[6]Aldrich and Sawada (2015) selected crimes per 1000 residents as social capital proxies, while our study measured the level of social capital using the number of social network members.

[7]Location-specific capital also includes noneconomic attachments, like the attachment to place. In other words, it is used as both economic and noneconomic capital.

Several comparative statics results emerge straightforwardly from the model. First, the likelihood of relocating decreases with income received (Y^S) in the current location and increases with income obtainable elsewhere (Y^O), conditional on the level of location-specific capital C. Our results endorse the compensating wage hypothesis (Harris & Todaro, 1970). Second, the probability of relocating increases with the level of destruction of location-specific capital (λ), holding the predisaster level of location-specific capital (C) fixed. Third, the probability of relocating decreases with C, holding the fraction of damage (λ) fixed. Our formulation allows interactions between the magnitude of damage (λ) and location-specific capital (C).

Location-specific capital is measured by four variables in our analysis. The first is an indicator of whether a respondent was born in the village. This variable is included because it directly represents an attachment to hometown. Landry et al. (2007) call it "connection to place" or "sense of place." The second variable is a dummy indicating whether respondents owned their residences before the disaster. Home ownership has been included in the previous studies because it can have a costly effect on the migration decision.

The third variable captures the number of people in social networks, which were subjectively measured numbers of people sorted into three categories: family and relatives, local associates, and workplace colleagues. It was obtained by asking the question, "How many (a) family members and relatives, (b) local associates, and (c) workplace colleagues are living within the walking distance now?[8] And how many were they before the earthquake, respectively?" We treat decreases in number from pre- to postdisaster as the damage to social networks, which are also measured in three categories.

Family and relatives are connected by blood relationship and they generally include relatives within the fourth degree of kinship (for example, cousins, including mother's brother's children) in Japan. Local associates stand for the relationship within the area of living place, not with blood relationship. For example, fire company and neighborhood association are organized locally as civic activities in Japan. Business colleagues are formed in the workplace but do not generally live so close, because the workplace is remote from the home.

The fourth variable indicates agriculture and fishing industries because those industries are location specific. Furthermore, fishing licenses granted to respondents are valid only for a specific region.

Our study measures individual utility using levels of income and location-specific capital. Income is based on the February 2013 survey, which took the whole household income into account. Income is divided into nine categories from zero to 10 million yen (approximately $111,000 as of February 2013).

SITE AND SAMPLE

Our data are from the *survey of resident's sentiment and current situation* conducted by the authors in Noda village, Iwate prefecture, one of the stricken areas, from February to March 2013. Noda's population was 4835 when the disaster occurred, according to the 2010 national census. The center of the village was hit by a tsunami 37.2 m high, killing 37 persons and obliterating or badly damaging 810 buildings and houses. Damage was concentrated on the village center, causing extensive losses to the village and basic industries such as fishery and agriculture. In this survey, excluding the nonrespondents, 36% (402) people are confirmed to have been injured.

---

[8]This survey was conducted in February 2013, which was almost two years after the earthquake.

Our sample encompassed all 2853 inhabitants aged 18 to 69 in Noda village in February 2013. The survey was conducted by mail. Respondents were not rewarded and spent 20 min on average to answer the questionnaire. The authors received 1142 responses, a 40.2% response rate. Our analyses are based on 971 residents (excluding nonrespondents): 53.2% were women and 46.8% were people who suffered heavy damage from the disaster.

The questionnaire design used for the research is based on previous interviews conducted in tandem with the support activity immediately following the disaster. This questionnaire survey was conducted in collaboration with the village office to establish the economic condition of the victims, damage to social capital, intention to migrate, and sense of life restoration. Our samples are drawn from people who remained within the stricken areas, not from evacuees, as in previous studies. There will be a sampling bias by excluding samples that have already relocated. However, this problem is left to be addressed in future studies.

Noda village is representative of stricken coastal areas. There are three villages and four towns along the coast in Iwate prefecture. These areas have similar geographic and economic characteristics and share the same challenges such as depopulation and aging. According to the 2010 national census, population density of Noda village is 57.3 persons per square kilometer, while the average of these areas is 49.0. The proportion of the residents aged 65 or more is 30.0%, while the average for these areas is 32.6%. The share of people engaged in primary industry in Noda village is 17.7%, while the average for these areas is 20.0%.

RESULTS

Table 15.1 shows sample means and standard deviations arranged according to whether respondents wished to remain in Noda or to relocate. The intention to relocate was subjectively measured by asking the question, "From now onwards, are you planning to live in Noda village?" with three choices; (a) "want to keep living in Noda village," (b) "want to live outside the village," and (c) "probably live outside the village once, but want to come back sometime." In all, 83.0% indicated they wanted to remain in Noda, 11.1% indicated they wanted to live elsewhere, and 5.9% said they probably would live elsewhere for a time but would return eventually. In the following analysis, we disregard the last group of respondents because the number of observations is insufficient to estimate separately.

From Table 15.1 we observe that primarily women and relatively young people tend to have a desire to relocate. Respondents who are married or have children were comparatively more interested in remaining in the village. Analyzing the average values for each group in Table 15.1, it is clear that incomes of respondents willing to remain in Noda were high, whereas incomes among groups wishing to relocate were low.

Table 15.2 shows the results of logistic regressions to determine factors behind respondents' intentions to relocate. The dependent variable is an indicator for hoping to live outside the village. We used logistic regression incorporating the idea of local-specific capital into Landry et al.'s (2007) evacuation-migration decision model. The merit of this model is that it allows both numerical and categorical independent variables to estimate a binary choice. The drawback of this model is that the parameter estimates are not intuitive, unlike the linear regression model, because they are interpreted as the rate of change in the logarithmic odds of intentions to relocate while the characteristics are changing.

Table 15.1 Descriptive Statistics

Variable	Remain	Relocate
Male (1 = male, zero otherwise.)	0.5 (0.5)	0.31 (0.47)*
Age (= years at last birthday)	53 (12)	41 (15)*
Married (1 = married, zero otherwise)	0.81 (0.4)	0.62 (0.5)*
Child (1 = child(ren), zero otherwise)	0.8 (0.4)	0.58 (0.50)*
Household income (= real annual income, 10,000 yen)	351 (234)	308 (228)
Hometown (1 = Noda, zero otherwise)	0.77 (0.42)	0.59 (0.49)*
House ownership (1 = yes, zero otherwise)	0.88 (0.33)	0.64 (0.48)*
Family and relatives network (= number of family and relatives)	9.8 (9.53)	9.2 (31.11)*
Local human network (= number of friends in Noda, not include family and relatives)	10.8 (20.33)	7.4 (30.11)*
Agriculture (1 = farms, zero otherwise)	0.09 (0.29)	0.02 (0.15)*
Fishery (1 = fishes, zero otherwise)	0.54 (0.23)	0.03 (0.18)
Damage to residence (1 = suffered damage beyond a specified extent, zero otherwise)	0.36 (0.48)	0.34 (0.48)
Damage to household income (1 = real income declined, zero otherwise)	−0.34 (0.62)	−0.30 (0.66)
Damage to job (1 = lost job due to the disaster, zero otherwise)	0.05 (0.21)	0.04 (0.19)
Damage to family and relatives network (= number of predisaster family and relatives minus postdisaster)	0.5 (3.6)	−1.7 (24.9)
Damage to local human network (= number of predisaster friends minus postdisaster)	1.4 (5.7)	−1.8 (26.3)

Notes: Standard errors are provided in parentheses. The sample contains 971 respondents of whom 846 indicated a desire to remain in the village and 146 indicated a desire to relocate. Asterisk means that values for those wishing to remain and to relocate differ at the 5% level of significance.

Model 1 indicates the estimated results of how individuals' characteristics affect the intentions to relocate. The coefficients of male, age, and marriage were negative in the model 1. Men, the elderly, and married respondents indicated strong willingness to remain in Noda. We added the household income as one of the variables of model 2. The variables that indicate individuals' characteristics in model 1 and 2 are equally significant.

Though there are natural differences between hurricanes and earthquakes,[9] we find some similarities between the determinants of returning to the home after Hurricane Katrina and those of staying at the home after the Great East Japan earthquake. Jeffrey Groen and Anne Polivka (2010) found that male and older evacuees were more likely to return than female and younger evacuees.

Our study also suggests that male and elderly people tend to stay in the home town.[10] The expected gains from staying is so high for men because men tend to have more opportunities to work than women.

[9]Concerning about migration associated with the disaster, we cannot evacuate prior to earthquake, though evacuation prior to hurricane is possible.

[10]The variables such as age, sex, and marital status are control variables to identify the effect of human networks in the regression analysis. So we hesitate to interpret the results by connecting such variables with the role of social networks. The meanings of the results can be explained by the economic reasons.

Table 15.2 Factors Contributing to Intentions to Relocate

Variable	Model 1			Model 2			Model 3		
Male	−0.85	***	(−0.22)	−0.75	***	(−0.23)	−0.55	*	(0.28)
Age	−0.06	***	(−0.09)	−0.06	***	(−0.01)	−0.06	***	(0.01)
Married	−0.72	**	(−0.29)	−0.59	*	(−0.31)	−0.89	**	(0.38)
Child	0.10		(−0.32)	0.14		(−0.33)	−0.01		(0.39)
Household income				−0.001	*	(0.001)	−0.001		(0.001)
Hometown							−1.10	***	(0.30)
House ownership							−0.67	**	(0.30)
Family and relatives network							0.01		(0.01)
Local human network							−0.01		(0.01)
Agriculture							−0.52		(0.75)
Fishery							−0.14		(0.59)
Number of observations	971			912			710		
Pseudo R^2	0.14			0.13			0.18		
Log L	−321.55			−294.01			−213.74		

Notes: The dependent variable is an indicator for hoping to live outside the village. Standard errors are provided in parentheses. ***, ** and * denote significance at 1%, 5%, and 10%, respectively.

Traditionally, men are expected to inherit family businesses or family estates, which is a kind of local-specific capital. Elderly people find it difficult to prepare sufficient money for moving or to build a new human network after relocating.

As Groen and Polivka found, our results show that respondents from low-income families indicated a reduced likelihood of remaining, which accords with the compensating wage hypothesis. Moreover, the variables of location-specific capitals are included in model 3. Our results suggest that the effect of "connection to place" or "sense of place" have a negative effect on propensity to relocate, because the sign of hometown dummy is negative. It accords with Landry et al.'s (2007) original conjecture, though they obtained an opposite result.

Landry et al. hypothesized that sense of place would be stronger among people who had roots in a certain region that Hurricane Katrina hit, and thus the likelihood of return after evacuation would be greater, but their data did not support this contention. To give a possible explanation for it, they suggested that the extent of damage tends to cause more distress to people with deep roots in a particular environment, which is called "solastalgia" (Albrecht, 2006). In other words, people with deeper connections to place might be more highly traumatized and less likely to return. Solastalgia is the distress that is produced by environmental change impacting on people, while they are directly connected to their home environment (Albrecht et al., 2007).

Table 15.3 Effects of the 2011 Disaster on Intentions to Relocate

Variable			
Damage to residence	0.25		(0.25)
Damage to household income	0.04		(0.20)
Damage to job	−0.15		(0.63)
Damage to family and relatives network	0.03		(0.03)
Damage to local human network	−0.05	**	(0.02)
Number of observations	669		
Pseudo R^2	0.018		
Log L	−238.56		

Notes: The dependent variable is an indicator for hoping to live outside the village. Standard errors are provided in parentheses. ** denotes significance at 5%.

The home ownership dummy also displays a negative effect on respondents' desire to relocate. However, the negative effect is smaller than we had expected, as the absolute value of coefficient of home ownership is less than that of hometown. In Japan, home ownership is a very important factor to determine the place of residence, because the liquidity of residential house is low while the house prices are high. However, in the coastal area that is stricken by the Great East Japan earthquake, home ownership can be thought as less important to determine the place of residence because the home prices are relatively low compared to the urban areas. On the other hand, hometown is an essential factor to determine where to live because hometown is the base of blood relationship and connection to place.

It is commonly thought that the intention to stay is stronger among persons who have a network of family and relatives. However, the presence of family and relatives did not display the expected results in the model 3. Furthermore, we could not obtain statistically sound results from dummies for location-specific industries.

We examine how the disaster influenced respondents' intentions to relocate. Using the logit model shown in Table 15.3, we regressed the effect of the 2011 disaster by employing five variables. The first variable describes whether a respondent's residence suffered damage beyond a specified extent. The second variable indicates whether respondents' real income declined, remained unchanged, or increased. The third variable is a dummy indicating whether respondents lost their jobs due to the disaster. We asked how many family members, relatives, and local associates the respondent could communicate with compared to the predisaster period and calculated the numerical difference between post- and predisaster. The fourth and fifth variables are concerning family and relatives and local associates, respectively.

Results were not significant for variables except changes in the number of local associates (variable Damage to Local Human Network). Although depopulation is a concern in the stricken area, our results indicate that the occurrence of the disaster was not the main factor strengthening respondents' intentions to relocate. As Table 15.3 indicates, loss of local human networks was more likely to drive respondents' intention to relocate in the village. Note that loss of local human networks is captured by a negative value in "Damage to Local Human Networks." The results of Table 15.3 mean that loss of human networks tends to raise the propensity to relocate. When local human network is lost, support

Table 15.4 Extent of Damage Suffered and its Effect on Intentions to Relocate

Variable	Damaged			Nondamaged		
Male	−0.42		(0.44)	−0.81	**	(0.39)
Age	−0.06	***	(0.02)	−0.05	***	(0.02)
Married	−0.94		(0.64)	−0.92	*	(0.49)
Child	0.75		(0.64)	−0.25		(0.53)
Household income	−0.001		(0.001)	−0.001		(0.001)
Hometown	−1.08	**	(0.49)	−0.93	**	(0.39)
House ownership	0.28		(0.48)	−1.10	***	(0.42)
Family and relatives network	−0.13	**	(0.05)	0.02		(0.02)
Local human network	−0.03		(0.05)	−0.01		(0.02)
Agriculture	−14.96		(1809)	−0.66		(1.06)
Fishery	−16.19		(1626)	0.65		(0.62)
Number of observations	266			440		
Pseudo R^2	0.253			0.215		
Log L	−78.79			−120.63		

Notes: The dependent variable is an indicator for hoping to live outside the village. Standard errors are provided in parentheses. ***, ** and * denote significance at 1%, 5%, and 10%, respectively.

for living cannot be expected. This is similar to Thiede and Brown's (2013) findings that local ties served as a "binding" force to lower respondents' propensity or ability to evacuate when Hurricane Katrina approached.

The Great East Japan earthquake created massive physical damage and altered the socioeconomic environment of daily life. Table 15.4 shows the results of estimation that clarify how intentions to relocate varied between two groups—those who suffered damage to residence from the 2011 disaster and those who did not. Among both groups, only age presents statistically stable results. Older respondents in both groups showed a strong intention to remain in place.

The coefficient for the hometown dummy is statistically negative for both groups. It suggests that there is a strong effect of "connection to place" or "sense of place" because hometown dummy has a negative effect on intention to relocate, irrespective of the severity of damage to residence. The coefficient for home ownership is significantly negative for the group that suffered no damage from the disaster, while home ownership does not strengthen the model for people who suffered damage from the disaster. As Paxson and Rouse (2008) and Groen and Polivka (2010) suggest, the predisaster stock of location-specific capital may not influence the migration decision for people who experienced high levels of damage. It is noteworthy that this contrasts with the hometown effect mentioned earlier.

Furthermore, the effect of losing their network of family and relatives on intentions to remain was profoundly negative and significant among the group that suffered damage. Respondents who lost much of material assets value nonmaterial assets like human networks especially with family and relatives. Our results do not find a significant local network effect, contrary to the results of Thiede and Brown (2013). This is partly because we differentiate networks of local associates from that of family and relatives, though local associates live near respondents' residences.

It is because of the specificity of social network in Japan why family and relatives are significant, while local associates are not. In general, loan transactions are rarely carried out with local associates, while it is possible to make a loan transaction with family and relatives. Transactions of land and house with relatives are also observed in rural areas. We think that such differences between relatives and local associates are reflected in our results.

Jacob Vigdor (2008) argues whether the destructed city of New Orleans should be fully rebuilt after Hurricane Katrina's attack, by comparing with other recovered cities from disaster or war and examining Katrina's impacts on the population, housing and labor markets of New Orleans. He states that many of the cities that have recovered successfully have regained somewhat through economic booms. On the other hand, New Orleans has lost its original economic rationale and been declining for quite a long time. As Vigdor pointed out, whether the city will come back is fundamentally an economic problem. However, we think that it is important to evaluate the value of local-specific capital to make cost-benefit comparison for rebuilding. If people do not come back, the city also does not come back.

We think that human networks are important to think about when considering reconstruction from the disaster. Rebuilding human networks raises the villager's utility of staying, leads to prevention of out-migration, and contributes to reconstruction of the village. Our findings echo Cathrine Brun's insight that networks help to maintain cultural continuity and everyday practices, thus developing "translocal" connections. Migrants can rely on these connections both during and after displacement as they can act as "a coping strategy and starting point for life in a new place" (Brun, 2005, p. 18).

CONCLUSIONS

Theories that emphasize purely economic factors fail to capture the broader social framework in which decisions to migrate are taken. This chapter has clarified several factors affecting the intention to relocate among the people of Noda village in northern7 Japan following the Great East Japan earthquake. Our survey results show that young, female, and unmarried respondents expressed greater intentions to relocate and that respondents from high-income families indicated greater likelihood of remaining, a finding that accords with the compensating wage hypothesis.[11] The home ownership and hometown dummy variables display significant and strong effects.

Direct damage to residence, job, or income did not significantly affect respondents' intentions to relocate. Unlike the case of Hurricane Katrina, residential damage did not exert a presiding influence on Noda residents' decisions to relocate. One reason is that they developed strong ties by helping each other overcome difficulties following the earthquake and in doing so formed location-specific capital.

---

[11]The variables such as age, sex, and marital status are control variables to identify the effect of human networks in the regression analysis. We think that it is not appropriate to interpret the results by connecting such variables with the role of social networks.

Another reason is that, as the compensating wage hypothesis suggests, the expected benefit from relocating is so small that people tend to remain in the village.

However, those who suffered greater devastation to personal networks expressed stronger intentions to leave the village. We analyzed how the disaster affected respondents' intentions to relocate. We found that home ownership was negatively correlated to the decision to relocate among respondents whose homes were not damaged. However, among the group harmed by the 2011 disaster, loss of personal networks had a negative and significant effect on intentions to remain in the area. In other words, the loss of their network of family and relatives rather than material capital stands out as a main motivation to leave the area.

Our results suggest an important message for authorities who create and implement reconstruction policy following natural disasters—that is, it is highly important to make initiatives based on residential ownership as location-specific capital. Also, for residents suffering from disaster, reconstructing the network of family and relatives is more essential than residence rebuilding. As one of the reconstruction processes of the Great East Japan earthquake, residential districts are being transferred to new places. This study shows that reconstruction authorities must consider residential reconstruction and reconstruction of human networks simultaneously.

In creating recovery policies that consider population mobility, it is crucial for policy makers to understand what drives peoples' intention to relocate following a natural disaster. In examining relocation following the 2011 earthquake and tsunami, we clarify how extensively loss of economic activity such as employment or of human networks affects Noda villagers' intentions to relocate. Policies intended to financially bolster locales devastated by disasters can perhaps restore economic activity, but they cannot necessarily restore human networks.

PERSONAL NETWORKS AND LONG-TERM GENDERED POSTDISASTER WELL-BEING IN MEXICO AND ECUADOR

Graham A. Tobin[1], Christopher McCarty[2], Arthur D. Murphy[3], Linda M. Whiteford[1], Eric C. Jones[4]

[1]*University of South Florida, Tampa, FL, United States;* [2]*University of Florida, Gainesville, FL, United States;*
[3]*University of North Carolina-Greensboro, Greensboro, NC, United States;* [4]*University of Texas Health Science*
Center at Houston, El Paso, TX, United States

CHAPTER OUTLINE

INTRODUCTION

This study looks at gendered differences in personal network characteristics and associated mental health status in postdisaster contexts in Mexico and Ecuador. Previous research has repeatedly demonstrated that disasters are associated with increased symptoms of stress and depression, that impacts of disaster are often distributed unevenly between men and women, and that the exchange (provision and receipt) of social support is frequently essential to mitigation, recovery, and well-being. A review of many disaster studies revealed that women are twice as likely to exhibit symptoms of posttraumatic stress following disasters (Tolin & Foa, 2006), along with potentially lower levels of social support in comparison with men (Faas, Jones, Whiteford, Tobin, & Murphy, 2014). Because women often bear the responsibility of providing care for family and others in a variety of cultural contexts (Cutter, 1995; Enarson, 2001; Steady, 1993)—and they are often important sources of social support for others during crises (Faas et al., 2014)—we direct our attention in this study to the possibility that inequities in the distribution of postdisaster mental health problems will mirror the gender-based distributions of other disaster impacts (Enarson & Morrow, 1998; Fothergill, 1996) and therefore place heavier mental health burdens on women.

In this chapter, we first establish how gender relates to postdisaster mental health and social support, then explain our study's methodology using analysis of personal networks to understand variation in social support. We present results on gendered differences in mental health, the extent to which these differences occur between the seven research sites considered in this study, and then examine the relationship of personal networks to posttraumatic stress and depression. We conclude the chapter with a summary of these findings and a discussion of the importance of understanding that regardless of the presence or absence of gendered differences in mental health outcomes in disasters, there remain gendered differences in the role of various facets of social support and social networks, including in nondisaster settings (Kawachi & Berkman, 2001).

GENDERED FACETS OF POSTDISASTER MENTAL HEALTH AND SOCIAL SUPPORT

Residents of hazard-prone communities are generally at high risk of mental illnesses such as posttraumatic stress (Baisden, 1979; Canino, Bravo, Rubio-Stipec, & Woodbury, 1990; Norris, Murphy, Baker, & Perilla, 2004; Tobin & Ollenburger, 1996; Warheit, 1985; Wood & Cowan, 1991; Wright, Ursano, Bartone, & Ingraham, 1990). In disasters, women are typically more adversely affected than men in terms of various aspects of mental health, with some of the greatest differences occurring in measures of posttraumatic stress (Norris, Perilla, Ibañez, & Murphy, 2001; Ollenburger & Tobin, 2008). David Tolin and Edna Foa's (2006) review of 31 disaster/fire studies from around the world found women to be over two times more likely than men to exhibit symptoms of posttraumatic stress.

In terms of *levels* of posttraumatic stress, in addition to the gendered *rates* of posttraumatic stress, Fran Norris and colleagues (2002) found in a review of 177 disaster studies that 65% of respondents recorded high levels of stress. They also reported that females had higher levels of posttraumatic stress symptoms in 42 of 45 disaster samples that tested for *gender differences*, and that these differences were more pronounced in non-Western populations. Norris and colleagues' (2001) study of white and black people in the United States and Mexico found gendered differences in posttraumatic stress symptoms where women had scores typically 15% to 30% higher than men for each of the populations and for each of the subscales of intrusion (unwelcome sensations), avoidance (staying away from possible experiences), and arousal (anxiety, panic, and heightened sensitivities). These findings prompted us to examine postdisaster gender differences in mental health in Ecuador and Mexico for comparison, where network differences seemed to be implicated (Jones et al., 2014).

An earlier study of mestizo agropastoralists in disaster-induced resettlements in the Andean highlands of Ecuador found evidence of relationships between social networks and well-being, in which women generally receive less support than men (Faas et al., 2014). A study comparing urban landslides in Mexico—what later became a research site we consider in the present study—with another flood site in Mexico found that women perceived less social support (Norris, Baker, Murphy, & Kaniasty, 2005). Gender-based differences in the distribution of mental health symptoms are of concern in disaster-prone areas, as women tend to be the primary caregivers and mental health can affect the way care is provided. Studies in disaster-induced resettlements in Ecuador (Faas et al., 2014) and central Asia

(Halvorson & Hamilton, 2007) found that labor patterns often lead to men being absent from disaster-prone communities, and those men who remain are typically unemployed and may represent a further burden on women as caregivers.

Our focus on *gendered network differences* comes from Valerie Haines and Jeanne Hurlbert's (1992) findings for the general population in Northern California that there were gender-specific differences in how a person's network density, diversity of ties, and size of network impacted stress exposure, social support access, and distress levels. They found that men who were embedded in relatively dense networks had better access to companionship and emotional support than men in less-dense networks, independent of the prevalence of strong ties, but men with a predominance of multiplex ties—multiple kinds of ties with any individual—also experienced higher levels of stress. Rather, men were better off seeking instrumental (i.e., tangible or practical) support from kin, while women tended to partition their network and ask for instrumental support from people upon whom they were not already relying for other forms of support. Later, in a related study, they found that the degree of gender diversity in personal networks and the degree to which a personal network has a higher proportion of men were both found to be significantly associated with the ability to activate ties in response to Hurricane Andrew in Louisiana in 1992 (Hurlbert, Haines, & Beggs, 2000).

Getting help or support from one's network members vastly increases the likelihood of reciprocating in some fashion to those specific individuals (Plickert, Côté, & Wellman, 2007). However, reciprocating relationships can be seen as protective factors, risk factors, or both (Noh & Avison, 1996). Such ties can present risk factors for increased mental health problems, especially for women with few social and economic resources who are required by social obligation to give support to others (Kawachi & Berkman, 2001). Several studies have also examined gendered social support in disaster settings. Norris and colleagues (2005) discovered that women perceive they have less social support available to them than do men, particularly following traumatic events. In their study of gender dynamics in the exchange of social support in disaster-induced resettlements in highland Ecuador, A.J. Faas and colleagues (2014) ascertained that women often provided more social support than they received.

Based on the previous empirical work just cited, in this study we expected to find a significant difference in the number of symptoms of depression and posttraumatic stress between men and women in several disaster contexts in Mexico and Ecuador. We likewise expected that these differences would in part be evident by study site (as explained by cultural and historical context and level of impact) and by personal networks (composition and structure of the relationships in a person's social world). The approach of social network analysis provides important details about the relative contribution of the various facets of social ties that other approaches to social support may lack.

RESEARCH DESIGN AND CONTEXT
SITES AND SAMPLING

Data for this study were collected between 2007 and 2009 in seven communities, five in Ecuador and two in Mexico (Table 16.1), all in close proximity to active volcanoes or subject to flooding and landslides (Fig. 16.1). In order to examine gender and networks, we attempted to select communities that maximized variability in traits that the literature suggested would affect mental health and well-being. Thus, the communities included in the study varied in size, experience with environmental crises, and whether they were newly resettled or had been evacuated.

Table 16.1 Study Sites and Characteristics

Site	Impact	Settlement Pattern	Population	Last Evacuation (years)
Mexico				
San Pedro	Evacuated (ash, explosions)	Rural village	3512	7
Ayotzingo	Resettled (landslide)	Urban neighborhood	1609	9
Ecuador				
Penipe Viejo	Continued ashfall	Urban village	710	NA
Penipe Nuevo	Resettled (ash)	Rural village	1405	3
Pusuca	Resettled (lahars, explosions)	Rural village	161	3
Pillate	Evacuated, continued ashfall	Rural village	193	3
San Juan	Evacuated, continued ashfall	Rural village	172	3

Adapted from Jones, E. C., Faas, A. J., Murphy, A. D., Tobin, G. A., Whiteford, L. M., & McCarty, C. (2013). Cross-cultural and site-based influences on demographic, well-being, and social network predictors of risk perception in hazard and disaster settings in Ecuador and Mexico. Human Nature, 24, 5–32.

2006 eruption of Mt Tungurahua in Ecuador; affected sites lie on volcano and across river from it.

1999 landslides from tropical depression in mountains of Puebla, Mexico (and all along Caribbean coast of Mexico).

FIGURE 16.1

Images from the Ecuador and Mexico field sites.

Our first Mexican study site, San Pedro Benito Juárez, sits at the base of Mt. Popocatepetl, an active stratovolcano located approximately 70 km southeast of Mexico City and 40 km west of Puebla, Mexico's fourth largest city. The volcano is surrounded by a number of small mestizo and indigenous Nahuatl communities, which for centuries have taken advantage of the rich volcanic soil for traditional milpa agriculture.[1] After lying dormant for nearly a century, Mt. Popocatépetl became active again in 1994 and since then has experienced periodic eruptions. Concern over possible major eruptions has triggered the government of Mexico to evacuate several communities on the Puebla side of the volcano on various occasions (Marcías Medrano, 2005). These evacuations lasted from a few days to up to several weeks. The volcano continued to be active at the time of writing, although major evacuations thus far occurred only in 1994 and 2000. Volcanologists in Mexico at the Center for the Prevention of Disasters indicated that an explosive eruption remained possible in 2016, though less likely than activity similar to or less than the 1994 and 2000 eruptions, and emergency evacuation procedures remained in place. The community of San Pedro Benito Juárez is an agricultural community with a population of 3500 that had evacuated several times, most recently in 2000, or seven years before the study was undertaken.

Our second Mexican research site, Ayotzingo, is a community of approximately 1600 and now a suburb of Teziutlán. Extensive flooding in 1999 led to landslides in Teziutlán, which caused a number of deaths and displaced several thousand families (Olazo García, 2009). After some delay, many of these households were resettled in the community of Ayotzingo. In many cases, families received a home that was much smaller than their previous dwelling and that was no longer near their work but approximately 10 km from the center of town. Our research 9 years later indicated that the displacement of communities was a more significant felt experience for those who relocated. Many residents no longer lived close to relatives and neighbors; in resettlement, they were now spatially removed from individuals that they had known for most of their lives.

In Ecuador, the focus of our attention was on communities around the active stratovolcano Mt. Tungurahua in the highland provinces of Tungurahua and Chimborazo. Tungurahua became active in 1999 after about an 80-year period of relative quiet. Since then, it has been highly active with numerous small-scale events interspersed with some major eruptions. The authors have been working in the area since 2000 and have built up a close association with various community leaders and organizations. These communities suffered devastating impacts from the eruptions of Mt. Tungurahua and many communities had serious social and economic damage. The communities in the high-risk zone (i.e., the areas closest to the volcano) in the two provinces were the most severely affected by the eruptions and were subjected to several mandatory evacuations. Ashfall damaged and obstructed roads, schools, and health centers, as well as crops, animals, and irrigation systems. In addition to the devastating effects the volcano had on household and regional health and economies, these eruptions also resulted in the permanent displacement and resettlement of thousands of former residents of the high-risk zone around the volcano. Next, we describe each of the Ecuador sites.

Penipe Viejo[2] is a small urban township (population ~700 at the time of research) that serves as the county administrative seat of the greater municipality of Penipe, in Chimborazo Province. At 10 km south of the volcano, Penipe Viejo sustained moderate ashfall during the 1999 and 2006 eruptions, light ashfall

---

[1] Milpa agriculture refers to the traditional Mesoamerican method of production where maize, beans, and squash are planted together in a symbiotic mix.

[2] Penipe Viejo and Penipe Nuevo are names of convenience that we employed to identify the two distinct communities, Penipe Nuevo having been established on the periphery of the old community and eventually exceeding Penipe Viejo in population size. The two communities constituted one administrative unit.

in the interim, and periodically in the chronic ash emissions that continue at the time of writing. Penipe Viejo was never evacuated, though it served as a base of emergency response operations and is the site of the Penipe Nuevo resettlement, so we consider this site affected—mildly by the direct impacts of volcanic emissions, yet more significantly by emergency operations and resettlement and recovery initiatives.

We also collected data in two resettlements built for those displaced by the eruptions of Mt. Tungurahua—Penipe Nuevo and Pusuca. Penipe Nuevo is a resettlement community built as an extension of the small urban center of Penipe Viejo. The resettlement consists of 287 houses (population ~1400 at the time of research) constructed by the Ministry of Housing and Urban Development and a US-based multinational, faith-based disaster relief organization, Samaritan's Purse. The resettlement is a landless, urban resettlement populated by smallholding rural agriculturalists displaced from the northern parishes of Bilbao, Puela, and El Altar in the wake of the eruptions. The second resettlement, La Victoria de Pusuca (hereafter, Pusuca), is a small, land-based rural resettlement community, built by the Ecuadorian nongovernmental organization Fundación Esquel. The hilltop resettlement consists of 45 houses (population 161 at the time of research) occupied by smallholding rural agriculturalists primarily displaced from the northern parish of Puela, with a few from Bilbao and El Altar. It sits 5 km south of the Penipe Nuevo resettlement.

Pillate and San Juan are two adjacent villages in the Cotaló parish of the municipality of Pelileo in Tungurahua Province, across the Chambo River valley, directly west of Penipe's northernmost parish, Bilbao. The two communities of approximately 35 households each (populations ~193 and 172, respectively, at the time of research) are 3 km west of the volcano and within the high-risk zone. They were evacuated for eruptions in 1999 and 2006, and the villages suffered damages from ashfall, incandescent material, and tremor-induced landslides, though later three-fourths of the residents returned.

We developed a stratified random sampling protocol for each community based on streets and housing numbers so that the geography of each community would be fully covered. However, we also controlled the sample for gender to achieve approximately equal numbers of male and female respondents. We could not control for sampling of outcome variables, such as depression and posttraumatic stress disorder, or the personal network covariates, as values for these were only known after data collection. In all, we conducted a total of 460 interviews across all seven sites (Table 16.2). This involved a lengthy process with individual surveys taking anywhere from 45 min to 2 h.

DEPENDENT VARIABLES

We used a 20-item Center for Epidemiologic Studies-Depression scale developed by Lenore Radloff (1977) to indicate levels of potential depression, and a 17-item scale and a related four-item functioning

Table 16.2 Sample Sizes for Dependent Variables, by Site

	Posttraumatic Stress Symptom Sample Size	Depression Symptom Sample Size
San Pedro	12	58
Ayotzingo	114	139
Penipe Viejo	26	46
Penipe Nuevo	86	99
Pusuca	37	40
Pillate	36	48
San Juan	21	30

scale from the Composite International Diagnostic Interview (World Health Organization, 1997) to determine symptoms of posttraumatic stress.[3] In addition, we conducted personal network interviews to capture personal network composition and structure. The posttraumatic stress disorder survey reflects smaller sample sizes because the application of that scale required respondents to report a traumatic event that was life threatening, and not all respondents had such an experience with the disasters.

Both the depression scale and the posttraumatic stress scale were normally distributed and required no transformation. The correlation between the depression and posttraumatic stress scales was $R = 0.483$, significant at $p<.005$, indicating some covariance in our two dependent variables.

INDEPENDENT VARIABLES

The procedure for collecting the network data involved a network name generator asking each person to list 45 people whom they knew by sight or by name, with whom they had had contact in the past 2 years or with whom they could they could have had contact if they wanted to do so (after McCarty, 2002a). We randomly subsampled 25 out of that list of 45 to reduce the respondent burden of answering questions on all 45 (thus requiring respondents to indicate ties between 600 unique pairs, rather than 1980). Research has shown that a random selection of as few as 25 alters from a larger list will yield reliable estimates of personal network structure (McCarty, Killworth, & Rennell, 2007). We then asked respondents to indicate whether or not they had exchanged (given or received) material, informational, or emotional support with each of the 25 alters. We also asked them to provide a variety of characteristics about each of the 25 people (network interpreters), including aspects of their relationship with each of them. These data were used to calculate the network composition and network support variables. Finally, we asked for the respondent's perception of the degree to which each unique pair of alters in the network interacted (little to none, some, a lot). These data were used to calculate three sets of independent variables:

1. **Network composition** included: percent of alters from same family as ego since family is among the most important category of relationships; percent of alters who were of the same sex since having more women in a personal network postdisaster can be a risk factor and is a form of homophily or similarity; and average age difference of alters from ego to serve as another measure of homophily.
2. **Network support** included the primary areas of received support: percent of alters providing any emotional support related to the disaster or recovery; percent of alters providing any material support ever in the past; and percent of alters providing any informational support related to the disaster or recovery.
3. **Network structure** measures were selected to determine how cohesive and connected the personal networks were, and included: average betweenness centrality of alters, which captures the connectedness in a personal network and thus the robustness of the network (even for networks that are not dense); number of components in the personal network to measure the amount of the network that is fully disconnected from other parts; and density of personal network to measure how much everyone is tied to everyone in the network.

---

[3]These were the same scales used by Norris and Murphy in their study of ambient stress and trauma in four Mexican cities RO1 MH51278 (Baker et al., 2005; Norris, Murphy, Baker, & Perilla, 2003; Norris et al., 2003; Slone et al., 2006).

RESULTS
TEST OF GENDER DIFFERENCES IN REPORTED SYMPTOMS OF DEPRESSION AND POSTTRAUMATIC STRESS

Our first analysis was to determine whether there were any significant differences between men and women on the two scales, posttraumatic stress and depression. There are two ways to ascertain differences: (1) using counts of males versus females with certain scores and (2) using average scale scores for each gender. First, using a count of symptoms of posttraumatic stress, women were not statistically more likely to have eight or more of the 17 possible clinical posttraumatic stress symptoms in any of the seven sites, but more likely to have depression symptoms in two of the sites. Table 16.3 shows the entire sample divided roughly in half to distinguish between high versus low symptomology (45% with high posttraumatic stress symptomology; 51% with high depression symptomology). The distribution of that dichotomization is not even across sites, but the important feature of Table 16.3 is when and if there are gendered differences in number of people experiencing a given level of symptoms. Results are similar to the scale scores approach presented in Table 16.5.

It is important to distinguish that, in the results in Table 16.4, we are presenting analyses concerning average levels of reported posttraumatic stress symptoms rather than the percentages of each gender experiencing a certain number of these symptoms, as shown in Table 16.3.

Table 16.4 shows that there were significant differences between men and women, with women scoring higher than men on both scales, thus supporting our first hypothesis. However, other research has additionally typically found many more women to have posttraumatic stress than do men. For example, Tolin and Foa (2006) reviewed the literature and noted that in 31 disaster/fire studies of adults, women were just over two times more likely to have posttraumatic stress. It jumps to over two and a half times when considering only life-threatening disasters. Levels, as opposed to rates, appear in Table 16.6.

T-tests of these variables (Table 16.5) demonstrate variability of these results by site. In both Ayotzingo, Mexico, and Pillate, Ecuador, women scored significantly higher than men on both scales despite being in different countries, one community being resettled and the other evacuated, and having

Table 16.3 Percentages of Men and Women Experiencing Higher Level of Stress or Depression Symptoms

	Posttraumatic Stress (8 or More of 17 Symptoms)		Depression (9 or More of 40 Points)	
	Men	**Women**	**Men**	**Women**
San Pedro (MX)	13	17	19	38
Ayotzingo (MX)	34+	51+	29***	60***
Penipe Viejo (EC)	24	22	47	44
Penipe Nuevo (EC)	53	65	56+	70+
Pusuca (EC)	59	61	59+	83+
Pillate (EC)	31	50	12**	46**
San Juan (EC)	63	71	75	50

*For gendered difference +p < .1, *p < .05, **p < .01, ***p < .001.*

passed different amounts of time since the stressful event. In Ecuador, women in both the Penipe Nuevo and Pusuca resettlements scored significantly higher on depression, but not on posttraumatic stress. For both communities, men scored nearly as high as women on the posttraumatic stress scale.

The data reported in Table 16.6 separate the sample (see Table 16.4) by level of impact and indicate that the difference in the number of stress and depression symptoms between men and women only appears in high-impact areas and not in relatively low-impact areas. The difference between men and women in *rates* of posttraumatic stress, when examined by sites in Table 16.5, is really the consequence of gendered differences occurring in only two of the seven sites—Ayotzingo and Pillate. The difference between men and women in *levels* of depression occurs in four of the seven sites.

We wanted to control for sample size and thus calculated the effect size using Cohen's *d*—which depends on sample size and standard deviation—and r^2, or the amount of variation in the dependent variable that is accounted for in the independent variable.[4] Our analysis summarized in Table 16.7 suggests that the general effect of gender on mental health, controlling only for impact level, is relatively small. Table 16.7 shows the same pattern as does Table 16.6, but adds information on the relationship between the means of each gender (see footnote), and that high-impact sites have about 15–30% of males and females that fall outside the distribution of the other gender for the two mental health scales.

As can be seen in Table 16.8, the global analysis in Table 16.7 can be misleading, since the sites vary considerably in effect sizes for gendered differences in mental health symptoms. Cohen's *d* suggests that there is more impact than was suggested by sample-size-dependent Pearson's *R*. Three instead of two sites showed gendered differences for symptoms associated with posttraumatic stress, and five instead of four sites showed gendered differences for depression. In the gray cells are the effects that are of possible

Table 16.4 Differences in Scales by Gender

Scale	Possible	Range Exhibited	Men	Women	Difference
Posttraumatic stress	0–17	0–17	7.4	8.3	$p=.067$
Depression	0–40	0–37	8.7	11.9	$p=.000$

[4]Typically, psychologists analyzing depression and posttraumatic stress scales consider the effect size, that is, how large the differences are compared to the distribution of responses across possible values, rather than calculating only absolute differences in averages. This approach controls for circumstances where subsamples may be statistically different based on chance or large sample size, but the actual effect size indicates a trivial finding. There are no firm conventions like those used with *p*-values. Cohen (1992) showed that effect size indicates the relationship of the mean of one sample to the mean of the other sample, as in the following table.

Effect Size (Cohen's *d*)	Where 50th Percentile of Sample 1 Meets Sample 2 (%)	Nonoverlap in Distributions (%)
0.8	79	47.4
0.5	69	33.0
0.2	58	14.7
0.0	50	0

Table 16.5 *T*-tests Between Women and Men for Number of Mental Health Symptoms, by Site

Site (Country)	Posttraumatic Stress Symptoms Mean (SD)		Depression Symptoms Scale Mean (SD)	
	Men	Women	Men	Women
San Pedro (MX)	7.5 (2.6)	5.4 (5.6)	13.1 (3.3)	15.0 (4.6)
Ayotzingo (MX)	5.5 (4.2)*	8.0 (4.3)*	15.0 (3.6)**	17.5 (5.8)**
Penipe Viejo (EC)	5.6 (2.0)	5.6 (3.0)	9.2 (8.3)	9.7 (7.2)
Penipe Nuevo (EC)	8.5 (4.5)	9.0 (3.8)	10.4 (6.9)*	14.9 (9.6)*
Pusuca (EC)	8.7 (3.4)	9.0 (3.8)	10.7 (6.1)*	15.4 (8.1)*
Pillate (EC)	6.0 (4.0)*	8.6 (4.0)*	4.6 (4.3)**	8.4 (4.4)**
San Juan (EC)	10.8 (3.9)	11.9 (4.0)	15.6 (9.5)	13.6 (10.3)

*Significant at p < .05 level.
**Significant at p < .005 level.

Table 16.6 Differences in Mental Health for Sites of Different Levels of Impact, by Gender

Site	Posttraumatic Stress Symptoms Mean (SD)		Depression Symptoms Mean (SD)	
	Men	Women	Men	Women
Low-impact sites	6.1 (2.3)	5.5 (3.9)	7.3 (6.9)	8.5 (6.6)
High-impact sites	7.6 (4.3)*	8.7 (4.1)*	9.1 (7.4)**	12.8 (8.7)**

*Significant at p < .05 level.
**Significant at p < .005 level.

Table 16.7 Gender Effect Size by Level of Impact: Cohen's *d* and *r*²

Site	Posttraumatic Stress Symptoms		Depression Symptoms	
	Cohen's *d*	*r*²	Cohen's *d*	*r*²
Low-impact sites	0.20	0.01	−0.10	0.002
High-impact sites	−0.26	0.02	−0.48	0.05

Table 16.8 Gender Effect Size by Site: Cohen's *d* and *r*²

Site (Country)	Posttraumatic Stress Symptoms		Depression Symptoms	
	Cohen's *d*	*r*²	Cohen's *d*	*r*²
San Pedro (MX)	0.48	0.05	−0.47	0.05
Ayotzingo (MX)	0.59	0.08	−0.52	0.06
Penipe Viejo (EC)	na	na	−0.06	<0.01
Penipe Nuevo (EC)	−0.12	<0.01	−0.54	0.07
Pusuca (EC)	−0.08	<0.01	−0.66	0.10
Pillate (EC)	−0.65	0.10	−0.87	0.16
San Juan (EC)	−0.28	0.02	0.20	0.01

substance. However, only Pillate exhibits a large effect size, and some caution should be taken when interpreting data from the sites that have only moderate levels of effect.

TEST OF DIFFERENCES BASED ON PERSONAL NETWORK CHARACTERISTICS

Our second hypothesis was that, given differences we have found in our data by gender, personal network variables would explain a significant proportion of the variance—after dividing the sample based on level of impact—in the number of stress and depression symptoms reported by respondents. We extracted the data for this analysis from the Egonet interviews and generated nine variables—six variables reflecting personal network composition and three variables representing personal network structure. We tested each variable to see if it was normally distributed and found the network composition and network support variables to be normally distributed, while the networks structure variables required a log transformation to reduce the skewness of their distribution.

We present the results of correlation analysis between the depression and posttraumatic stress scales with these variables in Tables 16.9–16.11. The contrast between men and women is evident. Composition variables have the most associations with mental health, followed by network support and network structure, both of which show about half as many associations with the outcomes of depression and posttraumatic stress. The significance of personal network composition suggests that for men in communities with low impacts, symptoms of depression and stress are associated with percent of

Table 16.9 Correlation of Depression and Posttraumatic Stress With Personal Network Composition, by Impact and Gender (Pearson's _R_)

		% Family		% Same Gender		Age Difference From Ego	
Sites	Symptoms	Men	Women	Men	Women	Men	Women
Low Impact	Depression	0.36*				0.29+	
	Posttraumatic stress			0.59*			
High impact	Depression		0.18*	−0.15+	−0.12+	0.20*	
	Posttraumatic stress		0.14+		−0.16*		

+p < .1, *p < .05.

Table 16.10 Correlation of Depression and Posttraumatic Stress With Personal Network Support, by Impact and Gender (Pearson's _R_)

		% Material Support		% Informational Support		% Emotional Support	
Sites	Symptoms	Men	Women	Men	Women	Men	Women
Low Impact	Depression						
	Posttraumatic stress			0.16+		0.43+	
High impact	Depression						
	Posttraumatic stress		0.14+		0.19*		0.20*

+p < .1, *p < .05.

Table 16.11 Correlation of Depression and Posttraumatic Stress With Personal Network Structure, by Impact and Gender (Pearson's *R*)

Sites	Symptoms	Avg. Betweenness (natural log)		Components (natural log)		Density or Mean Degree (natural log)	
		Men	Women	Men	Women	Men	Women
Low Impact	Depression						
	Posttraumatic stress						
High impact	Depression						−0.17*
	Posttraumatic stress	−0.18+				−0.18+	−0.15+

+p < .1, *p < .05.

family and percent of same gender in personal networks, whereas gender and age are associated with high impact (Table 16.9). In contrast, women show more symptoms of depression and stress associated with high impact experience.

In Table 16.10, our results indicate that the relationship between receiving network support (material, information, emotional) and posttraumatic stress is gendered by level of impact experienced from an event. Men in low-impact areas who received support from a larger portion of their personal network had higher levels of posttraumatic stress. However, it was women in high-impact areas who experienced the association between network support and posttraumatic stress.

Network structure—operationalized as average betweenness centrality, components, and density—elicited fewer correlations (Table 16.11). However, once again, there are discernible patterns. Men in high-impact areas experience associations between centrality measures—average betweenness and average degree—and posttraumatic stress. For women in high-impact sites, both posttraumatic stress and depression were associated with average degree centrality or density of personal networks.

DISCUSSION AND CONCLUSIONS

Personal networks in hazard-impacted communities do not appear to operate the same way for both men and women, and even behave differently based on the level of impact the hazard had on their village. We were able to test these hypotheses in two Latin American countries in order to build upon the work of other scholars on gendered postdisaster support that had primarily focused on the United States. Three sets of findings regarding stress and depression raise further questions with respect to composition of personal networks, support received, and network structure.

1. Composition (i.e., family, age, gender) of personal networks was more associated with symptoms of depression than with symptoms of posttraumatic stress, suggesting that the kind of people in one's life—or the characteristics of these people—may also be associated with depression. In terms of thinking about recovery from disasters, it is worth asking: In the processes of recovery, with what kind of individuals is one engaging? If networks and mental health are bidirectionality intertwined, then you select certain people because you are depressed, and the people in your life contribute to your depression.

2. Actual support received (material, information, emotional) from a personal network tended to be associated with more symptoms of posttraumatic stress in high-impact sites, which probably indicates that individuals with greater need were getting higher levels of support. It does not necessarily denote that receiving more support increases the strain on people. However, based on prior findings about the double-edged sword of reciprocity (Noh & Avison, 1996), it is possible that those receiving more support are also giving more support and are thus under stress due to high levels of social responsibility. In terms of thinking about recovery from disasters, it is worth asking: Are the people with greater need getting support from their network? And are people with greater need involved in high levels of reciprocation that causes further stress?

3. Structure (average betweenness centrality, components, density) of a personal network was implicated more with posttraumatic stress than with depression, but at similar levels of association for both men and women. Density was protective of both men and women in high-impact sites. In terms of thinking about recovery from disasters, it is worth asking: Are there optimal personal network structures that mitigate the effects of living under the threat of disaster or increase the likelihood of surviving a disaster?

Additionally, it is apparent that all three of the just-discussed sets of findings tend to occur more in the high-impact sites than low-impact sites. More specifically, in terms of the relevance of level of disaster impact for predicting the relationship between networks and well-being, the low-impact sites saw men's well-being (but not women's) predicted to a degree by what kinds of people were in their personal network (family, age, gender). In the high-impact sites, similarity in the age of network members was relevant for men, more family meant lower well-being for women, and both men and women were better off with more dense/closed networks and gender homophily.

Research in the general population on personality, mental health, and personal networks suggests that when people perceive themselves as susceptible to outside forces, their networks are more likely to be closed/dense but consisting of relatively weak connections—and they tend to see themselves as a part of a group or groups (Kalish & Robins, 2006). And, people who feel that they are more in control of their lives and thus less susceptible to external forces tend to generate structural holes or to keep some strong tie partners from being connected while also showing greater individualism. The implication for disaster networks research is that people who have gone through disasters might begin to adopt the outlook of vulnerability and group membership. This suggests the possibility that tight networks might be a result of disaster. If they are caused by disaster, and if these tighter networks are protective in disaster, then there is a natural tendency that can be built upon.

As men and women often have different kinds of support networks (see also Faas et al., 2014; Halvorson & Hamilton, 2007), how to help men and women seek and access support probably will need to differ in many disaster contexts. Additionally, the disruption of social networks by disaster and resettlement, although not examined here, may be gendered, such that rebuilding a network may on average look different for women than for men. Here, context can be important, and different sites may produce the inverse of an expected gendered facet of personal networks and postdisaster well-being.

Our findings support other research that attention should be given to mental health in postdisaster environments and perhaps also to mental health training in predisaster environments so that people are experienced with or thoughtful about how they react. Higher levels of stress and depression may compromise caregiving, for example, and thus exacerbate vulnerabilities. Similarly, the resettlement of people disrupts personal networks and presents new challenges as residents try to rebuild their personal networks. It was interesting that for the two resettled Ecuador sites, women had higher average levels of depressive symptoms than did men, but similar levels of posttraumatic stress symptoms. Usually these two mental health conditions covary. It is clear that the resettlement setting produces yet other challenges for the relationship between social support and postdisaster mental health. In addition to disruption by resettlement, networks are also compromised when members (usually men in these cases) are away from home, often traveling for work. The reduced support may lead to greater vulnerability.

ACKNOWLEDGMENTS

Thank you to our colleagues at CUPREDER in Puebla, Mexico, and the Instituto Geofísico at the Escuela Politécnica Nacional in Quito, Ecuador. Grants from the National Science Foundation (BCS-CMMI grants 075124/0751265 and BCS grant 0620213/0620264) funded the data collection in Ecuador and Mexico. We are indebted to field research assistants Fabiola Juárez Guevara and Isabel Pérez Vargas for their work in both countries.

CONCLUSIONS

THE PRACTICAL AND POLICY RELEVANCE OF SOCIAL NETWORK ANALYSIS FOR DISASTER RESPONSE, RECOVERY, AND ADAPTATION

17

Julie K. Maldonado
Livelihoods Knowledge Exchange Network, Santa Barbara, CA, United States

CHAPTER OUTLINE

Social Network Analysis of Disaster Response, Recovery, and Adaptation. http://dx.doi.org/10.1016/B978-0-12-805196-2.00017-0

Disasters often send shock waves through social systems that can inhibit responses to the very challenges the disaster presents, with few examples of people recovering with increased capacity. When disasters strike, survivors and responders are faced with challenges of receiving and conveying information; assessing impacts, damages, losses, and needs; and coordinating action and recovery efforts both in the short and long term. In particular, disaster survivors report losses of livelihood, health, access to education and services, resources and resource exchange practices and reciprocity, diminished cultural practices, and fragmented and disrupted social networks (Peterson & Maldonado, 2016). Survivors' capacities to cope with disaster impacts can hinge on the strength and range of their social relations—networks, kinship, and other patterns of group relationships that provide both immediate and long-term support independent of disaster aid and agency and organizational support. Yet, government- and agency-led disaster planning, responses, and recovery processes generally ignore community structures and relationships that facilitate information flows and tangible and nontangible recovery needs, which call for interorganizational collaboration (Harris & Doerfel, Chapter 6).

Social science research that uncovers the roles and consequences of networks before and after disasters can inform and improve disaster responses, recovery efforts, reduction of disaster risks, and performance of involved organizational structures and networks, from predisaster to postdisaster recovery and reconstruction. Levels of responses range from the individual to the community to the interagency or interorganizational, considering the ties and relationship patterns that enable (or inhibit) these various levels to deal with the risk or reality of hazards and disasters (Faas & Jones, Chapter 2; Kapucu & Demiroz, Chapter 3). Here I review the implication of arguments put forward in the theories, methodologies, and results of social network analysis presented in this book. I reflect on such questions as: What patterns emerge from social network analysis that could be useful for policy and decision makers? Can social network analysis suggest practical means of improving flows of information through a social system to build capacity?

The chapter considers the opportunities, benefits, and challenges of how social network analysis informs the different stages of disasters covered in this book—response, recovery, and adaptation—and how the empirical findings in the previous chapters can inform disaster-related policy and practice. In doing so, I refer back to the themes generated in Chapter 1: linking networks (or at least results) together across different levels of analysis, disciplines, and boundaries; the results when existing ties actually are activated; the ethics of social network analysis in disaster settings; and figuring out how to link local capacity with external resources. This chapter is not organized by those themes but comments on each of them in order to integrate insights in these areas into policy and practice.

Enhanced disaster policy is ever more critical under a new climate system with more frequent and intense disasters putting communities—particularly those that are already marginalized—at ever greater risk (Laska et al., 2015; Maldonado et al., 2013). It concludes with a parallel call to action in two parts: (1) to incorporate the ways social network analysis of communities, agencies, and organizations can inform disaster response, recovery, and adaptation into policy guidelines; and (2) a call for disaster researchers and practitioners to create a tool kit where we pool together our collective knowledge and map out our own social capital and networks, finding strategic inroads we can navigate to reach policy and decision makers to inform disaster policies and practices from the local to national to international level.

SOCIAL NETWORK ANALYSIS IN THE DISASTER CONTEXT: INFORMING POLICY AND PRACTICE

In disaster research, social network analysis provides key insight into how information and resources travel through a social system, who is reached and who is missed, who is most at risk, and mechanisms of survivors to both recover from disasters and adapt to future impacts of both past and future disasters. This is in part a result of seeing who is socially isolated and who and what are the key nodes through which information and resources flow (or do not flow) in a social system (Varda, Chapter 4). These key people or organizations can often appear in unexpected ways or in ways that disaster-related policies do not take into account. For example, it might be an elder sitting on their porch who is the center of information flowing to community members, as opposed to a local leader or organization announcing themselves during disaster response and recovery. Social network analysis can be useful to understand how networks that connect actors across various boundaries (e.g., sectoral, geographical, ideological, technological) can reduce vulnerability and build capacity when circumstances—and the networks themselves—shift and transform to varying needs and structures following a rapid disaster-induced shock. Analyzing social networks in communities exposed to hazards and at risk of disasters provides a marker around which to orient people's relational contacts, ties, and connections, even while they shift and reorient themselves following a disaster event.

Disaster research literature and the material and case studies in this book demonstrate that one's social network aids response and recovery efforts and long-term adaptation through both tangible (e.g., shelter) and nontangible (e.g., emotional) resources. Such support is often through informal exchanges at the community level that are hidden from direct outside view. Network analysis can unveil who (individuals or organizations) are the core nodes in a community or region, where they are creating solidarity and empowering collective action, and where people turn to find resources (e.g., churches, kinship networks). It can also aid in understanding how people who have been displaced following a disaster, or who have already moved (e.g., rural to urban migration, international migration), access information and obtain resources when at a distance from their community and kin networks. For example, following the 2010 British Petroleum Deepwater Horizon oil disaster and subsequent loss of fishing resources, Gulf Coast residents lost the ability to stockpile and send food to family members who had already moved away, resulting in people living outside the region losing access to local seafood supplies and the social connection of sharing and resource exchange, as well as cultural connection to place and community (Maldonado, 2014a). Thus, it is critical to see who is isolated or missed in the response, recovery, and adaptation process to provide the most effective support and create policies that enable the provision of such support (see Marcum et al., Chapter 8).

Through network analysis we can better understand who has a range of connections, which help postdisaster coping abilities. Likewise, this can point to who has fewer resources or relies on only a few close, strong ties without connections beyond their core network, also reducing coping abilities and well-being (Faas & Jones, Chapter 2; Meyer, Chapter 9). Such analysis can demonstrate how a disaster can strengthen or weaken ties, how such bonding can also produce and maintain hierarchies of race, class, gender, and ethnicity (Faas, 2015b), and the ways in which noneconomic exchanges either hinder or enhance postdisaster coping ability (Laska, 2012; Laska et al., 2010; Peterson & Maldonado, 2016). Network analysis could provide guidance to disaster practitioners and policy makers on the types of programs and policies that are most needed to support people following a disaster, what is needed to reduce disaster risk, and particularly who within a society or community is most vulnerable and needing to be reached. For example, network analysis could illuminate who people rely on for reciprocity and resource exchange, so if people are forced to move following a disaster event, it becomes clear what noneconomic resources they might now be severed from, and, in the reverse, for people who stay, what resources they might have lost. Such understanding is crucial when providing disaster support both in receiving communities for those displaced—as well as for people who are able to stay in the affected community—as it explicates what people have *really* lost, beyond economic means, that enables them to survive and thrive.

There is a distinct evolution in the international policy discourse about key issues of concern for disaster risk reduction that gives some cause for optimism about how policies are being enhanced and taking into consideration aspects that such tools as network analysis illuminate. For example, the Sendai Framework for Disaster Risk Reduction 2015–30, which was adopted at the 2015 Third United Nations World Conference on Disaster Risk Reduction as the successor instrument to the Hyogo Framework for Action, was the first major agreement of the post-2015 development agenda. It outlines seven targets and four priorities for action to prevent new and reduce existing disaster risks and explicitly calls for better understanding of disaster risk, which requires the need to "enhance the scientific and technical work on disaster risk reduction and its mobilization through the coordination of existing networks and scientific research institutions at all levels and in all regions," and that investing in disaster risk reduction for resilience involves "promot[ing] cooperation between academic, scientific and research entities and networks and the private sector to develop new products and services to help to reduce disaster risk" (UNISDR, 2015b).

The Sendai Framework highlights a shift from viewing human mobility as a driver of risk or consequence of disasters to seeing the potential of human movements acting as a social support mechanism to build the capacity and resilience of people affected by disaster events (Guadagno, 2016; see also Meyer, Chapter 9). Similar trends of paying attention to the nuances of social structures and significance such structures play in determining one's ability to cope following a disaster are seen in other international governance frameworks. For example, the Post-disaster Needs Assessment collaboration initiated in 2008 through a joint declaration between the United Nations Development Group, World Bank, and the European Union includes both a Damage and Loss Assessment plus a Human Recovery Needs Assessment in its approach, discussed in more detail in Section "Networks and Disaster Recovery". This is a substantial step forward, as previously, postdisaster assessments only included economic and asset loss, ignoring losses such as social networks, which can be included in human recovery needs assessments.

Despite the progress, such policy frameworks, as well as the disaster literature and the development and resettlement literature, largely frame networks in binary terms (e.g., fractured or not fractured), in which the myopic focus on network on/off binaries tends to be the only (and narrow) recognition patterns of social relations (Faas et al., 2015). However, human groups and social systems associate and flow in various levels, densities, and structures. The following sections build from arguments in the previous chapters about ways social network analysis demonstrates the need to move beyond a binary focus and help inform the call put forward by the Sendai Framework and international policy guidelines addressing response, recovery, and adaptation to disasters.

NETWORKS IN DISASTER RESPONSE
INTERORGANIZATIONAL COLLABORATION AND ASSESSING THE COLLECTIVE NETWORK

Disaster response understood as a network phenomenon informs how the resulting outcomes of the response phase are dictated by success or failures in the flow of information, coordinated action, and overall effect of the responding network of actors—what Branda Nowell and colleagues (Chapter 5) call an *incident response network*. Networks can reveal the various actors involved in disaster response, and what kinds of tensions or conflicts could arise from varied interests. For example, there is an inherent conflict of interest when different actors who do not share a common vision of response and recovery—and are located in distant locations from the disaster site—are charged with handling disaster aid and response. To adequately assess the functioning and effectiveness of response by the network of actors, it is critical to assess the collective network as a whole, which can be done through social network analysis, such as Nowell and colleagues (Chapter 5) did in assessing the performance of the response network to wildfire events in the United States.

No one agency or organization is equipped to handle disaster response and recovery efforts on its own (Almquist et al., Chapter 7; Nowell et al., Chapter 5). Interorganizational collaboration across geographic and organizational boundaries and varied communication pathways guided by stakeholder networks is critical to the response and recovery process. Assessing the performance of the network is both a key opportunity and a challenge, as networks do not exist in a vacuum but rather under messy conditions and circumstances (Nowell et al., Chapter 5). It is instead helpful to envision network performance as based on collaborative webs of relationships (Kapucu & Demiroz, Chapter 3), which serve to produce and maintain meaning and shape the networks' collective worldviews (see Geertz, 1973). Future work on social network analysis could inform practical guidelines for developing measures of network performance that account for the multilayered complexity in incident response (Nowell et al., Chapter 5).

EXTENDED KIN AS A SOCIAL SUPPORT MECHANISM

The shock or jolt of a disaster can lead survivors to activate social network ties to extended kin that are normally latent, as well as disengage friendship ties, as Christopher Marcum and colleagues (Chapter 8) demonstrated in their research following Hurricane Ike in the US Gulf Coast in 2008. They found that the role of family during the response and recovery phase moves beyond the core family unit and outward to extended kin, although such exchanges of reciprocity can be uneven. Thus, this helps understand the need for disaster risk reduction and response programs to consider how extended kin are activated as an instrument of social support that survivors rely on as a coping mechanism.

SOCIAL MEDIA COMMUNICATION TO EXPOSE KEY SOURCES

As disaster events send shock waves through social systems, we can readily see how attentional relationships shift during emergency response to organizations in the immediately affected area and lead to a reorganization of the attentional network, as Zack Almquist and colleagues (Chapter 7) demonstrated by examining Twitter accounts among US emergency management–related organizations. Analyzing social media communication during and after disasters (Almquist et al., Chapter 7) or mobile phone location and activity following disaster events (Lu, Chapter 10; Lu et al., 2016), helps to expose where the key sources of information and resources are located and where and to whom people turn for support. Such understanding shows where response and recovery efforts should be channeled to help bolster support, as these are often the more hidden pathways.

IMPORTANCE OF PREEXISTING RELATIONSHIPS AND LOCAL STAKEHOLDER NETWORKS

Existing policy frameworks (e.g., the National Response Framework in the United States) do not take into account the sweep of organizational-level activities that unfold during disaster response and recovery, particularly for community stakeholder networks, which are most efficient among the organizational stakeholders in a network to relay information because of their geographical and organizational positions (Harris & Doerfel, Chapter 6). Local stakeholder networks connected with regional and national response efforts provide a built-in flexibility for communicating and brokering new information and resources to be exchanged as the response phase progresses and the network transitions toward recovery. Thus, formal disaster planning and response needs to consider the preexisting community connections and civil and grassroots relationships in communities (Harris & Doerfel, Chapter 6).

CAPTURING PUBLIC–PRIVATE SYNERGIES

Social network analysis could offer guidance on better targeting relief efforts, especially for socially isolated disaster survivors, capturing public–private synergies, which could be done by investigating network ties and behaviors that may inform governmental and nongovernmental emergency management teams and help planners to improve the coordination and extent of coverage of disaster response, recovery, and adaptation (Varda, Chapter 4).

NETWORKS IN DISASTER RECOVERY
POSTDISASTER SUPPORT NETWORKS AND CIVIL SOCIETY

By understanding what kinds of support organizations mobilize post disaster, what kinds of relationships are found in postdisaster support networks, and how networks enable organizations to meet their needs during recovery, Joanne Stevenson and David Conradson's (Chapter 11) study of organizations affected by the 2010/2011 earthquakes in Canterbury, New Zealand, highlighted the significance of preexisting relationships stemming from trust and strong communication pathways. Network analysis can also reveal the ways grassroots participation supports disaster recovery through both formal and informal organizational networks, and how civil society actions evolve through relationship building and long-term maintenance, as Jia Lu (Chapter 10) demonstrated in research 3 years after China's Wenchuan earthquake in 2008.

Some key elements of the Post-disaster Needs Assessment collaboration, including damage and loss assessment and a human recovery needs assessment, focus on the distinct needs and priorities by gender and age group of the affected populations, the participation of affected stakeholders in their own recovery process, and support to the spontaneous recovery efforts of the affected population. Social network analysis can help inform a participatory and inclusive recovery process and interinstitutional coordination by pointing to the local and community stakeholder networks that need to be supported and incorporated in broader planning and operational activities (Harris & Doerfel, Chapter 6).

KIN EMBEDDEDNESS AS A SOCIAL SUPPORT MECHANISM FOR POSTDISASTER RECOVERY

Size and composition of networks, as well as the kinds of available resources that flow through networks, inform how supported one is during the recovery process and able to cope with the disaster effects. Michelle Meyer's (Chapter 9) research on people's perceptions of their social networks in a future disaster event provides information about the ways in which people determine which network ties to activate—or not activate—for support following a disaster. Family members are central to what Meyer calls disaster support networks, recognizing that kin embeddedness—the extent to which an individual is connected to family members—is a key signal for coping ability following a disaster, becoming more so as local structures and institutions can rupture when a disaster hits. This relates to how people cope with crises in their everyday lives, such as the reciprocity of informal exchange through child care or money lending.

Understanding social networks can provide insight about the various assets and capacities individuals and communities have access to during and following a disaster event; for example, for those with fewer kin or close family relations, increased support is needed for nonfamilial network mechanisms and both financial and nonfinancial institutional assistance (Meyer, Chapter 9; see also Casagrande, McIlvaine-Newsad, & Jones, 2015). However, personal network ties do not guarantee resources, which highlights the distinction between potential and realized ties. The types and availability of resources are often unclear until a disaster unfolds. Thus, an overreliance or assumption about resource access through personal ties can, at times, be a hindrance to recovery.

Further work that considers how social networks are used within the context of other available assets during a disaster would help inform how resources are allocated and where the resources come from (e.g., core family unit, extended kin, unexpected actors). Understanding the importance of kin embeddedness in disaster recovery informs consideration for other avenues (e.g., government support) people find support and the ways in which such support is limited and inhibits recovery. This helps inform postdisaster assessment work by considering various measures of social vulnerability (e.g., measuring family ties) (Meyer, Chapter 9). However, it can be both insensitive and unethical to research networks during and immediately following a disaster event when people are still in crisis mode. Such analysis should respectfully be done and possibly only once people are further along in the recovery process and can look back at what unfolded. Having completed predisaster network research, such as the case with Michelle Meyer's (Chapter 9) work, can aid and inform this process.

Practical guidelines that inform measuring and promoting how kin and extended kin are utilized and accessed as a social support mechanism for postdisaster recovery are needed, especially for extended kin at a distance (e.g., immigrant populations) (Marcum et al., Chapter 8). And policies are needed that include social capital—at the personal and community level—and promote network support (Meyer, Chapter 9), which is particularly crucial in the context of a changing climate and with already occurring and future predicted increased frequency and intensity of disasters.

BOUNDARY ORGANIZATIONS TO FACILITATE COLLABORATION

Some key elements interfere with or obstruct effective collaborations during disaster recovery efforts, such as cultural conflicts and lack of guiding plans or organizational development and shared understanding of decision-making processes (Kapucu & Demiroz, Chapter 3). Differing perspectives on networks and how such analysis can inform the recovery process can conflict to the extent that they inhibit the full realization of the utility of the network analysis tool. One way to address these issues is through boundary organizations, which act as hubs to facilitate networking and collaboration between diverse groups, individuals or even disciplines, particularly when there are high levels of distrust, disproportionate degrees of power, or unfamiliarity with each other (Rising Voices, 2015). Effective boundary organizations build accountability, trust, understanding, collaboration, and participation between individuals and organizations on both sides of the boundary (Cash et al., 2002; Maldonado et al., 2016), plus act as negotiators and translators between levels of organization, social or political structures, and multiple audiences (e.g., communities and policy makers) to broker information during disaster recovery (Harris & Doerfel, Chapter 6).

Networks that cross boundaries can help facilitate the recovery period, as well as long-term adaptation, as an opportunity to build capacity and connect to resources. As such, network-of-networks (e.g., disaster aid organizations), or "link-tanks" (Maldonado, Taylor, & Hufford, 2016), which cross geographic, cultural, organizational, and disciplinary boundaries, are key points of access to learn from them who their members are and the most trusted, influential conduits of information within them. Boundary organizations can foster connections with a range of resources, support, and actors in diverse geographic, economic, and social locations to mitigate loss or damage following a disaster event (Stevenson & Conradson, Chapter 11). Such organizations can translate across boundaries, conveying different meanings that disaster survivors and disaster agencies have in talking about preparedness, needs, losses, and recovery.

INTERSECTIONS OF BOUNDARY ORGANIZATIONS AND BRIDGING ACTORS

Boundary organizations in organizational networks and bridging actors in social networks are intimately related. Disaster and network researchers often focus exclusively on the translational and brokerage potential of boundary organizations, which essentially span network groups in the same way as do bridging actors. Thus, similar to the way bridging actors in networks can support their personal network subgroups—and can also increase their own power and prestige—this is also true for boundary organizations. Likewise, both boundary organizations and bridging actors are also vulnerable (see Faas & Jones, Chapter 2; Faas et al., 2015). The intersection of boundary organizations and bridging actors should be considered both in the policy design process and in disaster response, recovery, and adaptation efforts, as the vulnerability of both these entities needs to be mitigated to help bolster the support they can provide.

ROLE OF BRIDGERS TO CREATE DIVERSITY IN THE NETWORK

Ideas emerge throughout this book of how social scientists, particularly anthropologists, who have worked and lived with communities and have developed long-standing relationships built around trust and mutual respect, can act as a bridge between agencies and local communities, helping policy makers understand local people's perspectives, needs, and capacities. Bridging individuals—those trusted coming from either within or outside the community—may connect people and organizations who are not already connected, creating diversity in the network (Faas & Jones, Chapter 2). This is particularly effective when negotiated by people who cross boundaries and engage in both groups and create structural diversity to remove constraints that more closed networks limit in the disaster relief, recovery, and reconstruction processes (Marino & Lazrus, 2015).

Because of long-standing racial, ethnic, linguistic, and cultural biases against historically and politically marginalized communities, social scientists and other disaster researchers and practitioners can be privileged with policy-related opportunities not often provided to community members. Thus, they need to work *with* communities to understand how to most appropriately, effectively, and carefully carry the community's message forward while working to ensure that culturally sensitive information is protected. Therefore, a key consideration is the researchers' responsibility as a bridging entity between disaster survivors and disaster agencies and organizations (Maldonado, 2014b). A major challenge in bringing diverse voices, knowledge, understandings, and perspectives to the same table is collaborating on a shared understanding of decision-making—particularly as diverse stakeholders bring a range of cultures, guidelines, and limiting functions (Kapucu & Demiroz, Chapter 3). A key way to navigate this terrain is through the development of more effective communication and information pathways, as discussed in the following section.

NETWORKS IN HAZARD ADAPTATION
EMERGENT EFFECTS OF RISK MANAGEMENT AND HAZARD ADAPTATION

Social networks that underlie informal risk-sharing institutions play a significant role in disaster-related adaptation for underserved communities; thus, it is critical to understand how disaster events can alter communities' informal risk-sharing institutions against subsequent nondisaster shocks. For example, Yoshito Takasaki (Chapter 14) demonstrated how, in the reconstruction phase in rural Fiji 2 years after a tropical cyclone struck, people who suffer loss or damage from a disaster have a weakened reciprocity-based safety net against future nondisaster-related shocks, which household networks can partly mitigate. Further, emergent effects of hazard adaptation practices can occur in unintended ways, as Mark Moritz (Chapter 13) showed in research with pastoralists' communities in Cameroon, considering why livestock transfers provide short-term support but do not allow pastoralists to rebuild their herds. Pastoralists do not think of these exchanges or practice them as a type of risk management or conscious adaptation, yet these practices have risk management effects (Moritz, Chapter 13). Therefore, it is essential that these practices be reflected in the creation of policies focusing on risk management and hazard adaptation, particularly for underserved individuals and communities.

BUILT-IN FLEXIBILITY TO REDUCE SOCIAL DISPARITIES

Understanding the role that social networks play in risk sharing and disaster-related adaptation can inform policies to support strengthening network-based risk sharing and lowering constraints within the information institutions. The effectiveness of long-term adaptation and risk-reduction strategies would be boosted and sustained if they were implemented at the network level (Stevenson & Conradson, Chapter 11). However, policy creation and guidelines need to be crafted with the utmost sensitivity due to context-dependent variation in the roles of networks (Tobin et al., Chapter 16). Flexibility needs to be built into guidelines to account for context and differences in varying access points and types of support needed depending on race, class, and gender understood within a historically specific context, while also considering the ways in which such practices provide opportunities for risk reduction (Faas, 2015a). This is particularly relevant for issues related to mental health following a disaster event, as Rangel and colleagues (Chapter 12) found the participation in networks associated with protesting and raising concerns about justice may influence outcomes such as posttraumatic stress and depression.

UNDERSTANDING THE DRIVERS OF RELOCATION

Networks also play a key role in migration decisionmaking following a disaster event (Fussell, 2006), as demonstrated by Lee Young-Jun and colleagues (Chapter 15) in their study that considered the factors that influence whether residents of Noda in Iwate prefecture remained in or relocated from their village following the 2011 Great East Japan earthquake. They found that the loss of human networks, and not material capital, was the main driver to relocate. Considering disaster recovery and adaptation policies and processes to support human mobility needs, policy makers must understand the drivers of relocation post disaster (e.g., loss of human networks or social capital).

The sheer number of people displaced by disasters every year demands social science attention to this issue. In 2014 alone, nearly 20 million people were displaced in 100 countries by disasters. From 2007 to 2014, an average of 22.5 million people—or 62,000 people every day—were displaced each year by climate or weather-related disasters (NRC & IDMC, 2015). These numbers will surely rise as the intensity of extreme weather and sudden-onset events are predicted to become more severe with a changing climate (IPCC, 2014; Melillo, Richmond, & Yohe, 2014). The number on record of disasters emanating out of natural hazards doubled from about 200 to 400 in the 1990s and 2000s, with 9 out of 10 of these disasters being considered climate-related (Kolmannskog, 2008).

RECONSTRUCTING HUMAN NETWORKS

Research has also shown how people who are displaced by a disaster and have weaker or disrupted social networks are left in a state of limbo, staying longer in shelters provided by agencies or organizations, unable to either return to their communities or to relocate, as Elizabeth Fussell (2006) found in New Orleans following Hurricane Katrina in 2005. Lee and colleagues' (Chapter 15) work demonstrates the key role social networks play during the relief, recovery, and long-term adaptation processes for those internally displaced by a disaster and the need for disaster-recovery policies to include both residential rebuilding and reconstruction of human networks. Policies that ignore relationships and connections might work to restore economic activity, but they will most often fail at restoring human networks (Lee et al., Chapter 15) and will thus fail at building and supporting disaster survivors' capacity and resilience.

SUSTAINABLE COMMUNITY DEVELOPMENT-FOCUSED RELOCATION

The resettlement of people disrupts personal networks and presents new challenges as residents try to rebuild their personal networks. This disruption results in varying consequences by factors such as gender, leading to even greater vulnerability to future disaster and nondisaster shocks (Tobin et al., Chapter 16). For individuals, households, or entire communities displaced by disaster events and needing to relocate either temporarily or permanently, there are important elements to consider beyond just physically relocating, such as choosing a site location, housing configuration, maintaining social networks, livelihood opportunities, access to culturally important resources, and creating a plan for sustainable community development (e.g., see Displacement Solutions, 2013). Simply regrouping people without the thought of these other components in mind can leave people still feeling dispersed and unsure of what to do (Fullilove, 2005). Understanding the role social networks play in this process can help inform effective policy guidelines to support individuals and communities going through this and other postdisaster recovery and adaptations.

COMMUNITY-WIDE RELOCATION SUPPORT

Despite the disaster and resettlement literature that demonstrates the impacts on communities as they are scattered and as social networks and resource exchanges are fragmented and disrupted, there is a lack of government policies and measures supporting community-wide (as opposed to individual) relocation. For example, the US Federal Emergency Management Agency's buyouts support individual, voluntary relocation, as opposed to community-wide relocation. Offering support only individually further scatters communities and risks loss of community, culture, and sovereignty. A policy framework addressing disaster-induced displacement needs to provide *community-wide* support, not just *individual* or a binary focus of patterns of social relations, while still being attuned to existing inequities.

ADDRESSING THE SOCIAL ROOTS OF DISASTERS TO STEM FURTHER DISPLACEMENT

The Sendai Framework on Disaster Risk Reduction includes root causes and risk drivers as one of its basic principles, and formal institutional contexts have recognized the social roots of disasters, yet there has been minimal result in the actual practice and implementation of disaster risk reduction that addresses the social roots. Disaster risks are largely being produced by ill-informed and ill-equipped policies and practices that are framed more to protect development than actually reduce disaster risk (Oliver-Smith, 2016; UNISDR, 2015a). This is seen most readily by practices of the international financial institutions that further the spread of displacement in the name of disaster risk reduction and climate mitigation. Disaster risk reduction cannot become an excuse to move people out of harm's way in the name of the greater good, since the people being moved are often those who have already been pushed aside and marginalized—with consequences for their social networks—in the name of development and progress (Roy, 1999).

SUPPORT FOR CONTINUING AND REESTABLISHING SUBSISTENCE PRACTICES

Disaster policies are often designed to push displaced people from rural to urban locales and into a capitalist-based economic marketplace. This distinctly ignores people's right to be rural (Marino & Lazrus, 2015), and the significance of subsistence practices and informal exchange practices not only in economic terms, but also social and cultural value, with such traditional practices often tied to identity, place connection, community structure, and cultural traditions. Therefore, disaster policies need to include support for the relationships or social facets that people require for continuing their subsistence practices in place.

INCLUSION OF DIVERSE TYPES OF KNOWLEDGE

Diverse types of knowledge and the distribution of such knowledge across networks need to be included in the long-term adaptation process and decision-making. Communities understand the effects of disasters based on their relationships to the environment and generational dwelling in place; policy makers and disaster responders must be ready to engage with these communities and make sense of the impacts being experienced from local people's knowledge and perspectives. Thus, there needs to be a continued shift from the current economic-based framework for adaptation needs to a people-centered framework that focuses on human rights and local participation in decision-making. This is starting to be seen in frameworks such as the UN–World Bank–EU Post-disaster Needs Assessments, but it is yet to be seen in full implementation.

MORE EFFECTIVE COMMUNICATION AND INFORMATION PATHWAYS

Postdisaster needs assessments have called for establishing a network of "robust communication and dissemination systems" to strengthen things such as flood forecasting and warning systems (Government of Samoa, 2013; United Nations Development Group, World Bank, & European Union, 2013), and developing an Emergency Communications Policy Framework (Government of Fiji, 2013). However, such recommendations tend to focus on channels of communication such as mobile phone text messages, radio-telephone systems, and satellite phones, but often fail to include the human-power ways that communication is often transferred and flows through a region, communities, and households. Network analysis can help in this regard by demonstrating who and what are the key nodes for information flow and resource access; informing how information moves (or does not move) through a social system can help improve disaster response and recovery efforts, as well as long-term adaptation.

EFFECTIVE FLOW OF INFORMATION AND RESOURCES TO REDUCE DISASTER RISK

Network analysis can inform institutional and interagency guidelines for developing practical ways to ensure effective flow of information and resources. For example, with the support of partners at Michigan State University, the United States National Climate Assessment conducted a preliminary social network analysis to diagram the interactions between people and organizations engaged by the National Climate Assessment and show how the network of stakeholders changed over the course of the Third National Climate Assessment (2010–14). Social network analysis was used to inform gaps in the types and locations of stakeholders and networks engaged with and involved in the Third National Climate Assessment and that were missed, thus informing ways to improve future engagement, reach, and flow of information and, in turn, improve understanding of stakeholders' informational needs and improve the usefulness and usability of information products (Cloyd, Moser, Maibach, Maldonado, & Chen, 2015). This translates to the ways disaster risk can be reduced by understanding who is missed, ways to more effectively engage with people at risk along with appropriate avenues to reach them, and what kinds of information are needed to be most useful for recovery, adaptation, and reducing risk.

CONCLUSIONS: GUIDANCE FOR INTERNATIONAL POLICY FRAMEWORKS ON DISASTERS

Social network analysis can push aside the binary framing of networks and help disaster practitioners and policy makers think through the social-related disaster response, recovery, and adaptation by actually measuring these dynamics to better understand not only the flows of information but also the implications of what this means for people in varying dimensions of vulnerability, so as to not further produce or maintain hierarchies of race, class, gender, and ethnicity in the postdisaster context. This is where it becomes the responsibility of the disaster researchers to communicate that story. The visuals and graphics they create are only meaningful if there is an accompanying simple, pointed, well-articulated story that is crafted and directed for practitioners and policy makers in such a way that is readily digestible and useful for program and policy development.

Social scientists have been beating down the disaster policy door for over three decades with the similar rhetoric of needing to include local communities in decision-making processes, having local knowledge guide adaptation and disaster risk reduction, and needing to address the social roots of disasters to reduce disaster risk. I repeat this mantra here because while it might appear that such repetition has not changed policy, it has certainly shifted the conversation in subtle, yet critical ways. For example, in disaster recovery, a framework that calls for accounting for human needs, and not just economic and infrastructural damages and losses, now guides United Nations agencies' efforts. In the past, the international postdisaster recovery and reconstruction process was focused on valuating physical damages and economic losses, known as the Damage and Loss Assessment. As a result, the total effects of disasters were largely underestimated and many social needs left unaddressed. In part with the continued persistence and presence of researchers and practitioners beating down that door, this led to a shift in focusing on the impacts on human development, including households and communities, social, cultural, and governance issues, including crosscutting issues such as gender and human rights, and a shift from economic recovery to broader restoration of functions, systems, and capacities. A similar story has occurred with the Hyogo Framework, and subsequent Sendai Framework for Disaster Risk Reduction, which recognize the need to address root causes and risk drivers. While these might not yet be seen as monumentally changing practice on the ground, it has certainly shifted the policy framework and guidelines, which is a critical start. Therefore, it is essential to continue the pounding on the policy door making these demands.

It is equally critical that we—as disaster researchers and practitioners—consider whatever system we are working within and the ways in which we can utilize our ties within that system to effect change in disaster response and recovery. At this critical moment, we need to do so with urgency, as more and more communities are facing extreme conditions where the very land on which they depend for their lives, livelihoods, and cultures is disappearing underneath their feet, and they might be one storm away from absolute loss.

This chapter is not only a call for incorporating the ways social network analysis of communities, agencies, and organizations can inform disaster response, recovery, and adaptation into policy guidelines but a call for us to turn this analysis on ourselves, to link together our collective knowledge and map out our own social capital and networks, and strategically navigate pathways to reach policy and decision makers to inform disaster policies and practices from the local to national to international level. It is within our collective reach, and we know what to do, but sometimes, much like the elder on the porch who is the key point of information, the challenge is knowing where to turn to make the most use of our collective energy and see where change can most readily be made. It is one thing to say get away from policies that maintain binaries and produce and reproduce social inequalities, and that is part of the objective here. But the other, perhaps more important one, is a challenge to the disaster research community to map out your social network and see where and to whom you are bonded and tied—however strong or weak—as the pathways to change might arise in the most surprising places. And that is the story we must be ready to tell.

References

An asterisk (*) indicates a work covering networks and extreme events.

Aakhus, M. (2007). Communication as design. *Communication Monographs, 74*(1), 112–117.

Abe, Y., & Tamada, K. (2007). Regional patterns of employment changes of less-educated men in Japan: 1990–2007. *Japan and the World Economy, 22*, 69–79.

* Acock, A. C., & Hulbert, J. S. (1993). Social networks, marital status, and well-being. *Social Networks, 15*(3), 309–334.

Adger, W. N., Hughes, T. P., Folke, C., Carpenter, S. R., & Rockström, J. (2005). Social-ecological resilience to coastal disasters. *Science, 309*(5737), 1036–1039.

* Ahuja, G. (2000). Collaboration networks, structural holes, and innovation: a longitudinal study. *Administrative Science Quarterly, 45*(3), 425–455.

* Aktipis, C. A., Cronk, L., & Aguiar, R. (2011). Risk-pooling and herd survival: an agent-based model of a Maasai gift-giving system. *Human Ecology, 39*(2), 131–140. http://dx.doi.org/10.1007/s10745-010-9364-9.

Albrecht, G. (2006). Solastalgia. *Alternatives Journal, 32*(4/5), 33–36.

Albrecht, G., Sartoreb, G., Connorc, L., Higginbothamd, N., Freemane, S., Kellyb, B., et al. (2007). Solastalgia: the distress caused by environmental change. *Australasian Psychiatry, 15*(S1), S95–S98.

* Aldrich, D. P. (2008). The crucial role of civil society in disaster recovery and Japan's preparedness for emergencies. *Japan Aktuell, 3*, 81–96.

* Aldrich, D. P. (2011). The power of people: social capital's role in recovery from the 1995 Kobe earthquake. *Natural Hazards, 56*(3), 595–611.

Aldrich, D. (2012). *Building resilience: Social capital and post-disaster recovery.* Chicago: University of Chicago Press.

* Aldrich, D. P., & Meyer, M. (2015). Social capital and community resilience. *American Behavioral Scientist, 59*(2), 254–269.

Aldrich, D., & Sawada, Y. (2015). The physical and social determinants of mortality in the 3.11 Tsunami. *Social Science & Medicine, 124*, 66–75.

Almquist, Z. W. (2012). Random errors in egocentric networks. *Social Networks, 34*, 493–505.

Almquist, Z. W., & Butts, C. T. (2013). Dynamic network logistic regression: a logistic choice analysis of inter-and intra-group blog citation dynamics in the 2004 US presidential election. *Political Analysis, 21*, 430–448.

Almquist, Z. W., & Butts, C. T. (2014a). Bayesian analysis of dynamic network regression with joint edge/vertex dynamics. In *Bayesian inference in the social sciences* (pp. 1–26). John Wiley & Sons.

Almquist, Z. W., & Butts, C. T. (2014b). Logistic network regression for scalable analysis of networks with joint edge/vertex dynamics. *Sociological Methodology, 44*, 273–321.

Amin, S., & Goldstein, M. (2008). In *Data against natural disasters: Establishing effective systems for relief, recovery, and reconstruction.* Washington, DC: World Bank.

Anderson, B. (1983). *Imagined communities: Reflections on the origin and spread of nationalism.* New York: Verso.

Aranda, M. P. (2008). Relationship between religious involvement and psychological well-being: a social justice perspective. *Health & Social Work, 33*(1), 9–21.

Arif, A., Shanahan, K., Chou, F.-J., Dosouto, Y., Starbird, K., & Spiro, E. S. (2016). How information snowballs: exploring the role of exposure in online rumor propagation. In *Proceedings of the 19th ACM conference on computer-supported cooperative work and social computing (CSCW).*

Arnberg, F. K., Hultman, C. M., Michel, P.-O., & Lundin, T. (2012). Social support moderates posttraumatic stress and general distress after disaster. *Journal of Traumatic Stress, 25*(6), 721–727.

Arnold, M. (2006). The globalization of disaster: disaster reconstruction and risk management for poverty reduction. *Journal of International Affairs, 59*(2), 269–279.

Associated Press. (2011). *Costs to fight duckett fire reach $4m*. CBS Denver.

Axinn, W. G., Ghimire, D. J., Williams, N. E., & Scott, K. M. (2015). Association between the social organization of communities and psychiatric disorder in rural Asia. *Social Psychiatry and Psychiatric Epidemiology*, *50*(10), 1537–1545.

Baisden, B. (1979). Crisis intervention in smaller communities. In E. J. Miller, & R. P. Wolensky (Eds.), *The small city and regional community: Proceedings of their 1979 conference* (pp. 107–123). Stevens Point, WI: University of Wisconsin.

Baker, W. E. (1990). Market networks and corporate behavior. *The American Journal of Sociology*, *96*(3), 589–625.

Baker, C. K., Norris, F. H., Diaz, D. M. V., Perilla, J. L., Murphy, A. D., & Hill, E. G. (2005). Violence and PTSD in Mexico: gender and regional differences. *Social Psychiatry & Psychiatric Epidemiology*, *40*(7), 519–528.

Bales, R. F., & Parsons, T. (1956). *Family: Socialization and interaction process*. Routledge.

Banks, D., & Carley, K. (1996). Models for network evolution. *Journal of Mathematical Sociology*, *21*(1–2), 173–196.

Barnett, P. A., & Gotlib, I. H. (1998). Psychosocial functioning and depression: distinguishing among antecedents, concomitants, and consequences. *Psychological Bulletin*, *104*, 97–126.

Barrett, C. B. (Ed.). (2005). *The social economics of poverty: On identities, Communities, Groups, and Networks*. London and New York: Routledge, Taylor and Francis.

Bateson, R. (2012). Crime victimization and political participation. *American Political Science Review*, *106*(3), 570–587.

Bearman, P., Moody, J., & Stovel, K. (2004). Chains of affection: the structure of adolescent romantic and sexual networks. *American Journal of Sociology*, *110*(1), 44–91.

Bejarano, C. L. (2002). Las super madres de Latino America: transforming motherhood by challenging violence in Mexico, Argentina, and El Salvador. *Frontiers: A Journal of Women Studies*, *23*(1), 126–150.

Bergenholtz, C., & Waldstrøøm, C. (2011). Inter-organizational network studies: a literature review. *Industry and Innovation*, *18*(6), 539–562.

Bertram, G. (1986). "Sustainable development" in Pacific micro-economies. *World Development*, *14*(7), 809–822.

Blake, N., & Stevenson, K. (2009). Reunification: keeping families together in crisis. *Journal of Trauma & Acute Care Surgery*, *67*(2), S147–S151.

Bloch, B. F., Genicot, G., & Ray, D. (2007). Reciprocity in groups and the limits to social capital. *American Economic Review*, *97*(2), 65–69.

Boin, A., & Lagadec, P. (2000). Preparing for the future: critical challenges in crisis management. *Journal of Contingencies & Crisis Management*, *8*(4), 185–191.

Bolin, R. (1976). Family recovery from natural disaster: a preliminary model. *Mass Emergencies*, *1*(4), 267–277.

Bolin, R. (1985). Disasters and long-term recovery policy: a focus on housing and families. *Review of Policy Research*, *4*, 709–715.

Bolin, R., Jackson, M., & Crist, A. (1998). Gender inequality, vulnerability, and disaster: issues in theory and research. In E. Enarson, & B. H. Morrow (Eds.), *The gendered terrain of disaster: Through Women's Eyes* (pp. 27–44). Westport, CT: Greenwood.

* Bollig, M. (1998). Moral economy and self-interest: kinship, friendship, and exchange among the Pokot (N.W. Kenya). In T. Schweizer, & D. R. White (Eds.), *Kinship, networks, and exchange* (pp. 137–157). Cambridge, UK: Cambridge University Press.

* Bollig, M. (2006). *Risk management in a hazardous environment: A comparative study of two pastoral societies*. New York, NY: Springer.

Borgatti, S. P. (2005). Centrality and network flow. *Social Networks*, *27*(2005), 55–71. http://doi.org/10.1016/j.socnet.2004.11.008.

Borgatti, S. P., & Everett, M. G. (2006). A Graph-theoretic perspective on centrality. *Social Networks*, *28*(4), 466–484. http://doi.org/10.1016/j.socnet.2005.11.005.

Borgatti, S. P., Everett, M. G., & Freeman, L. C. (1999). *Ucinet 5.0 for Windows*. Natick: Analytic Technologies.

Borgatti, S. P., Everett, M. G., & Freeman, L. C. (2002). *Ucinet for windows: Software for social network analysis*.

Borgatti, S. P., & Freeman, L. C. (2002). *Ucinet for windows: Software for social network analysis*. Harvard, MA: Analytic Technologies.

Borgatti, S. P., Mehra, A., Brass, D., & Labianca, G. (2009). Network analysis in the social sciences. *Science, 323*(5916), 892–895.

Borgerhoff Mulder, M., Fazzio, I., Irons, W., McElreath, R. L., Bowles, S., Bell, A., & Hazzah, L. (2010). Pastoralism and wealth inequality. *Current Anthropology, 51*(1), 35–48. http://dx.doi.org/10.1086/648561.

Bourdieu, P. (1985). The forms of capital. In J. Richardson (Ed.), *Handbook of theory and research for the sociology of education* (pp. 241–258). New York: Greenwood.

Boutrais, J. (2008). La vache d'attache chez les Peuls pasteurs (Niger et Centrafrique). *Journal des Africanistes, 78*(1–2), 71–104.

Bowden, S. (2011). Aftershock: business relocation decisions in the wake of the February 2011 Christchurch earthquake. *Journal of Management & Organization, 17*(201), 857–863.

Bowman, D. M., Balch, J., Artaxo, P., Bond, W. J., Cochrane, M. A., D'Antonio, C. M., et al. (2011). The human dimension of fire regimes on Earth. *Journal of Biogeography, 38*(12), 2223–2236.

Bradburd, D. A. (1982). Volatility of animal wealth among Southwest Asian pastoralists. *Human Ecology, 10*(1), 85–106.

* Brettell, C. B. (2008). Theorizing migration in atropology: the social construction of networks, identities, communities, and globalscapes. In C. B. Bretell, & J. F. Hollifield (Eds.), *Migration theory: Talking Across Disciplines*. New York, London: Routledge.

Brewin, C. R., Andrews, B., & Valentine, J. D. (2000). Meta-analysis of risk factors for posttraumatic stress disorder in trauma-exposed adults. *Journal of Consulting and Clinical Psychology, 68*(5), 748–766.

Brouwer, A. (2004). The inert firm; why old firms show a stickiness to their location. In *European regional science association conference* (pp. 1–23) (Porto, Portugal).

* Brun, C. (2005). *Research guide on internal displacement*. Retrieved from http://www.forcedmigration.org/research-resources/expert-guides/internal-displacement/fmo041.pdf.

Bruneau, M., Chang, S. E., Eguchi, R. T., Lee, G. C., O'Rourke, T. D., Reinhorn, A. M., et al. (2003). A framework to quantitatively assess and enhance the seismic resilience of communities. *Earthquake Spectra, 19*(4), 733–752.

Bruns, A., Burgess, J., Crawford, K., & Shaw, F. (2011). *#qldfloods and @QPS-Media: Crisis communication on twitter in the 2011 South East Queensland floods* Technical Report Cci ARC Centre of Excellence for Creative Industries & Innovation.

Burnard, K., & Bhamra, R. (2011). Organisational resilience: development of a conceptual framework for organisational responses. *International Journal of Production Research, 49*(18), 5581–5599. http://dx.doi.org/10.1080/00207543.2011.563827.

Burt, R. S. (1992). *Structural holes*. Cambridge, MA: Harvard University Press.

Butts, C. T. (2008a). A relational event framework for social action. *Sociological Methodology, 38*, 155–200.

Butts, C. T. (2008b). Social network analysis with sna. *Journal of Statistical Software, 24*.

* Butts, C. T., Acton, R. M., & Marcum, C. S. (2012). Interorganizational collaboration in the Hurricane Katrina response. *Journal of Social Structure, 13*(1), 1–36.

Butts, C. T., & Cross, B. R. (2009). Change and external events in computer-mediated citation networks: english language weblogs and the 2004 U.S. electoral cycle. *Journal of Social Structure, 10*.

Butts, C. T., Sutton, J., Spiro, E. S., Johnson, B., & Fitzhugh, S. (2011). *Hazards, emergency response, and online informal communication project data*.

Buzzanell, P. M. (2010). Resilience: talking, resisting, and imagining new normalcies into being. *Journal of Communication, 60*(1), 1–14.

Canino, G., Bravo, M., Rubio-Stipec, M., & Woodbury, M. (1990). The impact of disaster on mental health: prospect and retrospect analysis. *Journal of Mental Health, 19*(1), 51–69.

Carley, K. M. (2002). Smart agents and organizations of the future. In L. Lievrouw, & S. Livingstone (Eds.), *The handbook of new media* (pp. 206–220). Thousand Oaks, CA: Sage.

Carlson, E. J., Poole, M. S., Lambert, N. J., & Lammers, J. C. (2016). *A Study of organizational reponses to dilemmas in Interorganizational emergency management*. Communication Research, 1–29. http://doi.org/10.1177/0093650215621775.

* Casagrande, D. G., McIlvaine-Newsad, H., & Jones, E. C. (2015). Social networks of help-seeking in different types of disaster responses to the 2008 Mississippi river floods. *Human Organization, 74*(4), 351–361.

Cash, D., Clark, W. C., Alcock, F., Dickson, N. M., Eckley, N., & Jäger, J. (2002). *Salience, credibility, legitimacy and boundaries: Linking research, assessment and decision making* KSG Working papers series RWP02–046. http://dx.doi.org/10.2139/ssrn.372280.

Castles, S. (2003). Towards a sociology of forced migration and social transformation. *Sociology, 77*(1), 13–34.

CERA. (2012). *Economic recovery programme for greater Christchurch* (Christchurch, NZ).

Cernea, M. (1996). Understanding and Preventing impoverishment from displacement: reflections on the state of knowledge. In C. McDowell (Ed.), *Understanding impoverishment: The Consequences of Development-Induced Displacement* (pp. 13–32). Oxford: Berghahn Books.

Cernea, M. (1997). The risks and reconstruction model for resettling displaced populations. *World Development, 25*(10), 1569–1587.

Charuvastra, A., & Cloitre, M. (2008). Social bonds and posttraumatic stress disorder. *Annual Review of Psychology, 59*, 321–328.

Chatters, L. M., Taylor, R. J., & Neighbors, H. W. (1989). Size of informal helper network mobilized during a serious personal problem among black Americans. *Journal of Marriage & the Family, 51*(3), 667–676.

* Cheong, S. (2012). Community adaptation to the Hebei-Spirit oil spill. *Ecology and Society, 17*(3), 26. http://dx.doi.org/10.5751/ES-05079-170326.

Chewning, L. V., Lai, C.-H., & Doerfel, M. L. (2013). Communication technologies to rebuild communication structures. *Management Communication Quarterly, 27*(2), 237–263. http://doi.org/10.1177/0893318912465815.

* Choi, S. O., & Brower, R. (2006). When practice matters more than government plans: a network analysis of local emergency management. *Administration & Society, 37*(6), 651–678.

Chow, W. S., & Chan, L. S. (2008). Social network, social trust and shared goals in organizational knowledge sharing. *Information & Management, 45*(7), 458–465. http://dx.doi.org/10.1016/j.im.2008.06.007.

Chow, W.-H., Chrisman, M., Ye, Y., Gomez, H., Dong, Q., Anderson, C., et al. (2015). Cohort profile: The Mexican American mano-a-mano Cohort. *International Journal of Epidemiology* (Available on-line only).

Chua, V., Madej, J., & Wellman, B. (2011). Personal communities: the world according to me. In J. Scott, & P. J. Carrington (Eds.), *The Sage handbook of social network analysis* (pp. 101–115). Thousand Oaks, CA: Sage.

Clark, D., & Cosgrave, J. (1991). Amenities versus labor market opportunities: choosing the optimal distance to move. *Journal of Regional Science, 31*(3), 311–328.

Cloyd, E., Moser, S., Maibach, E., Maldonado, J., & Chen, T. (2015). Engagement in the third US national climate assessment: commitment, capacity, and communication for impact. *Climatic Change, 135*(1), 39–54.

Cobb, S. (1976). Social support as a moderator of life stress. *Psychosomatic Medicine, 78*(5), 300–314.

Cohen, J. (1992). A power primer. *Psychological Bulletin, 112*(1), 155–159.

Cohen, J. L., & Arato, A. (1992). *Civil society and political theory*. Cambridge, MA: MIT Press.

Cohen, S., Underwood, L., & Gottlieb, B. (Eds.). (2000). *Social support measurement and intervention: A Guide for Health and Social Scientists*. New York: Oxford University Press.

Cohen, S., & Wills, T. A. (1985). Stress, social support, and the buffering hypothesis. *Psychological Bulletin, 98*(2), 310–357.

Coleman, J. S. (1988). Social capital in the creation of human capital. *American Journal of Sociology, 94*(s1), S95. http://dx.doi.org/10.1086/228943.

Colten, C. E., Hay, J., & Giancarlo, A. (2012). Community resilience and oil spills in coastal Louisiana. *Ecology and Society, 17*(3), 5.

* Comfort, L. K. (2007). Crisis management in hindsight: cognition, communication, coordination, and control. *Public Administration Review, 67*(s1), 189–197.

* Comfort, L. K., & Haase, T. W. (2006). Communication, coherence, and collective action the impact of Hurricane Katrina on communications infrastructure. *Public Works Management & Policy, 10*(4), 328–343.

* Comfort, L. K., & Kapucu, N. (2006). Interorganizational coordination in extreme events: the world trade center attack, September 11, 2001. *Natural Hazards, 39*(2), 309–327.

Comfort, L. K., Ko, K., & Zagorecki, A. (2004). Coordination in rapidly evolving disaster response systems: the role of information. *American Behavioral Scientist, 48*, 295–313.

Contractor, N., Wasserman, S., & Faust, K. (2006). Testing multi-theoretical multilevel hypotheses about organizational networks: an analytic framework and empirical example. *Academy of Management Review, 31*(3), 681–703.

Cook, J. D., & Bickman, L. (1990). Social support and psychological symptomatology following a natural disaster. *Journal of Traumatic Stress, 3*(3), 541–556.

Coser, R. (1975). The complexity of roles as seedbed of individual autonomy. In L. Coser (Ed.), *The idea of social structure: Essays in honor of Robert Merton* (pp. 237–263). New York: Harcourt Brace Jovanovich.

Cox, D., & Fafchamps, M. (2008). Extended family and kinship networks: economic insights and evolutionary directions. In T. P. Schultz, & J. A. Strauss (Eds.), *Handbook of development economics* (Vol. 4) (pp. 3713–3784). Amsterdam: Elsevier.

Cranmer, S. J., & Desmarais, B. A. (2011). Inferential network analysis with exponential random graph models. *Political Analysis, 19*, 66–86.

Creswick, N., Westbrook, J., & Braithwaite, J. (2009). Understanding communication networks in the emergency department. *BMC Health Services Research, 9*(1), 247.

Cutter, S. (1995). The forgotten casualties: women, children, and environmental change. *Global Environmental Change: Human & Policy Dimensions, 5*(3), 181–194.

Cutter, S. L., & Smith, M. M. (2009). Fleeing from the Hurricane's wrath: evacuation and the two Americas. *Environment: Science & Policy for Sustainable Development, 51*(2), 26–36.

Dahl, G., & Hjort, A. (1976). *Having herds: Pastoral herd growth and household economy.* Stockholm, Sweden: Department of Social Anthropology, University of Stockholm.

Danieli, Y. (2009). Massive trauma and the healing role of reparative justice. *Journal of Traumatic Stress, 22*(5), 351–357.

De Haas, H. (2010). Migration and development: a theoretical perspective. *International Migration Review, 44*(1), 227–264.

De Weerdt, J., & Dercon, S. (2006). Risk-sharing networks and insurance against illness. *Journal of Development Economics, 81*(2), 337–356.

De Wever, S., Martens, R., & Vandenbempt, K. (2005). The impact of trust on strategic resource acquisition through Interorganizational networks: towards a conceptual model. *Human Relations, 58*(12), 1523–1543. http://dx.doi.org/10.1177/0018726705061316.

Deb, P., Okten, C., & Osili, U. O. (2010). Giving to family versus giving to the community within and across generations. *Journal of Population Economics, 23*(3), 963–987.

Department of Homeland Security (DHS). (January 2008). *National response framework.* Retrieved from https://www.fema.gov/pdf/emergency/nrf/nrf-core.pdf.

Department of Homeland Security (DHS). (May, 2013a). *National response framework* (2nd ed.). Retrieved from http://www.fema.gov/media-library-data/20130726-1914-25045-1246/final_national_response_framework_20130501.pdf.

Department of Homeland Security (DHS). (September 30, 2013b). *The resilient social network: @Occupysandy #superstormsandy.* Science and Technology Directorate. Retrieved from http://homelandsecurity.org/docs/the%20resilient%20social%20network.pdf.

Department of Homeland Security (DHS). (2015). *State and major urban area fusion centers.* Accessed November 9, 2015 via http://www.dhs.gov/state-and-major-urban-area-fusion-centers.

Dercon, S., & Krishnan, P. (2000). In sickness and in health: risk sharing within households in Ethiopia. *Journal of Political Economy*, *108*(4), 688–727.

Dercon, S., & Krishnan, P. (2005). Food aid and informal insurance. In S. Dercon (Ed.), *Insurance against poverty* (pp. 305–329). Oxford and New York: Oxford University Press.

Descola, P. (1994). *In the society of nature: A native ecology in Amazonia*. Cambridge, UK: Cambridge University Press.

Diani, M. (2015). *The cement of civil society: Studying networks in localities*. New York: Cambridge University Press.

Diesner, J., & Carley, K. M. (2005). Revealing social structure from texts: meta-matrix text analysis as a novel method for network text analysis. In V. K. Narayanan, & D. J. Armstrong (Eds.), *Causal mapping for information systems and technology research* (pp. 81–108). Harrisburg, PA: Idea Group Publishing.

Dilworth-Anderson, P., & Marshall, S. (1996). Social support in its cultural context. In G. R. Pierce, B. R. Sarason, & I. G. Sarason (Eds.), *Handbook of social support and the family* (pp. 67–79). New York: Springer.

DiMaggio, P. J., & Powell, W. W. (1983). The iron cage revisited: institutional isomorphism and collective rationality in organizational fields. *American Sociological Review*, *48*, 147–160.

Displacement Solutions. (2013). *The peninsula principles on climate displacement within states*. Geneva: Displacement Solutions. Retrieved from http://displacementsolutions.org/wp-content/uploads/2014/12/Peninsula-Principles.pdf.

* Doerfel, M. L. (2016). Networked forms of organizing, disaster-related disruptions, and public health. In T. R. Harrison, & E. A. Williams (Eds.), *Organizations, health, and communication* (pp. 365–383). New York: Routledge.

* Doerfel, M. L., Chewning, L. V., & Lai, C.-H. (2013). The evolution of networks and the resilience of interorganizational relationships after disaster. *Communication Monographs*, *80*(4), 533–559. http://doi.org/10.1080/03637751.2013.828157.

* Doerfel, M. L., & Haseki, M. (2013). Networks, disrupted: media use as an organizing mechanism for rebuilding. *New Media & Society*, *17*, 432–452. http://doi.org/10.1177/1461444813505362 .

Doerfel, M. L., & Harris, J. L. (2016). Resilience processes. In C. R. Scott, & L. L. Lewis (Eds.), *The International Encyclopedia of Organizational Communication. Hoboken*. NJ: Wiley-Blackwell.

* Doerfel, M. L., Lai, C.-H., & Chewning, L. V. (2010). The evolutionary role of Interorganizational communication: modeling social capital in disaster contexts. *Human Communication Research*, *36*(2), 125–162. http://dx.doi.org/10.1111/j.1468-2958.2010.01371.x.

Doerfel, M. L., & Taylor, M. (2004). Network dynamics of Interorganizational cooperation: the Croatian civil society movement. *Communication Monographs*, *71*(4), 373–394. http://doi.org/10.1080/0363452042000307470.

Dolfin, S., & Genicot, G. (2010). What do networks do? The role of networks on migration and "Coyote" use. *Review of Development Economics*, *14*(2), 343–359.

Donner, W., & Rodríguez, H. (2008). Population composition, migration and inequality: the influence of demographic changes on disaster risk and vulnerability. *Social Forces*, *87*(2), 1089–1114.

Drabek, T. E. (1969). Social processes in disaster: family evacuation. *Social Problems*, *16*(3), 336–349.

* Drabek, T. E. (1986). *Human system responses to disaster: An inventory of sociological findings*. New York: Springer-Verlag.

* Drabek, T. E., & Boggs, K. S. (1968). Families in disaster: reactions and relatives. *Journal of Marriage & Family*, *30*(3), 443–451.

* Drabek, T. E., Key, W. H., Erickson, P. E., & Crowe, J. L. (1975). The impact of disaster on kin relationships. *Journal of Marriage & the Family*, *37*(3), 481–494.

* Drabek, T. E., & McEntire, D. A. (2002). Emergent phenomena and multiorganizational coordination in disasters: lessons from the research literature. *International Journal of Mass Emergencies and Disasters*, *20*, 197–224. Retrieved from http://www.ijmed.org/detailed_article.php?id=395.

Drabek, T. E., & McEntire, D. A. (2003). Emergent phenomena and the sociology of disaster: lessons, trends, and opportunities from the research literature. *Disaster Prevention & Management*, *12*(2), 97–112. http://doi.org/10.1108/09653560310474214.

Dunning, T. (2012). *Natural experiments in the social sciences: A design-based approach.* Cambridge: Cambridge University Press.

Dupire, M. (1962). *Peuls nomades: Étude Descriptive des Wodaabe du Sahel nigérien.* Paris: Institut d'ethnologie.

Dyson-Hudson, R., & McCabe, J. T. (1985). *South Turkana nomadism: Coping with an unpredictably varying environment.* New Haven (CT): Human Relations Area Files.

Earthquake Engineering Research Institute (EERI). (2008). *The Wenchuan, Sichuan province, China, earthquake of May 12, 2008. EERI special earthquake report.* Retrieved November 29, 2010, from http://www.eeri.org/site/images/eeri_newsletter/2008_pdf/Wenchuan_China_Recon_Rpt.pdf.

Edwards, F. (2012). All hazards, whole community: creating resiliency. In N. Kapucu, C. V. Hawkins, & F. I. Rivera (Eds.), *Disaster resiliency: Interdisciplinary perspectives* (pp. 21–47). New York: Routledge.

Edwards, M., & Gaventa, J. (2001). *Global citizen action.* Boulder, CO: Lynne Rienner.

Eisenman, D., Chandra, A., Fogleman, S., Magana, A., Hendricks, A., Wells, K., et al. (2014). The Los Angeles County community disaster resilience project–A community-level, public health initiative to build community disaster resilience. *International Journal of Environmental Research & Public Health, 11*(8), 8475–8490.

Elliot, J., & Sullivan, L. (June 3, 2015). *How the red cross raised half a billion dollars for haiti and built six homes.* New York: Pro Publica. Retrieved from https://www.propublica.org/article/how-the-red-cross-raised-half-a-billion-dollars-for-haiti-and-built-6-homes.

* Elliott, J. R., Haney, T. J., & Sams-Abiodun, P. (2010). Limits to social capital: comparing network assistance in two New Orleans neighborhoods devastated by Hurricane Katrina. *The Sociological Quarterly, 51*(4), 624–648.

Elliott, J. R., & Pais, J. (2006). Race, class, and Hurricane Katrina: social differences in human responses to disaster. *Social Science Research, 35*(2), 295–321.

Emmett, E. (2013). *Hurricane Ike five years later: A more resilient community. Report, Harris County Homeland Security & Emergency Management, Houston, TX.*

Enarson, E. (2001). What women do: gendered labor in the Red River Valley flood. *Environmental Hazards, 3*(1), 1–18.

Enarson, E., & Morrow, B. H. (Eds.). (1998). *The gendered terrain of disaster: Through Women's Eyes.* Westport, CT: Praeger.

Erikson, K. (1976). *Everything in its path: Destruction of community in the buffalo creek flood.* New York: Simon and Schuster.

Esnar, A.-M., & Sapat, A. (2014). *Displaced by disaster: Recovery and resilience in a globalizing world.* New York and London: Routledge.

Faas, A.J. (in press). Reciprocity and Vernacular Statecraft: Changing Practices of Andean Cooperation in Post-Disaster Highland Ecuador. *Journal of Latin American and Caribbean Anthropology.*

* Faas, A. J. (2012). *Reciprocity and political power in disaster-induced resettlements in the Ecuadorian Andes.* Unpublished Ph.D. Dissertation. Tampa: University of South Florida.

Faas, A. J. (2015a). Disaster resettlement organizations and the culture of cooperative labor in the Ecuadorian Andes. In M. Companion (Ed.), *Disasters' impact on livelihood and cultural survival: Losses, opportunities, and mitigation* (pp. 51–62). Boca Raton, FL: CRC Press.

Faas, A. J. (2015b). Metaphors, metrics, and ethnographic heuristics of social networks in disaster. *Paper presented at the annual meeting of the society for applied anthropology, Pittsburgh, Pa, March 23–28.*

* Faas, A. J., Jones, E. C., Whiteford, L. M., Tobin, G. A., & Murphy, A. D. (2014). Gendered paths to formal and informal resources in post-disaster development in the Ecuadorian Andes. *Mountain Research & Development, 34*(3), 233–234.

* Faas, A. J., Jones, E. C., Tobin, G. A., Whiteford, L. M., & Murphy, A. D. (2015). Critical aspects of social networks in a resettlement setting. *Development in Practice, 25*(2), 221–233.

* Faas, A. J., Velez, A., FitzGerald, C., Nowell, B., & Steelman, T. (2016). *Patterns of preference and practice: Bridging actors in wildfire response networks in the American Northwest.* Disasters (In press).

Fafchamps, M., & Gubert, F. (2007). The formation of risk sharing networks. *Journal of Development Economics*, *83*(2), 326–350.

Faist, T. (2000). Transnationalization in international migration: implications for the study of citizenship and culture. *Ethnic and Racial Studies*, *23*(2), 189–222.

Farmer, S. M., Tierney, P., & Kung-McIntyre, K. (2003). Employee creativity in Taiwan: an application of role identity theory. *Academy of Management Journal*, *46*(5), 618–630.

Ferguson, J., & Gupta, A. (2002). Spatializing states: toward an ethnography of neoliberal governmentality. *American Ethnologist*, *29*(4), 981–1002.

Feuer, A. (November 9, 2012). *Where FEMA fell short, Occupy Sandy was there*. New York: The New York Times. Retrieved from http://www.nytimes.com/2012/11/11/nyregion/where-fema-fell-short-occupy-sandy-was-there.html.

Fischetti, M. (2011). Terms of property damage? Global warming could send more money up in smoke. *Scientific America*.

Flannigan, M., Cantin, A. S., de Groot, W. J., Wotton, M., Newbery, A., & Gowman, L. M. (2013). Global wildland fire season severity in the 21st century. *Forest Ecology & Management*, *294*, 54–61.

Fleming, C. J., McCartha, E. B., & Steelman, T. A. (2015). Conflict and collaboration in wildfire management: the role of mission alignment. *Public Administration Review*, *75*(3), 445–454.

Fogleman, C. W., & Parenton, V. J. (1959). Disaster and aftermath: selected aspects of individual and group behavior in critical situations. *Social Forces*, *38*(2), 129–135.

Foner, N., & Dreby, J. (2011). Relations between the generations in immigrant families. *Annual Review of Sociology*, *37*, 545–564.

Form, W. H., Loomis, C. P., Clifford, R. A., Moore, H. E., Nosow, S., Stone, G. P., et al. (1956). The persistence and emergence of social and cultural systems in disasters. *American Sociological Review*, *21*(2), 180–185.

Fortun, K. (2001). *Advocacy after Bhopal: Environmentalism, disaster, new global orders*. Chicago: University of Chicago Press.

Foster, A. D., & Rosenzweig, M. R. (2001). Imperfect commitment, altruism and the family: evidence from transfer behavior in low-income rural areas. *Review of Economics & Statistics*, *83*(3), 389–407.

Fothergill, A. (1996). Gender, risk and disaster. *International Journal of Mass Emergencies & Disasters*, *14*(1), 33–56.

Fothergill, A., Maestas, E. G., & Darlington, J. (1999). Race, ethnicity and disasters in the United States: a review of the literature. *Disasters*, *23*(2), 156–173.

Fothergill, A., & Peek, L. A. (2004). Poverty and disasters in the United States: a review of recent sociological findings. *Natural Hazards*, *32*(1), 89–110.

Fothergill, A., & Peek, L. (2015). *Children of Katrina*. Austin, TX: University of Texas Press.

Frank, K. A. (2009). Quasi-Ties: directing resources to members of a collective. *American Behavioral Scientist*, *52*(12), 1613–1645.

Frank, O., & Strauss, D. (1986). Markov graphs. *Journal of the American Statistical Association*, *81*(395), 832–842.

Fratkin, E., & Roth, E. A. (1990). Drought and economic differentiation among Ariaal pastoralists of Kenya. *Human Ecology*, *18*(4), 385–402.

Fullilove, M. T. (2005). *Root shock: How tearing up city neighborhoods hurts America and what we can do about it*. New York, NY: One World Press.

Furman, W., & Buhrmester, D. (1992). Age and sex differences in perceptions of networks of personal relationships. *Child Development*, *63*(1), 103–115.

Furstenberg, F. F., & Kaplan, S. B. (2004). Social capital and the family. In J. Scott, J. Treas, & M. Richards (Eds.), *The Blackwell companion to sociology of families* (pp. 218–232). Malden, MA: Wiley-Blackwell.

* Fussell, E. (2006). *Leaving new Orleans: Social stratification, networks, and hurricane evacuation. Understanding Katrina: Perspectives from the social sciences*. Social Science Research Council. Retrieved from http://understandingkatrina.ssrc.org/Fussell/.

* Ganapati, E. (2005). *Rising from the rubble: Disaster victims, social capital, and public policy.* Turkey: Case of Golcuk. Retrieved from ProQuest Digital Dissertations. (UMI No. 3219806).

Gargiulo, M., & Benassi, M. (2000). Trapped in your own net? Network cohesion, structural holes, and the adaptation of social capital. *Organization Science*, *11*(2), 183–196. http://dx.doi.org/10.1287/orsc.11.2.183.12514.

Garnett, J. D., & Moore, M. (2010). Enhancing disaster recovery: lessons from exemplary international disaster management practices. *Journal of Homeland Security & Emergency Management*, *7*(1), 1–20.

Garrison, M. E. B., & Sasser, D. D. (2009). Families and disasters: making meaning out of adversity. In K. E. Cherry (Ed.), *Lifespan perspectives on natural disasters: Coping with Katrina, Rita, and other storms* (pp. 113–130). New York, NY: Springer Science + Business Media.

Geertz, C. (1973). *The interpretation of cultures.* New York, NY: Basic Books.

Gelman, A., Su, Y.-S., Yajima, M., Hill, J., Pittau, M. G., Kerman, J., et al. (2009). *Arm: Data Analysis using regression and multilevel/hierarchical models (R package, version 9.01).*

GeoNet. (2011). *Canterbury quakes.* Retrieved from http://info.geonet.org.nz/display/home/Canterbury+Quakes.

Gertler, P., & Gruber, J. (2002). Insuring consumption against illness. *American Economic Review*, *92*(1), 51–76.

Gilligan, M. J., Pasquale, B. J., & Samii, C. S. (2011). *Civil war and social capital: Behavioral-game evidence from Nepal.* New York University (unpublished manuscript).

Girvan, M., & Newman, M. E. J. (2002). Community structure in social and biological networks. *Proceedings of the National Academy of Sciences*, *99*, 7821–7826.

Goldschmidt, W. (1969). *Kambuya's cattle: The legacy of an African Herdsman.* Berkeley: University of California Press.

Goldschmidt, W. (1972). The operations of a Sebei capitalist: a contribution to economic anthropology. *Ethnology*, *11*(3), 187–201.

Goldschmidt, W. (1990). *The human career: The Self in the Symbolic World.* Cambridge, MA: Blackwell.

Gooty, J., & Yammarino, F. J. (2011). Dyads in organizational research: conceptual issues and multilevel analyses. *Organizational Research Methods*, *14*(3), 456–483.

Gordon, J. S., Luloff, A., & Stedman, R. C. (2012). A multisite qualitative comparison of community wildfire risk perceptions. *Journal of Forestry*, *110*(2), 74–78.

Gouldner, A. W. (1960). The norm of reciprocity: a preliminary statement. *American Sociological Review*, *25*(2), 161–178.

Government of Fiji. (2013). *Post-disaster needs assessment: Tropical cyclone Evan, 17th December 2012.* Suva, Fiji: Applied Geoscience and Technology Division, Secretariat of the Pacific Community. Retrieved from http://www.gfdrr.org/sites/gfdrr.org/files/Fiji_Cyclone_Evan_2012.pdf.

Government of Samoa. (2013). *Samoa post-disaster needs assessment: Cyclone Evan, 2012.* Apia, Samoa: Government of Samoa. Retrieved from http://www.gfdrr.org/sites/gfdrr/files/SAMOA_PDNA_Cyclone_Evan_2012.pdf.

Graham, L. T. (2007). Permanently failing organizations? Small business recovery after September 11, 2001. *Economic Development Quarterly*, *21*(4), 299–314. http://dx.doi.org/10.1177/0891242407306355.

Granovetter, M. S. (1973). The strength of weak ties. *American Journal of Sociology*, *78*(6), 1360–1380.

Granovetter, M. (1982). The strength of weak ties: a network theory revisited. In P. Marsden, & N. Lin (Eds.), *Social structure and network analysis* (pp. 105–130). Beverly Hills: Sage Publications.

* Gray, C. L., & Mueller, V. (2012). Natural disasters and population mobility in Bangladesh. *Proceedings of the National Academy of Sciences*, *109*(16), 6000–6005.

Grimard, F. (1997). Household consumption smoothing through ethnic ties: evidence from Cote d'Ivoire. *Journal of Development Economics*, *53*(2), 391–422.

* Groen, J. A., & Polivka, A. E. (2010). Going home after Hurricane Katrina: determinants of return migration and changes in affected areas. *Demography*, *47*(4), 821–844 .

Guadagno, L. (2016). Human mobility in the Sendai framework for disaster risk reduction. *International Journal of Disaster Risk Science*, *7*(1), 30–40.

Guo, X., & Kapucu, N. (2015). Examining coordination in disaster response using simulation methods. *Journal of Homeland Security & Emergency Management*, *12*(4), 891–914.

Habermas, J. (1989). *The structural transformation of the public sphere: An Inquiry into a Category of Bourgeois Society*. Cambridge, MA: The MIT Press.

Haden, S. C., Scarpa, A., Jones, R. T., & Ollendick, T. H. (2007). Posttraumatic stress disorder symptoms and injury: the moderating role of perceived social support and coping for young adults. *Personality & Individual Differences, 42*(7), 1187–1198.

Haines, V. A., & Hurlbert, J. S. (1992). Network range and health. *Journal of Health & Social Behavior, 33*, 254–266.

* Haines, V. A., Hurlbert, J. S., & Beggs, J. J. (1996). Exploring the determinants of support provision: provider characteristics, personal networks, community contexts, and support following life events. *Journal of Health & Social Behavior, 37*(3), 252–264.

Halpern, D. (2005). *Social capital*. Cambridge: Polity Press.

Halstead, P., & O'Shea, J. (1989). Introduction: cultural responses to risk and uncertainty. In P. Halstead, & J. O'Shea (Eds.), *Bad year economics: Cultural responses to risk and uncertainty* (pp. 1–7). Cambridge, UK: Cambridge University Press.

* Halvorson, S. J., & Hamilton, J. P. (2007). Vulnerability and the erosion of seismic culture in mountainous Central Asia. *Mountain Research & Development, 27*(4), 322–330.

Hamra, J., Hossain, L., Owen, C., & Abbasi, A. (2012). Effects of networks on learning during emergency events. *Disaster Prevention and Management: An International Journal, 21*, 584–598. http://dx.doi.org/10.1108/09653561211278716. http://www.emeraldinsight.com/doi/abs/10.1108/09653561211278716.

Han, Q. (2002). Contemporary social transformation of Chinese society. *Modern Philosophy, 3*, 27–35.

Handcock, M. S., Hunter, D. R., Butts, C. T., Goodreau, S. M., & Morris, M. (2008). statnet: software tools for the representation, visualization, analysis and simulation of network data. *Journal of Statistical Software, 24*(1), 1–11. http://www.jstatsoft.org/v24/i01/.

Hanneke, S., Fu, W., & Xing, E. P. (2010). Discrete temporal models of social networks. *Electronic Journal of Statistics, 4*, 585–605.

Hanneke, S., & Xing, E. P. (2007). Statistical network analysis: models, issues, and new directions: ICML 2006 workshop on statistical network analysis, Pittsburgh, PA, June 29, 2006, revised selected papers. In E. M. Airoldi, D. M. Blei, S. E. Fienberg, A. Goldenberg, E. P. Xing & A. X. Zheng (Eds.), *Lecture notes in computer science: Vol. 4503*. TBA chapter discrete temporal models of social networks (pp. 115–125). Springer-Verlag.

Hanneman, R. A., & Riddle, M. (2005). *Introduction to social network methods*. Riverside, CA: University of California, Riverside. Retrieved from http://faculty.ucr.edu/~hanneman/.

Harris, J. L., & Doerfel, M. L. (2016). Resilient social networks and disaster-struck Communities. In J. P. Fyke, J. Faris, & P. Buzzanell (Eds.), *Cases in organizational and managerial communication: Stretching boundaries*. New York, NY: Routledge. in press.

Harris, J., & Todaro, M. (1970). Migration, unemployment and development: a two-sector analysis. *American Economic Review, 60*(1), 126–142.

Hasegawa, R. (2013). *Disaster evacuation from Japan's 2011 Tsunami disaster and the Fukushima nuclear accident*. Study no. 05/13. IDDRI/Sciences Po.

Hatfield, E., Walster, G. W., & Berscheid, E. (1978). *Equity: Theory and research*. Boston: Allyn & Bacon.

* Hawkins, R. L., & Maurer, K. (2010). Bonding, bridging and linking: how social capital operated in New Orleans following Hurricane Katrina. *British Journal of Social Work, 40*(6), 1777–1793. http://dx.doi.org/10.1093/bjsw/bcp087.

Helleringer, S., & Kohler, H. (2005). Social networks, perceptions of risk, and changing attitudes towards HIV/AIDS: new evidence from a longitudinal study using fixed-effects analysis. *Population Studies, 59*(3), 265–282.

Helslott, I., & Ruitenberg, A. (2004). Citizen response to disasters: a survey of literature and some practical implications. *Journal of Contigencies & Crisis Management, 12*(3), 98–111.

Hiltz, S. R., Kushma, J., & Plotnick, L. (2014). Use of social media by us public sector emergency managers: barriers and wish lists. *Proceedings of ISCRAM*.

Hoffman, S. (1999). The best of times, the worst of times: toward a model of cultural response to disaster. In A. Oliver-Smith, & S. Hoffman (Eds.), *The angry earth: Disaster in anthropological perspective* (pp. 134–155). New York: Routledge.

Horne, J. F., & Orr, J. E. (1998). Assessing behaviors that create resilient organizations. *Employment Relations Today, Winter*, 29–40.

Hox, J. (2010). *Multilevel Analysis: Techniques and Applications*. New York: Routledge.

Huang, S.-K., Lindell, M. K., Prater, C. S., Wu, H.-C., & Siebeneck, L. K. (2012). Household evacuation decision making in response to Hurricane Ike. *Natural Hazards Review, 13*(4), 283–296.

Hughes, A. L., St Denis, L. A., Palen, L., & Anderson, K. M. (2014). Online public communications by police & fire services during the 2012 Hurricane Sandy. In *CHI*. Toronto, Ontario, Canada: ACM. http://dl.acm.org/citation.cfm?id=2557227.

* Hu, Q., & Kapucu, N. (2014). Information communication technology utilization for effective emergency management networks. *Public Management Review*. http://dx.doi.org/10.1080/14719037.2014.969762 (first published online).

Humphrey, M., & Valverde, E. (2007). Human rights, victimhood and impunity: an anthropology of democracy in Argentina. *Social Analysis, 51*(1), 179–219.

Hunter, D. R., Handcock, M. S., Butts, C. T., Goodreau, S. M., & Morris, M. (2008). Ergm: a package to Fit, simulate and diagnose exponential-family models for networks. *Journal of Statistical Software, 24*.

* Hurlbert, J. S., Beggs, J. J., & Haines, V. A. (2001). Social networks and social capital in extreme environments. In N. Lin, K. Cook, & R. S. Burt (Eds.), *Social capital: Theory and Research* (pp. 209–232). New York: Aldine de Gruyter.

Hurlbert, J. S., Beggs, J. J., & Haines, V. (2005). *Bridges over troubled waters: What are the Optimal Networks for Katrina's Victims?* New York: Social Science Research Council. http://understandingkatrina.ssrc.org.

* Hurlbert, J. S., Haines, V. A., & Beggs, J. J. (2000). Core networks and tie activation: what kinds of routine networks allocate resources in nonroutine situations? *American Sociological Review, 65*(4), 598–618.

Hutter, G. (2011). Organizing social resilience in the context of natural hazards: a research note. *Natural Hazards, 67*(1), 47–60. http://dx.doi.org/10.1007/s11069-010-9705-4.

Ibañez, G. E., Buck, C., Khatchikian, N., & Norris, F. H. (2004). Qualitative analysis of coping strategies among Mexican disaster survivors. *Anxiety, Stress, & Coping, 17*(1), 69–85.

Inkpen, A. C., & Tsang, E. W. K. (2005). Social capital, networks, and knowledge transfer. *Academy of Management Review, 30*(1), 146–165.

Internal Displacement Monitoring Centre. (2015). *Global estimates 2015: People displaced by disasters*. Available online at http://www.internal-displacement.org/publications/2015/global-estimates-2015-people-displaced-by-disasters.

International Federation of Red Cross and Red Crescent Societies. (2014). *World disasters report 2014: Focus on culture and risk*. Geneva: International Federation of Red Cross and Red Crescent Societies.

IPCC (Intergovernmental Panel on Climate Change). (2014). Climate change 2014: impacts, adaptation, and vulnerability. Part A: global and sectoral aspects. In C. B. Field, V. R. Barros, D. J. Dokken, K. J. Mach, M. D. Mastrandrea, T. E. Bilir, et al. (Eds.), *Contribution of working group II to the fifth assessment report of the intergovernmental panel on climate change*. Cambridge, UK and New York, NY: Cambridge University Press.

Irwin, R. L. (1989). The incident command system (ICS). In E. Auf Der Heide (Ed.), *Disaster response: Principles of Preparation and Coordination* (pp. 133–163). St. Louis, MO: C.V. Mosby.

Islam, R., & Walkerden, G. (2014). How bonding and bridging networks contribute to disaster resilience and recovery on the Bangladeshi coast. *International Journal of Disaster Risk Reduction, 10*(A), 281–291.

Jackson, M. O. (2008). *Social and economic networks*. Princeton: Princeton University Press.

Jackson, M. O. (2010). An overview of social networks and economic applications. In J. Benhabib, M. Jackson, & A. Bisin (Eds.), *Handbook of social economics*. Amsterdam: Elsevier.

Jadacki, M. (2011). *Effectiveness and costs of FEMA's disaster housing assistance program* Report OIG-11–102. Washington, DC: Department of Homeland Security.

* Johnson, B. R. (1999). Social networks and exchange. In M. A. Little, & P. W. Leslie (Eds.), *Turkana herders of the dry savanna: Ecology and Biobehavioral Response of Nomads to an Uncertain Environment* (pp. 89–106). Oxford. UK: Oxford University Press.

Johnson, N., & Elliott, D. (2011). Using social capital to organise for success? A case study of public-private interface in the Uk Highways Agency. *Policy & Society, 30*(2011), 101–113.

Johnson, N., Elliott, D., & Drake, P. (2013). Exploring the role of social capital in facilitating supply chain resilience. *Supply Chain Management: An International Journal, 18*(3), 324–336.

* Jones, E. C., Faas, A. J., Murphy, A. D., Tobin, G. A., Whiteford, L. M., & McCarty, C. (2013). Cross-cultural and site-based influences on demographic, well-being, and social network predictors of risk perception in hazard and disaster settings in Ecuador and Mexico. *Human Nature, 24*(1), 5–32.

* Jones, E. C., Gupta, S., Murphy, A. D., & Norris, F. H. (2011). Inequality, socioeconomic status, and social support in post-disaster mental health in Mexico. *Human Organization, 70*(1), 33–43.

* Jones, E. C., & Murphy, A. D. (2015). Social organization of suffering and justice-seeking in a tragic day care fire disaster. In R. Anderson (Ed.), *World suffering and quality of life* (pp. 281–292). New York: Springer.

* Jones, E. C., Murphy, A. D., Faas, A. J., Tobin, G. A., McCarty, C., & Whiteford, L. M. (2015). Postdisaster reciprocity and the development of inequality in personal networks. *Economic Anthropology, 2*, 385–404.

* Jones, E. C., Tobin, G. A., McCarty, C., Faas, A. J., Yepes, H., Whiteford, L. M., et al. (2014). Articulation of personal network structure with gendered well-being in disaster and relocation settings. In L. Roeder (Ed.), *Issues of gender and sexual orientation in humanitarian emergencies* (pp. 50–119). New York: Springer.

* Jordan, A. E. (2010). Collaborative relationships resulting from the urban area security initiative. *Journal of Homeland Security & Emergency Management, 7*(1), 38.

* Juma, R. (2009). *Turkana livelihood strategies and adaptation to drought in Kenya.* Unpublished Ph.D. dissertation. School of Geography: Environment and Earth Sciences, Victoria University of Wellington.

Kachali, H., Stevenson, J. R., Whitman, Z., Seville, E., Vargo, J., & Wilson, T. (2012). Organisational resilience and recovery for Canterbury organisations after the 4 September 2010 earthquake. *Australasian Journal of Disaster & Trauma Studies, 1*, 11–19.

Kaelin, W. (2015). The Nansen initiative. In *Discussion paper on the relationship between climate change and human mobility* Retrieved from http://www.thecvf.org/wp-content/uploads/2015/05/migration.pdf.

Kage, R. (2011). *Civic engagement in postwar Japan: The Revival of a Defeated Society.* New York: Cambridge University Press.

Kalish, Y., & Robins, G. (2006). Psychological predispositions and network structure: the relationship between individual predispositions, structural holes and network closure. *Social Networks, 28*(1), 56–84.

Kana'Iaupuni, S. M., Donato, K. M., Thompson-Colon, T., & Stainback, M. (2005). Counting on kin: social networks, social support, and child health status. *Social Forces, 83*(3), 1137–1164.

* Kaniasty, K. (2012). Predicting social psychological well-being following trauma: the role of post-disaster social support. *Psychological trauma: Theory, research, practice, & policy, 4*(1), 22–33.

* Kaniasty, K., & Norris, F. H. (1995). In search of altruistic community: patterns of social support mobilization following Hurricane Hugo. *American Journal of Community Psychology, 23*(4), 447–477.

Kaniasty, K., & Norris, F. H. (2000). Help-seeking comfort and receiving social support: the role of ethnicity and context of need. *American Journal of Community Psychology, 28*(4), 545–581.

Kaniasty, K., & Norris, F. H. (2008). Longitudinal linkages between perceived social support and posttraumatic stress symptoms: sequential roles of social causation and social selection. *Journal of Traumatic Stress, 21*(3), 274–281.

Kapucu, N. (2004). Public non-profit partnerships in emergency and crisis management: September 11, 2001 crisis coordination. *PA (Public Administration) Times, 27*(5) (May).

* Kapucu, N. (2005). Interorganizational coordination in dynamic context: networks in emergency response management. *Connections, 26*(2), 33–48.

* Kapucu, N. (2006a). Interagency communication networks during emergencies boundary spanners in multiagency coordination. *The American Review of Public Administration, 36*(2), 207–225.

* Kapucu, N. (2006b). Examining national response plan in response to a catastrophic disaster: hurricane Katrina in 2005. *International Journal of Mass Emergencies & Disasters, 24*(2), 271–299.

* Kapucu, N. (2009). Interorganizational coordination in complex environments of disasters: the evolution of intergovernmental disaster response system. *Journal of Homeland Security and Emergency Management, 6*(1) Article 47.

* Kapucu, N., Arslan, T., & Collins, M. L. (2010). Examining intergovernmental and Interorganizational response to catastrophic disasters: toward a network-centered approach. *Administration & Society, 42*(2), 222–247.

Kapucu, N., Arslan, T., & Demiroz, F. (2010). Collaborative emergency management and national emergency management network. *Disaster Prevention & Management, 19*(4), 452–468.

* Kapucu, N., Augustin, M. E., & Garayev, V. (2009). Interstate partnerships in emergency management: emergency Management Assistance Compact (EMAC) in response to catastrophic disasters. *Public Administration Review, 69*(2), 297–313.

* Kapucu, N., & Demiroz, F. (2011). Measuring performance for collaborative public management using network analysis methods and tools. *Public Performance & Management Review, 34*(4), 549–579.

* Kapucu, N., & Garayev, V. (2014). Structure and network performance: horizontal and vertical networks in emergency management. *Administration and Society*, 1–31. http://dx.doi.org/10.1177/0095399714541270 (first published online).

* Kapucu, N., & Hu, Q. (2014). Understanding multiplexity of collaborative emergency management networks. *American Review of Public Administration*, 1–19. http://dx.doi.org/10.1177/0275074014555645 (first published online).

Kapucu, N., Hu, Q., & Khosa, S. (2014). The state of network research in public administration. *Administration & Society*, 1–34. http://dx.doi.org/10.1177/0095399714555752.

Kass, R. E., & Wasserman, L. (1995). A reference Bayesian test for nested hypotheses and its relationship to the Schwarz criterion. *Journal of the American Statistical Association, 90*, 928–934.

Kavanaugh, A., Fox, E. A., Sheetz, S., Yang, S., Li, L. T., Whalen, T., et al. (2011). Social media use by government: from the routine to the critical. In *Proceedings of the 12th annual international conference on Digital Government Research* (pp. 121–130). College Park, MD: ACM.

* Kawachi, I., & Berkman, L. F. (2001). Social ties and mental health. *Journal of Urban Health: Bulletin of the New York Academy of Medicine, 78*(3), 458–467.

* Kendra, J. M., & Wachtendorf, T. (2003). Elements of resilience after the world trade center disaster: reconstituting New York City's emergency operations centre. *Disasters, 27*(1), 37–53. http://dx.doi.org/10.1111/1467-7717.00218.

Kenis, P., & Provan, K. G. (2009). Towards an exogenous theory of public network performance. *Public Administration, 87*(3), 440–456.

Kessler, R. C., Berglund, P., Demler, O., et al. (2005). Lifetime prevalence and age- of-onset distributions of DSM-IV disorders in the national Comorbidity survey Replication. *Archives of General Psychiatry, 62*, 592–602.

Kessler, R. C., & McLeod, J. (1984). Social support and psychological distress in community surveys. In S. Cohen, & S. L. Syme (Eds.), *Social support and health* (pp. 19–40). New York: Academic Press.

Kessler, R. C., Sonnega, A., Bromet, E., et al. (1995). Posttraumatic stress disorder in the national Comorbidity survey. *Archives of General Psychiatry, 52*, 1048–1060.

Kessler, R. C., & Ustun, T. B. (2008). *The WHO world mental health Surveys: Global perspectives on the epidemiology of mental disorders*. New York: Cambridge University Press.

Khosa, S. (2013). *Examining multi-level and inter-organizational collaborative response to disasters: The case of Pakistan floods in 2010*. Unpublished Dissertation. Orlando, FL: Universtiy of Central Florida.

* Killian, L. M. (1952). The significance of multiple-group membership in disaster. *American Journal of Sociology*, *57*(4), 309–314.

Klinenberg, E. (2002). *Heat wave: A social autopsy of disaster in Chicago*. Chicago: University of Chicago Press.

Knobben, J., & Oerlemans, L. (2006). Proximity and inter-organizational collaboration: a literature review. *International Journal of Management Reviews*, *8*(2), 71–89.

Knoke, D., & Yang, S. (2008). (2nd ed.). *Social network analysis* (Vol. 2). Thousand Oaks, CA: Sage.

Kochar, A. (1995). Explaining household vulnerability to idiosyncratic income shocks. *American Economic Review*, *85*(2), 159–164.

Koehly, L. M., Ashida, S., Goergen, A. F., Skapinsky, K. F., & Hadley, D. W. (2011). Willingness of Mexican-American adults to share family health history with healthcare providers. *American Journal of Preventive Medicine*, *40*(6), 633–636.

Koliba, C. (2014). Governance network performance: a complex adaptive systems approach. In R. Keast, M. Mandell, & R. Agranoff (Eds.), *Network theory in the public sector: Building New Theoretical Frameworks* (pp. 184–192). New York: Routledge.

Kolmannskog, V. (2008). *Climate of displacement, climate for protection? Migration scenarios, DIIS brief*. Copenhagen: Danish Institute for International Studies.

Kordon, D. R., Edelman, L. I., & Lagos, N. B. (1988). *Psychological effects of political repression*. Buenos Aires: Sudamericana/Planeta.

Kramer, R. M., & Tyler, T. R. (1996). *Trust in organizations: Frontiers of theory and research*. Thousand Oaks: SAGE Publications, Inc.

Kroll-Smith, S. (In press). *Recovering inequality: A tale of two American disasters*. Austin, TX: University of Texas Press.

Kryvasheyeu, Y., Chen, H., Obradovich, N., Moro, E., Van Hentenryck, P., Fowler, J., et al. (2016). Rapid assessment of disaster damage using social media activity. *Science Advances*, *2*, e1500779.

Kunreuther, H. (1967). The peculiar economics of disaster. *Papers on Non-Market Decision Making*, *3*(1), 67–83.

Kurekova, L. (2011). Theories of migration: conceptual review and empirical testing the context of the EU East-West flows. In *Paper prepared for interdisciplinary conference on Migration, Economic change, social Challenge, April 6–9*. University College London. Retrieved from https://cream.conference-services.net/resources/952/2371/pdf/MECSC2011_0139_paper.pdf.

Kwak, H., Lee, C., Park, H., & Moon, S. (2010). What is twitter, a social network or a news media? In *Proceedings of the 19th international conference on World Wide Web WWW'10* (pp. 591–600). New York, NY: ACM. http://dx.doi.org/10.1145/1772690.1772751. http://doi.acm.org/10.1145/1772690.1772751.

La Ferrara, E. (2003). Kin groups and reciprocity: a model of credit transactions in Ghana. *American Economic Review*, *93*(5), 1730–1751.

* Landry, C., Bin, O., Hindsley, P., Whitehead, J., & Wilson, K. (2007). Going home: evacuation-migration decisions of hurricane Katrina survivors. *Southern Economic Journal*, *74*(2), 326–343.

Lapierre, D., & Moro, J. (2002). *Five past midnight in Bhopal: The Epic Story of the World's Deadliest Industrial Disaster*. New York, NY: Warner Books.

Laska, S. (2012). Dimensions of resiliency: essential, exceptional, and scale. *International Journal of Critical Infrastructure*, *6*(3), 246–276.

Laska, S., Peterson, K., Alcina, M. E., West, J., Volion, B. Tranchina, A., & Krajeski, R. (2010). *Enhancing Gulf of Mexico coastal communities' resiliency through participatory community engagement* CHART Publications, Paper 21 . Retrieved from http://scholarworks.uno.edu/chart_pubs/21.

Laska, S., Peterson, K., Rodrigue, C., Cosse, T., Philippe, R., Burchett, O., et al. (2015). 'Layering' of natural and human caused disasters in the context of anticipated climate change disasters: the coastal Louisiana experience. In M. Companion (Ed.), *Disasters' impact on livelihood and cultural survival: Losses, Opportunities, and Mitigation* (pp. 225–238). Boca Raton, FL: CRC Press.

Laumann, E. O., & Knoke, D. (1987). *The organizational state: Social Choice in National Policy Domains.* Madison, WI: University of Wisconsin Press.

Lawrence, A. R., & Schigelone, A. R. S. (2002). Reciprocity beyond dyadic relationships aging-related communal coping. *Research On Aging*, 24(6), 684–704.

* Lee, T. R. (1980). The resilience of social networks to changes in mobility and propinquity. *Social Networks*, 2(4), 423–435.

Lee, Y., & Sugiura, H. (2014). Impact of the great East Japan earthquake on intentions to relocate. *Journal of Integrated Disaster Risk Management*, 4(2), 156–165.

Lee, A. V., Vargo, J., & Seville, E. (2013). Developing a tool to measure and compare organizations' resilience. *Natural Hazards Review*, 14(February), 29–41. http://dx.doi.org/10.1061/(ASCE)NH.1527-6996.0000075.

Levi-Strauss, C. (1969). *The raw and the cooked.* Chicago: University of Chicago Press.

Liang, J., Krause, N. M., & Bennett, J. M. (2001). Social exchange and well-being: is giving better than receiving? *Psychology & Aging*, 16(3), 511.

Ligon, E., Thomas, J. P., & Worrall, T. (2002). Informal insurance arrangements with limited commitment: theory and evidence from village economies. *Review of Economic Studies*, 69(1), 209–244.

Lin, N. (2000). Inequality in social capital. *Contemporary Sociology*, 29(6), 785–795.

Lindsay, B. R. (2011). *Social media and disasters: Current Uses, Future Options, and Policy Considerations.*

Lin, N., & Dumin, M. (1986). Access to occupations through social ties. *Social Networks*, 8(4), 365–385.

Linnenluecke, M., & Griffiths, A. (2010). Beyond adaptation: resilience for business in light of climate change and weather extremes. *Business & Society*, 49(3), 477–511. http://dx.doi.org/10.1177/0007650310368814.

Lin, N., Woelfel, M. W., & Light, S. C. (1985). The buffering effect of social support subsequent to an important life event. *Journal of Health & Social Behavior*, 26(3), 247–263.

Little, P. D., McPeak, J., Barrett, C. B., & Kristjanson, P. (2008). Challenging orthodoxies: understanding poverty in pastoral areas of East Africa. *Development & Change*, 39(4), 587–611.

Litwak, E., & Szelenyi, I. (1969). Primary group structures and their functions: kin, neighbors, and friends. *American Sociological Review*, 34(4), 465–481.

Longino, C. F., Jr., & Lipman, A. (1981). Married and spouseless men and women in planned retirement communities: support network differentials. *Journal of Marriage & the Family*, 43(1), 169–177.

Long, D. A., & Perkins, D. D. (2007). Community social and place predictors of sense of community: a multilevel and longitudinal analysis. *Journal of Community Psychology*, 35(5), 563–581.

Lowenstein, A., & Daatland, S. O. (2006). Filial norms and family support in a comparative cross-national context: evidence from the OASIS study. *Aging & Society*, 26(2), 203–223.

Lu, Y. (2009). *Non-governmental Organizations in China: The Rise of Dependent Autonomy.* New York, NY: Routledge.

* Lu, J. (2013). *The Wenchuan Earthquake recovery: Civil society, institutions, and planning.* Retrieved from ProQuest Digital Dissertations. (UMI No. 3609950).

* Lu, X., Bengtsson, L., & Holme, P. (2012). Predictability of population displacement after the 2010 Haiti earthquake. *Proceedings of the National Academy of Sciences*, 109(29), 11576–11581.

Lu, X., Wrathall, D. J., Sundsøy, P. R., Nadiruzzaman, M., Wetter, E., Iqbale, A., et al. (2016). Unveiling hidden migration and mobility patterns in climate stressed regions: a longitudinal study of six million anonymous mobile phone users in Bangladesh. *Global Environmental Change*, 38, 1–7.

Lybbert, T., Barrett, C., Desta, S., & Coppock, L. (2005). Stochastic wealth dynamics and risk management among a poor population. *Economic Journal*, 114, 750–777.

Lyons, R. F., Mickelson, K. D., Sullivan, M. J., & Coyne, J. C. (1998). Coping as a communal process. *Journal of Social & Personal Relationships*, 15(5), 579–605.

Ma, Q. (2006). *Non-governmental organizations in contemporary China: Paving the Way to Civil Society?* New York, NY: Routledge.

Macmillan, R. (2001). Violence and the life course: the consequences of victimization for personal and social development. *Annual Review of Sociology*, 27, 1–22.

* Magsino, S. L. (Ed.). (2009). *Applications of social network analysis for Building community disaster Resilience: Workshop Summary*. Washington, DC: National Academies Press.

Majchrzak, A., Jarvenpaa, S. L., & Hollingshead, A. B. (2007). Coordinating expertise among emergent groups responding to disasters. *Organization Science, 18*(1), 147–161.

Maldonado, J. K. (2014a). *Facing the rising tide: Co-occurring disasters, displacement and adaptation in coastal Louisiana's tribal communities* Dissertation. Department of Anthropology: American University.

Maldonado, J. K. (2014b). Translating environmental change: when local experiences and outside perspectives collide. *Anthropology News, 55*(7–8), 7–8.

Maldonado, J. K., Shearer, C., Bronen, R., Peterson, K., & Lazrus, H. (2013). The impact of climate change on tribal communities in the US: displacement, relocation, and human rights. *Climatic Change, 120*(3), 601–614.

Maldonado, J. K., Lazrus, H., Gough, B., Bennett, S. K., Chief, K., Dhillon, C., et al. (2016). The story of rising voices: facilitating collaboration between indigenous and western ways of knowing. In M. Companion, & M. Chaiken (Eds.), *Responses to disasters and climate change: Understanding Vulnerability and Fostering Resilience*. Boca Raton, FL: CRC Press.

Maldonado, J., Taylor, B., & Hufford, M. (2016). The livelihoods knowledge exchange network: grow where you are. *Practicing Anthropology, 38*(3) Summer 2016.

Mallak, L. (1998). Putting organizational resilience to work. *Industrial Management, 40*(6), 8.

Mandell, M. P., & Keast, R. (2008). Evaluating the effectiveness of Interorganizational relations through networks: developing a framework for revised performance measures. *Public Management Review, 10*(6), 715–731.

Manni, F., Guérard, E., & Heyer, E. (2004). Geographic patterns of (genetic, morphologic, linguistic) variation: how barriers can be detected by using Monmonier's algorithm. *Human Biology, 76*(2), 173–190.

Marcias Medrano, J. M. (2005). *La disputa por el riesgo en el volcán Popocatépetl*. Mexico, DF: CIESAS/La Casa Chata.

Marcum, C. S., Bevc, C. A., & Butts, C. T. (2012). Mechanisms of control in emergent Interorganizational networks. *Policy Studies Journal, 40*, 516–546.

Marcum, C. S., & Koehly, L. M. (2015). Inter-generational contact from a network perspective. *Advances in Life Course Research, 24*(2), 10–20.

Marino, E., & Lazrus, H. (2015). Migration or forced displacement?: The complex choices of climate change and disaster migrants in Shishmaref, Alaska and Nanumea, Tuvalu. *Human Organization, 74*(4), 341–350.

Marsden, P. V. (1987). Core discussion networks of Americans. *American Sociological Review, 52*(1), 122–131.

Marsden, P. V. (1990). Network data and measurement. *Annual Review of Sociology, 16*, 435–463.

Mathbor, G. M. (2007). Enhancement of community preparedness for natural disasters the role of social work in building social capital for sustainable disaster relief and management. *International Social Work, 50*, 357–369.

Mauss, M. (1925). *The gift: The form and reason for exchange in archaic societies (W. D. Halls, Trans. 1990 ed.)*. New York: W.W. Norton.

Mayton, J., & Kazem, H. (September 17, 2015). *California residents condemn red cross for slow response to wildfires*. New York: The Guardian. Retrieved from http://www.theguardian.com/us-news/2015/sep/17/california-residents-condemn-red-cross-wildfires?CMP=share_btn_fb.

McCarty, C. (2002a). Measuring structure in personal networks. *Journal of Social Structure, 3*, 1. Retrieved from http://www.cmu.edu/joss/content/articles/volume3/McCarty.html.

McCarty, C. (2002b). Structure in personal networks. *Journal of Social Structure, 3*(1), 20.

McCarty, C., Killworth, P. D., & Rennell, J. (2007). Impact of methods for reducing respondent burden on personal network structural measures. *Social Networks, 29*(2), 300–315.

McCormick, S. (2012). After the cap: Risk assessment, citizen science, and disaster recovery. *Ecology and Society, 17*(4), 31. http://dx.doi.org/10.5751/ES-05263-170431.

McGraw, S. (2008). Cumulative impacts and recovery considerations. In R. Perry (Ed.), *Hurricane ike impact report* (pp. 50–57). Austin, TX: Division of Emergency Management: Homeland Security. fema.gov.

McGuire, M., & Silvia, C. (2010). The effect of problem severity, managerial and organizational capacity, and agency structure on intergovernmental collaboration: evidence from local emergency management. *Public Administration Review, 70*(2), 279–288.

* McPeak, J. G. (2006). Confronting the risk of asset loss: what role do livestock transfers in northern Kenya play? *Journal of Developmental Economics, 81*, 415–437.

McPeak, J. G., Little, P. D., & Doss, C. R. (2011). *Risk and social change in an African Rural Economy: Livelihoods in Pastoralist Communities.* London: Routledge.

McPherson, J. M., & Smith-Lovin, L. (1987). Homophily in voluntary organizations: status distance and the composition of face-to-face groups. *American Sociological Review, 53*(3), 370–379.

McPherson, M., Smith-Lovin, L., & Brashears, M. E. (2001). Birds of a feather: homophily in social networks. *Annual Review of Sociology, 2*(1), 415–444.

McPherson, M., Smith-Lovin, L., & Brashears, M. E. (2006). Social isolation in America: changes in core discussion networks over two decades. *American Sociological Review, 71*, 353–375.

McPherson, M., Smith-Lovin, L., & Cook, J. (2001). Birds of a feather: homophily in social networks. *Annual Review of Sociology, 27*, 415–444.

Medical Decision Logic, Inc. (2014). *VisuaLyzer 2.2 user Manual.* Baltimore [MD]: Medical Decision Logic, Inc.

Mehrotra, S., Butts, C. T., Kalashnikov, D. V., Venkatasubramanian, N., Altintas, K., Hariharan, R., et al. (2004). Camas: a citizen awareness system for crisis mitigation. In *Proceedings of the ACM SIGMOD international conference on Management of Data (SIGMOD'04).*

Melillo, J. M., Richmond, T. C., & Yohe, G. W. (Eds.). (2014). *Climate change impacts in the United States: The third national climate assessment.* Washington, DC: U.S. Global Change Research Program. Retrieved from h ttp://nca2014.globalchange.gov/.

* Messias, D. K. H., Barrington, C., & Lacy, E. (2012). Latino social network dynamics and the Hurricane Katrina disaster. *Disasters, 36*(1), 101–121.

Miami-Dade County. (2015). *Local mitigation strategy.* Accessed August 25, 2015 via http://www.miamidade.gov/fire/library/OEM/local-mitigation-strategy-part-4-appendices.pdf.

Mintzberg, H. (2015). Time for the plural sector. *Stanford Social Innovation Review, Summer*, 6–10.

Mische, A. (2003). Cross-talk in movements: Reconceiving the culture-network link. In M. Diani, & D. McAdam (Eds.), *Social relations and networks: Relational approaches to collective action* (pp. 258–280). Oxford, UK: Oxford University Press.

Molm, L. D. (1994). Dependence and risk: Transforming the structure of social exchange. *Social Psychology Quarterly, 57*(3), 163–176.

Monge, P. R., & Contractor, N. S. (2001). Emergence of communication. In F. M. Jablin, & L. L. Putnam (Eds.), *The new handbook of organizational communication* (pp. 48–78). Thousand Oaks: SAGE Publications, Inc. http://dx.doi.org/10.4135/9781412986243.

Monge, P. R., & Contractor, N. S. (2003). *Theories of communication networks.* New York, NY: Oxford University Press.

Monge, P. R., Heiss, B. M., & Margolin, D. B. (2008). Communication network evolution in organizational communities. *Communication Theory, 18*(4), 449–477. http://doi.org/10.1111/j.1468-2885.2008.00330.

Moody, J., McFarland, D., & Bender-deMoll, S. (2005). Dynamic network visualization 1. *American Journal of Sociology, 110*(4), 1206–1241.

Moore, S., Eng, E., & Daniel, M. (2003). International NGOs and the role of network centrality in humanitarian aid operations: a case study of coordination during the 2000 Mozambique floods. *Disasters, 27*(4), 305–318.

Moore, M.-L., & Westley, F. (2011). Surmountable chasms: networks and social innovation for resilient systems. Art. 5 *Ecology & Society, 16*(1) Article.

Morduch, J. (2005). Consumption smoothing across space: testing theories of risk-sharing in the ICRISAT study region of South India. In S. Dercon (Ed.), *Insurance against poverty* (pp. 38–58). Oxford and New York: Oxford University Press.

Moritz, M. (2003). *Commoditization and the pursuit of piety: The transformation of an African pastoral system. (Dissertation).* Los Angeles: University of California at Los Angeles.

Moritz, M. (2012). Individualization of livestock ownership in Fulbe family herds: the effects of pastoral Intensification and Islamic Renewal. In A. Khazanov, & F. Schlee (Eds.), *Who owns the stock? Collective and multiple forms of property in animals* (pp. 193–214). Oxford, UK: Berghahn.

Moritz, M. (2013). Livestock transfers, risk management, and human careers in a West African pastoral system. *Human Ecology, 41*(2), 205–219. http://dx.doi.org/10.1007/s10745-012-9546-8.

Moritz, M. A., Batllori, E., Bradstock, R. A., Gill, A. M., Handmer, J., Hessburg, P. F., et al. (2014). Learning to coexist with wildfire. *Nature, 515*(7525), 58–66.

Moritz, M., Giblin, J., Ciccone, M., Davis, A., Fuhrman, J., Kimiaie, M., et al. (2011). Social risk-management strategies in pastoral systems: a qualitative comparative analysis. *Cross-Cultural Research, 45*(3), 286–317.

Moritz, M., Ritchey, K. K., & Kari, S. (2011). The social context of herding contracts in the Far North Region of Cameroon. *Journal of Modern African Studies, 49*(2), 263–285. http://dx.doi.org/10.1017/S0022278X11000048.

Morss, R. E., & Hayden, M. H. (2010). Storm surge and "certain death": interviews with Texas coastal residents following Hurricane Ike. *Weather, Climate, & Society, 2*(3), 174–189.

Moscovici, S., & Zavalloni, M. (1969). The group as a polarizer of attitudes. *Journal of Personality & Social Psychology, 12*(2), 125.

Mount, M. (2008). *Texans evacuate coast ahead of Ike. Online.* CNN.

Moynihan, D. P. (2008). Learning under uncertainty: networks in crisis management. *Public Administration Review, 68*(2), 350–361.

* Moynihan, D. P. (2009). The network governance of crisis response: case studies of incident command systems. *Journal of Public Administration Research & Theory, 19*(4), 895–915.

Mukherjee, S. (2010). *Surviving Bhopal: Dancing bodies, written texts, and oral testimonials of women in the wake of an industrial disaster.* New York: Palgrave Macmillan.

Mutch, R. W. (1970). Wildland fires and ecosystems—a hypothesis. *Ecology, 51*(6), 1046–1051.

Nahapiet, J., & Ghoshal, S. (1998). Social capital, intellectual capital, and the organizational advantage. *The Academy of Management Review, 23*(2), 242. http://dx.doi.org/10.2307/259373.

Nakagawa, Y., & Shaw, R. (2004). Social capital: a missing link to disaster recovery. *International Journal of Mass Emergencies & Disasters, 22*(1), 5–34.

Neal, D. M. (1997). Reconsidering the phases of disaster. *Journal of Mass Emergencies & Disasters, 15*(2), 239–264.

Neely, A., Richards, H., Mills, J., Platts, K., & Bourne, M. (1997). Designing performance measures: a structured approach. *International Journal of Operations & Production Management, 17*(11), 1131–1152.

Newman, L., & Dale, A. (2005). Network structure, diversity, and proactive resilience building: a response to Tompkins and Adger. *Ecology & Society, 10*(1), r2. (online) http://\www.ecologyandsodety.org/vol10/iss l/resp2/.

Ng, M., Diaz, R., & Behr, J. (2015). Departure time choice behavior for hurricane evacuation planning: the case of the understudied medically fragile population. *Transportation Research Part E: Logistics & Transportation Review, 77,* 215–226.

Nigg, J. M. (1995). *Disaster recovery as a social process* Technical Report 219. Disaster Research Center.

Noh, S., & Avison, W. R. (1996). Asian immigrants and the stress process: a study of Koreans in Canada. *Journal of Health & Social Behavior, 37*(2), 192–206.

Noji, E. K. (1997). *The public health consequences of disasters.* Oxford, UK: Oxford University Press.

* Nolte, M. I., & Boenigk, S. (2013). A study of ad hoc network performance in disaster response. *Nonprofit & Voluntary Sector Quarterly, 42*(1), 148–173.

Norris, F. H., Baker, C. K., Murphy, A. D., & Kaniasty, K. (2005). Social support mobilization and deterioration after Mexico's 1999 flood: effects of context, gender and time. *American Journal of Community Psychology, 36*(1–2), 15–28.

Norris, F. H., Friedman, M. J., Watson, P. J., Byrne, C. M., Diaz, E., & Kaniasty, K. (2002). 60,000 disaster victims speak: Part I. An empirical review of the empirical literature, 1981—2001. *Psychiatry, 65*(3), 207–239.

* Norris, F. H., & Kaniasty, K. (1996). Received and perceived social support in times of stress: a test of the social support deterioration deterrence model. *Journal of Personality & Social Psychology, 71*(3), 498–511.

Norris, F. H., Murphy, A. D., Baker, C. K., & Perilla, J. L. (2003). Severity, timing and duration of reactions to trauma in the population: an example from Mexico. *Biological Psychiatry, 53*(9), 769–778.

Norris, F. H., Murphy, A. D., Baker, C. K., & Perilla, J. L. (2004). Post-disaster PTSD over four waves of a panel study of Mexico's 1999 flood. *Journal of Traumatic Stress, 17*(4), 283–292.

Norris, F. H., Murphy, A. D., Baker, C. K., Perilla, J. L., Gutierrez, F., & Gutierrez, J. (2003). Epidemiology of trauma and posttraumatic stress disorder in Mexico. *Journal of Abnormal Psychology, 112*(4), 646–656.

Norris, F. H., Perilla, J. L., Ibañez, G. E., & Murphy, A. D. (2001). Sex differenc3es in symptoms of posttraumatic stress: does culture play a role? *Journal of Traumatic Stress, 14*(1), 7–28.

Norris, F., Stevens, S. P., Pfefferbaum, B., Wyche, K. F., & Pfefferbaum, R. L. (2008). Community disaster resiliency as a metaphor, theory, set of capacities, and strategy for disaster readiness. *American Journal of Community Psychology, 41*(1–2), 127–150.

Norwegian Refugee Council (NRC) and Internal Displacement Monitoring Centre (IDMC). (2015). *Global estimates 2015: People Displaced by Disasters*. Retrieved from http://www.internal-displacement.org/assets/publications/2015/20150713-global-estimates-2015-en.pdf.

* Nowell, B., & Steelman, T. (2013). The role of responder networks in promoting community resilience: toward a measurement framework of network capacity. In N. Kapucu, C. Hawkins, & F. Rivera (Eds.), *Disaster resiliency: Interdisciplinary Perspectives* (pp. 232–255). New York: Routledge.

* Nowell, B., & Steelman, T. (2014). Communication under fire: the role of embeddedness in emergence and efficacy of disaster response communication networks. *Journal of Public Administration Research & Theory, 25*(3), 929–952.

* Nowell, B., & Steelman, T. (2015). Communication under fire: the role of embeddedness in the emergence and efficacy of disaster response communication networks. *Journal of Public Administration Research & Theory, 25*(3), 929–952.

NZ Police. (2012). *Christchurch earthquake: List of deceased*. Retrieved January 1, 2013, from http://www.police.govt.nz/major-events/previous-major-events/christchurch-earthquake/list-deceased.

O'Toole, L. J. (2015). Networks and networking: the public administrative agendas. *Public Administration Review, 75*(3), 361–371.

O'Toole, L. J., Jr. (1997). Treating networks seriously: practical and research-based agendas in public administration. *Public Administration Review, 57*(1), 45–52.

Ohta, S. (2005). Chiiki no nakano jakunen koyo mondai (Regional characteristics of the Japanese youth labor market). *Japanese Journal of Labour Studies, 539*, 17–33 (in Japanese).

Olazo García, J. L. (2009). *Lo de ayer y lo de hoy: Teziutlán, a 10 años de la tragedia*. Puebla, Mexico: Benemérita Universidad Autónoma de Puebla.

Oliver-Smith, A. (1979). Post disaster consensus and conflict in a traditional society: the 1970 avalanche of Yungay, Peru. *Mass Emergencies, 4*(1), 39–52.

Oliver-Smith, A. (1992). *The martyred city: Death and Rebirth in the Peruvian Andes*. Prospect Heights, IL: Waveland Press.

Oliver-Smith, A. (2002). Theorizing disasters: Nature, power, and culture. In S. M. Hoffman, & A. Oliver-Smith (Eds.), *Catastrophe & culture: The Anthropology of disaster* (pp. 23–47). Santa Fe, NM: School of American Research.

Oliver-Smith, A. (2016). Disaster risk reduction and applied anthropology. *Annals of Anthropological Practice*.

Ollenburger, J. C., & Tobin, G. A. (2008). Women, aging and post-disaster stress: risk factors for psychological morbidity. In B. D. Phillips, & B. H. Morrow (Eds.), *Women and disasters: From Theory to Practice* (pp. 117–130). Bloomington, IN: Xlibris.

Ozer, E. J., Best, S. R., Lipsey, T. L., & Weiss, D. S. (2003). Predictors of posttraumatic stress disorder and symptoms in adults: a meta-analysis. *Psychological Bulletin, 129*(1), 52–73.

Palen, L., Vieweg, S., & Sutton, J. (2007). Crisis informatics: studying crisis in a networked world. In *Third International Conference on e-Social Science*. Michigan: Ann Arbor.

Palinkas, L. A., Downs, M., Petterson, J., & Russell, J. (1993). Social, cultural, and psychological impacts of the Exxon Valdez oil spill. *Human Organization, 52*(1), 1–13.

Pattison, P. E., & Wasserman, S. (1999). Logit models and logistic regressions for social networks: II. Multivariate relations. *British Journal of Mathematical & Statistical Psychology, 52*(2), 169–194.

Paxson, C., & Rouse, C. (2008). Returning to new Orleans after hurricane Katrina. *American Economic Review, 98*(2), 38–42.

Paxton, P. (1999). Is social capital declining in the united states? A multiple indicator assessment. *American Journal of Sociology, 105*(1), 88–127. http://dx.doi.org/10.1086/210268.

Pelling, M., & Dill, K. (2010). Disaster politics: Tipping points for change in the adaptation of sociopolitical regimes. *Progress in Human Geography, 34*(1), 21–37.

Perrow, C. (1999). Organizing to reduce the vulnerabilities of complexity. *Journal of Contingencies & Crisis Management, 7*(3), 150–155.

Perry, R. W., & Quarantelli, E. (Eds.). (2005). *What is a disaster? New answers to old questions*. Philadelphia, Pa: Xlibris Corp.

Perry, M., Williams, R. L., Wallerstein, N., & Waitzkin, H. (2008). Social capital and health care experiences among low-income individuals. *American Journal of Public Health, 98*(2), 330–336.

Peterson, K. J., & Maldonado, J. K. (2016). When adaptation is not enough: between now and then of community-led resettlement. In S. Crate, & M. Nuttall (Eds.), *Anthropology and climate change: From Actions to Transformations, 2nd edition* (pp. 336–353). New York, NY: Routledge.

Petev, I. D. (2013). The association of social class and lifestyle: persistence in American sociability, 1974-2010. *American Sociological Review, 78*(4), 633–661.

* Petrescu-Prahova, M., & Butts, C. (2005). *Emergent coordination in the world trade center disaster*. Institute for Mathematical Behavioral Sciences, 1–23. Retrieved from http://citeseerx.ist.psu.edu/viewdoc/download?doi=10.1.1.59.8310&rep=rep1&type=pdf.

* Pettigrew, O. (2011). Children's support networks after the 1999 landslides in Teziutlán, Mexico. *Explorations, 6*, 151–167.

Phan, T. Q., & Airoldi, E. M. (2014). A natural experiment of social network formation and dynamics. *PNAS, 112*(21), 6595–6600.

Phillips, B., Garza, L., & Neal, D. (1994). Issues of cultural diversity in time of disaster: the case of Hurricane Andrew. *Journal of Intergroup Relations, 21*(3), 18–27.

Pickering, A. (1995). *The mangle of practice: Time, Agency, & Science*. Chicago: University of Chicago Press.

Picou, J. S., Marshall, B. K., & Gill, D. A. (2005). Disaster, litigation, and the corrosive community. *Social Forces, 82*(4), 1493–1522.

Pierce, G. R., Sarason, I. G., & Sarason, B. R. (1991). General and relationship-based perceptions of social support: are two constructs better than one? *Journal of Personality & Social Psychology, 61*(6), 1028–1039.

Plickert, G., Côté, R. R., & Wellman, B. (2007). It's not who you know, it's how you know them: who exchanges what with whom? *Social Networks, 29*(3), 405–429.

Podolny, J. M., & Page, K. L. (1998). Network forms of organisation. *Annual Review of Sociology, 24*, 57–76.

* Poros, M. (2011). Migrant social networks: Vehicles for migration, integration, and development. *Migration information Source, the Online Journal of the Migration Policy Institute*. Retrieved from http://www.migrationpolicy.org/article/migrant-social-networks-vehicles-migration-integration-and-development.

Potkanski, T. (1999). Mutual assistance among the Ngorongoro maasai. In V. Broch-Due, & D. M. Anderson (Eds.), *The poor are not us: Poverty and Pastoralism* (pp. 199–217). Oxford, UK: James Currey.

Powell, J. W., Rayner, J., & Finesinger, J. E. (1953). Responses to disaster in American cultural groups. In *Symposium on stress*. Washington, D.C. Army Medical Service Graduate School.

Procidano, M. E., & Heller, K. (1983). Measures of perceived social support from friends and from family: Three validation studies. *American Journal of Community Psychology, 11*(1), 1–24.

Provan, K. G., Fish, A., & Sydow, J. (2007). Interorganizational networks at the network level: a review of the empirical literature on whole networks. *Journal of Management, 33*(3), 479–516.

Provan, K. G., & Kenis, P. (2008). Modes of network governance: structure, management, and effectiveness. *Journal of Public Administration Research & Theory, 18*(2), 229–252.

Provan, K. G., & Milward, H. B. (1995). A preliminary theory of network effectiveness: a comparative study of four community mental health systems. *Administrative Science Quarterly, 40*(1), 1–33.

Provan, K. G., & Sebastian, J. G. (1998). Networks within networks: service link overlap, organizational cliques, and network effectiveness. *Academy of Management Journal, 41*(4), 453–463.

PTSD Alliance. (n.d.). PTSD "What is PTSD." Available at http://www.ptsdalliance.org/about-ptsd/.

Quinn, P. M. (2015). *Qualitative research and evaluation methods.* California EU: Sage Publications Inc.

Quarantelli, E. L. (1960). A note on the protection function of the family in disasters. *Marriage and Family Living, 22*, 263–264.

Quarantelli, E. L., & Dynes, R. R. (1977). Response to social crisis and disaster. *Annual Review of Sociology, 3*, 23–49.

R Core Team. (2015). *R: A language and environment for statistical computing.* Vienna, Austria: R Foundation for Statistical Computing. URL: http://www.r-project.org/.

R Development Core Team. (2011). *R: A language and environment for statistical computing.* Vienna, Austria: The R Foundation for Statistical Computing. Available online at http://www.R-project.org/.

Raab, J., & Kenis, P. (2009). Heading toward a society of networks: empirical developments and theoretical challenges. *Journal of Management Inquiry, 18*(3), 198–210. http://dx.doi.org/10.1177/1056492609337493.

Raab, J., Mannak, R. S., & Cambré, B. (2015). Combining structure, governance, and context: a configurational approach to network effectiveness. *Journal of Public Administration Research & Theory, 25*(2), 479–511.

Radloff, L. S. (1977). The CES-D scale: a self-report depression scale for research in the general population. *Applied Psychological Measurement, 1*(3), 385–401.

Raudenbush, S. W., & Bryk, A. S. (2002). *Hierarchical linear models: Applications and Data Analysis Methods* (2nd ed.). Newbury Park, CA: Sage Publishing.

Ravuvu, A. (1983). *Vaka i taukei: The Fijian Way of Life.* Suva: Institute of Pacific Studies.

Rayner, S., & Malone, E. L. (1998). *Human choice and climate change, Volume 4: What Have We Learned?* Columbus, OH: Battelle Press.

Reconstruction Agency. (2013). *Hisaisyatou-no jokyo (The state of the sufferers) (in Japanese).* Retrieved from http://www.reconstruction.go.jp/topics/001169.html.

Redfield, R. (1941). *The folk culture of Yucatan.* Chicago: University of Chicago Press.

Reeder, H. T., McCormick, T. H., & Spiro, E. (2014). *Online information behaviors during disaster events: Roles, routines, and reactions.*

Riad, J. K., Norris, F. H., & Ruback, R. B. (1999). Predicting evacuation in two major disasters: risk perception, social influence, and access to resources. *Journal of Applied Social Psychology, 29*(5), 918–934.

Ripley, R. M., Snijders, T. A. B., & Preciado, P. (2012). *Manual for RSiena.* Oxford, UK: University of oxford: Department of Statistics; Nuffield College.

Rising Voices. (2015). *Learning and doing: Education and adaptation through diverse ways of knowing* Workshop report (June 29–July 1) Boulder, CO: National Center for Atmospheric Research. Retrieved from http://risingvoices.ucar.edu/.

* Ritchie, L. A. (2012). Individual stress, collective trauma, and social capital in the wake of the Exxon Valdez oil spill. *Sociological Inquiry, 82*(2), 187–211.

Roback, J. (1982). Wages, rents, and the quality of life. *Journal of Political Economy, 90*(6), 1257–1278.

Robinaugh, D. J., Marques, L., Traeger, L. N., Marks, E. H., Sung, S. C., Beck, J. G., et al. (2011). Understanding the relationship of perceived social support to post-trauma cognitions and posttraumatic stress disorder. *Journal of Anxiety Disorder, 25*(8), 1072–1078.

* Robinson, S. E., Berrett, B., & Stone, K. (2006). The development of collaboration of response to Hurricane Katrina in the Dallas area. *Public Works Management & Policy, 10*(4), 315–327.

Robins, G. L., Pattison, P. E., & Wasserman, S. (1999). Logit models and logistic regressions for social networks, III. Valued Relations. *Psychometrika, 64*(3), 371–394.

Roijakkers, N., & Hagedoorn, J. (2006). Inter-firm R&D partnering in pharmaceutical biotechnology since 1975: trends, patterns, and networks. *Research Policy, 35*(3), 431–446.

Rook, K. S. (1987). Reciprocity of social exchange and social satisfaction among older women. *Journal of Personality & Social Psychology, 52*(1), 145–154.

Rosen, S. (1974). Hedonic prices and implicit markets: product differentiation in perfect competition. *Journal of Political Economy, 82*(1), 34–55.

Rowlands, A., & Tan, N. T. (2008). Social redevelopment following the Indian Ocean Tsunami: an International social work response through the FAST Project. *Social Development Issues, 30*(1), 47–58.

Rowley, T. J. (1997). Moving beyond dyadic ties: a network theory of stakeholder influences. *Academy of Management, 22*(4), 887–910.

Roy, A. (1999). *The cost of living.* New York, NY: The Modern Library.

Saban, L. I. (2014). Entrepreneurial brokers in disaster response network in typhoon Haiyan in the Phillippines. *Public Management Review.* http://dx.doi.org/10.1080/14719037.2014.943271.

Sahlins, M. (1972). *Stone age economics.* Chicago: Aldine and Atherton, Inc.

Sampson, R. J. (1991). Linking the micro- and macrolevel dimensions of community social organization. *Social Forces, 70*(1), 43–64.

Sandström, A., & Carlsson, L. (2008). The performance of policy networks: the relation between network structure and network performance. *Policy Studies Journal, 36*(4), 497–524.

Sanil, A., Banks, D., & Carley, K. (1995). Models for evolving fixed node networks: model fitting and model testing. *Social Networks, 17*(1), 65–81.

Saramäki, J., Leicht, E., López, E., Roberts, S. G., Reed-Tsochas, F., & Dunbar, R. I. (2014). Persistence of social signatures in human communication. *Proceedings of the National Academy of Sciences, 111*(3), 942–947.

Schensul, J. J., LeCompte, M. D., Cromley, E. K., & Singer, M. (1999). *Mapping social networks, spatial data, and hidden populations.* Walnut Creek, CA: AltaMira Press.

* Scherer, C. W., & Cho, H. (2003). A social network contagion theory of risk perception. *Risk Analysis, 23*(2), 261–267.

Schokkaert, E. (2006). The empirical analysis of transfer motives. In S. C. Kolm, & J. M. Ythier (Eds.), *Handbook of the economics of giving, altruism and reciprocity* (Vol. 1). Amsterdam: Elsevier.

Schwarzer, R., Bowler, R. M., & Cone, J. E. (2014). Social integration buffers stress in New York police after the 9/11 terrorist attack. *Anxiety, Stress & Coping, 27*(1), 18–26.

Schwarzer, R., & Luszczynska, A. (2012). Stressful life events. In I. B. Weiner, A. M. Nezu, C. M. Nezu, & P. A. Geller (Eds.), *Handbook of psychology, volume 9 health psychology* (pp. 29–56). New York: Wiley.

Schweinberger, M. (2011). Instability, sensitivity, and degeneracy of discrete exponential families. *Journal of the American Statistical Association, 106,* 1361–1370.

Scott, M. F., & Gormley, B. (1980). The animal of friendship (Habbanaae): an indigenous model of Sahelian pastoral development in Niger. In D. Brokensha, D. M. Warren, & O. Werner (Eds.), *Indigenous knowledge systems and development* (pp. 92–110). Lanham, MD: University Press of America.

Seminole County. (n.d.). *Local mitigation strategy for Seminole county and its municipalities 2015–2020.* Accessed August 25, 2015 via http://www.seminolecountyfl.gov/core/fileparse.php/3333/urlt/LMS-2015-2020.pdf.

Seville, E., Van Opstal, D., & Vargo, J. (2015). A primer in resiliency: seven principles for managing the unexpected. *Global Business & Organizational Excellence, 34*(4), 6–18. http://dx.doi.org/10.1002/joe.

* Shavit, Y., Fischer, C., & Koresh, Y. (1994). Kin and non-kin under collective threat: Israeli networks during the Gulf War. *Social Forces, 72*(4), 1197–1215.

Shioji, E. (2001). Composition effect of migration and regional growth in Japan. *Journal of the Japanese and International Economies, 15*(1), 29–49.

Sieff, D. F. (1999). The effects of wealth on livestock dynamics among the Datoga pastoralists of Tanzania. *Agricultural Systems, 59*(1), 1–25.

* Siciliano, M. D., & Wukich, C. (2015). Network Features and Processes as Determinants of Organizational Interaction during Extreme Events. *Complexity, Governance & Networks, 2*(1), 23–44.

Simo, G., & Bies, A. L. (2007). The role of nonprofits in disaster response: an expanded model of cross-sector collaboration. *Public Administration Review, 67*(s1), 125–142. http://doi.org/10.1111/j.1540-36849062687.

Simon, H. A. (1957). *Models of man: Social and rational.* New York: John Wiley and Sons, Inc.

Simon, H. A. (1976). *Administrative behavior: a study of decision-making processes in administrative organization* (3rd ed.).

Sjaastad, L. (1962). The costs and returns of human migration. *Journal of Political Economy, 70*(5.2), 80–93.

Slone, L. B., Norris, F. H., Murphy, A. D., Baker, C. K., Perilla, J. L., Diaz, D., et al. (2006). Epidemiology of major depression in four cities in Mexico. *Depression & Anxiety, 23*(3), 158–167.

Smith, S. K., & McCarty, C. (1996). Demographic effects of natural disasters: a case study of Hurricane Andrew. *Demography, 33*(2), 265–275.

Smith, P. C., & Simpson, D. M. (2009). Technology and communications in an urban crisis: the role of mobile communications systems in disasters. *Journal of Urban Technology, 16*, 133–149. http://dx.doi.org/10.1080/10630730903076494. http://www.tandfonline.com/doi/abs/10.1080/10630730903076494.

Snijders, T. A. B. (1996). Stochastic actor-oriented dynamic network analysis. *Journal of Mathematical Sociology, 21*(1–2), 149–172.

Snijders, T. A. B., van de Bunt, G. G., & Steglich, C. E. G. (2010). Introduction to stochastic actor-based models for network dynamics. *Social Networks, 32*(1), 44–60.

Somers, S. (2009). Measuring resilience potential: an adaptive strategy for organizational crisis planning. *Journal of Contingencies & Crisis Management, 17*(1).

Spiro, E. S., Acton, R. M., & Butts, C. T. (2013). Extended structures of mediation: Re-examining brokerage in dynamic networks. *Social Networks, 35*(1), 130–143. http://dx.doi.org/10.1016/j.socnet.2013.02.001.

* Spiro, E. S., Dubois, C. L., & Butts, C. T. (2012). Waiting for a retweet: modeling waiting times in information propagation. In *Nips workshop on social media and social networks.* http://snap.stanford.edu/social2012/papers/spiro-dubois-butts.pdf.

Spiro, E. S., Fitzhugh, S., Sutton, J., Pierski, N., Greczek, M., & Butts, C. T. (2012). Rumoring during extreme events: a case study of Deepwater Horizon 2010. In *Proceedings of the 4th annual ACM web science conference* (pp. 275–283). ACM.

* Sprinkle, N. (2012). The role of formal and informal support in the decision to return to new orleans after Hurricane Katrina. *Explorations, 7*, 171–183.

Stack, C. B. (1975). *All our kin: Strategies for survival in a black community.* New York: Basic Books.

Stallings, R. A., & Quarantelli, E. L. (1985). Emergent citizen groups and emergency management. *Public Administration Review, 45*(special issue), 93–100.

Starbird, K., & Palen, L. (2011). "Voluntweeters": self-organizing by digital Volunteers in times of crisis. In *CHI* (pp. 1071–1080). Vancouver, Canada: ACM.

Starbird, K., Spiro, E. S., Edwards, I., Zhou, K., Maddock, J., & Narasimhan, S. (2016). Could this be true? i think so! expressed uncertainty in online rumoring. In *Proceedings of the ACM CHI conference on human factors in computing systems (CHI).*

Statistics New Zealand. (2013). *Canterbury's earthquake recovery progresses.* Retrieved from http://www.stats.govt.nz/browse_for_stats/snapshots-of-nz/yearbook/people/region/cera.aspx.

Staub, E., & Vollhardt, M. A. (2008). Altruism born of suffering: the roots of caring and helping after victimization and other trauma. *American Journal of Orthopsychiatry, 78*(3), 267–280.

Steady, F. C. (Ed.). (1993). *Women and children first: Environment, poverty, and sustainable development.* Rochester, NY: Schenkman Books.

Steelman, T. A., Nowell, B., Bayoumi, D., & McCaffrey, S. (2012). Understanding information exchange during disaster response: methodological insights from infocentric analysis. *Administration and Society*, *46*(6), 707–743.

* Steelman, T., Nowell, B., Bayoumi, D., & McCaffrey, S. (2014). Understanding information flows during disasters: methodological insights from social network analysis. *Administration & Society*, *46*(6), 707–743.

Steelman, T. A., Nowell, B. L., Velez, A. K., & Godette, S. K. (2016). Operationalizing performance measures in networked settings: lessons from large scale wildfires in the United States. In E. C. Jones, & A. J. Faas (Eds.), *Social networks and disasters.* Elsevier.

Steglich, C. E. G., Snijders, T. A. B., & Pearson, M. (2010). Dynamic networks and behavior: separating selection from influence. *Sociological Methodology*, *40*(1), 329–392.

* Stevenson, J. R., Chang-Richards, Y., Conradson, D., Wilkinson, S., Vargo, J., Seville, E., et al. (2014). Organizational networks and recovery following the Canterbury earthquakes. *Earthquake Spectra*, *30*(1), 555–575. http://dx.doi.org/10.1193/022013EQS041MR.

Stevenson, J. R., Seville, E., & Vargo, J. (2012). *The Canterbury earthquakes: Challenges and opportunities for central business district organisations (No. Report 2).* Retrived from www.resorgs.org.nz.

Strauss, J., & Thomas, D. (1998). Health, nutrition, and economic development. *Journal of Economic Literature*, *36*(2), 766–817.

* Sutton, J. N. (2010). Twittering Tennessee: distributed networks and collaboration following a Technological disaster. In *Proceedings of the 7th International ISCRAM Conference–Seattle* (Vol. 1). ISCRAM.

Sutton, J., Gibson, C. B., Phillips, N. E., Spiro, E. S., League, C., Johnson, B., et al. (2015a). A cross-hazard analysis of terse message retransmission on twitter. *Proceedings of the National Academy of Sciences*, *112*, 14793–14798.

Sutton, J., Gibson, C. B., Spiro, E. S., League, C., Fitzhugh, S. M., & Butts, C. T. (2015b). What it takes to get passed on: message content, style, and structure as predictors of retransmission in the boston marathon bombing response. *PLoS ONE*, *10*, e0134452. http://dx.doi.org/10.1371/journal.pone.0134452.

Sutton, J. N., Johnson, B., Greczek, M., Spiro, E. S., Fitzhugh, S. M., & Butts, C. T. (2012). Connected communication: network structures of Official communications in a Technological disaster. In L. Rothkrants, J. Ristvej, & Z. Franceo (Eds.), *Proceedings of the 9th International systems for Crisis response and Management conference* (pp. 1–10) (Vancouver, Canada).

* Sutton, J., Palen, L., & Shklovski, I. (May 2008). Backchannels on the front lines: emergent uses of social media in the 2007 southern California wildfires. In *Proceedings of the 5th international ISCRAM conference* (pp. 624–632) (Washington, DC).

Sutton, J., Spiro, E. S., Greczek, M., Johnson, B., Fitzhugh, S., & Butts, C. T. (2013). Tweeting the spill: online informal communications, social networks, and conversational microstructures during the Deepwater Horizon oil spill. *International Journal of Information Systems for Crisis Response and Management*, *5*, 58–76.

* Sutton, J., Spiro, E. S., Johnson, B., Fitzhugh, S., Gibson, B., & Butts, C. T. (2014). Warning tweets: serial transmission of messages during the warning phase of a disaster event. *Information, Communication & Society*, *17*(6), 765–787. http://dx.doi.org/10.1080/1369118X.2013.862561.

Swanson, D. A., Forgette, R., Van Boening, M., Kinnell, A. M., & Holley, C. (2007). Assessing Katrina's demographic and social impacts on the Mississippi gulf coast. *Journal of the Mississippi Academy of Sciences*, *52*(4), 228–242.

Takasaki, Y. (2011a). Do local elites capture natural disaster reconstruction funds? *Journal of Development Studies*, *47*(9), 1281–1298.

Takasaki, Y. (2011b). Groups, networks, and hierarchy in household private transfers: evidence from Fiji. *Oxford Development Studies*, *39*(1), 97–130.

Takasaki, Y. (2011c). *Post-disaster informal risk sharing.* Tokyo: Tokyo Center for Economic Research.

Takasaki, Y. (2011d). Targeting cyclone relief within the village: kinship, sharing, and capture. *Economic Development & Cultural Change, 59*(2), 387–416.

Takasaki, Y. (2012). *Do natural disasters decease the gender gap in schooling?* Tsukuba, Japan: University of Tsukuba.

Takasaki, Y. (2014). How is disaster aid allocated within poor communities? Risk sharing and social hierarchy. *Journal of International Development, 26*(8), 1097–1114.

Tang, S. (1996). Civil society, modern state and the relation between state and society in China. *Journal of Peking University: Philosophy & Social Sciences, 6,* 65–127.

Tao, W. (2009). On the government-led civil society in China. *Journal of Changshu Institute of Technology (Philosophy & Social Sciences), 7,* 31–34.

Tausig, M. (1992). Caregiver network structure, support and caregiver distress. *American Journal of Community Psychology, 20*(1), 81–96.

Taylor, M., & Doerfel, M. L. (2011). Evolving network roles in international aid efforts: evidence from Croatia's post-war transition. *Voluntas: International Journal of Voluntary and Non Profit Organizations, 22*(2), 311–334.

Tedeschi, R. G., & Calhoun, L. G. (2004). Posttraumatic growth: conceptual foundations and empirical evidence. *Psychological Inquiry, 15*(1), 1–18.

The Japan Association for Fire Science and Engineering. (1996). *Reports on fire in the 1995 South Hyogo earthquake. (in Japanese).*

* Thiede, B. C., & Brown, D. I. (2013). Hurricane Katrina: who stayed and why? *Population Research and Policy Reviewer, 32*(6), 803–824.

Thoits, P. A. (2011). Mechanisms linking social ties and support to physical and mental health. *Journal of Health and Social Behavior, 52*(2), 145–161.

Thoresen, S., Jensen, T. K., Wentzel-Larsen, T., & Dyb, G. (2014). Social support barriers and mental health in terrorist attack survivors. *Journal of Affective Disorder, 156,* 187–193.

Thornton, A. (2013). *Reading history sideways: The fallacy and enduring impact of the developmental paradigm on family life.* Chicago: University of Chicago Press.

Tierney, K. (2006). Social inequality, hazards, and disasters. In R. J. Daniels, D. F. Kettl, & H. Kunreuther (Eds.), *On risk and disaster: Lessons from hurricane Katrina* (pp. 109–128). Philadelphia: University of Pennsylvania Press.

Tierney, K. J. (2013). "Only connect!" Social capital, resilience, and recovery. *Risk, Hazards & Crisis in Public Policy, 4*(1), 1–5.

* Tierney, K. J., & Trainor, J. (2004). Networks and resilience in the world trade center disaster. *MCEER: Research Progress and Accomplishments, 2003–2004,* 157–172.

Tobin, G. A., & Ollenburger, J. C. (1996). Predicting levels of post-disaster stress in adults following the1993 floods in the upper Midwest. *Environment & Behavior, 28*(3), 340–357.

Tobin, G. A., Whiteford, L. M., Jones, E. C., & Murphy, A. D. (2007). Chronic hazard: weighing risk against the effects of emergency evacuation from Popocatepetl, Mexico. *Proceedings of the Applied Geography Conference, 30,* 288–297.

* Tobin, G. A., Whiteford, L. M., Jones, E. C., Murphy, A. D., Garren, S. J., & Vindrola Padros, C. (2011). The role of individual well-being in risk perception and evacuation for chronic vs. acute natural hazards in Mexico. *Journal of Applied Geography, 31*(2), 700–711.

* Tobin, G. A., Whiteford, L. M., Murphy, A. D., Jones, E. C., Faas, A. J., & Yepes, H. (2012). A social network analysis of resilience in chronic hazard settings. In *International Committee on Global Geological and Environmental change "GEOCHANGE." Proceedings: Natural cataclysm and global problems of the modern civilization, World forum–international Congress* (pp. 431–438). London: SWB.

* Tobin, G.A., Whiteford, L.M., Murphy, A.D., Jones, E.C., & McCarty, C., 2014. Modeling social networks and community resilience in chronic disasters: case studies from Volcanic areas in Ecuador and Mexico. In P. Gasparini, G. Manfredi, & D. Asprone (Eds.), Resilience and sustainability in relation to natural disasters: A Challenge for future Cities (pp. 13–24), Springer Briefs in Earth Sciences, http://doi.org/10.1007/978-3-319-04316-6_2.

Tolin, D. F., & Foa, E. B. (2006). Sex differences in trauma and posttraumatic stress disorder: a quantitative review of 25 years of research. *Psychological Bulletin, 132*(6), 959–992.

Treas, J. (2008). Transnational older adults and their families. *Family Relations, 57*(4), 468–478.

Turner, J. W. (1987). Blessed to give and receive: Ceremonial exchange in Fiji. *Ethnology, 26*(3), 209–219.

* Turrini, A., Cristofoli, D., Frosini, F., & Nasi, G. (2010). Networking literature about determinants of network effectiveness. *Public Administration, 88*(2), 528–550.

United Nations Development Group, World Bank, and European Union. (2013). Culture. In *Post-disaster needs assessments, Volume B*. New York, NY: United Nations Development Programme. Retrieved from https://gfdr r.org/sites/gfdrr/files/WB_UNDP_PDNA_Culture_FINAL.pdf.

United Nations Development Programme. (2009). *Human development report 2009: Overcoming barriers: Human mobility and development*. Available online at http://hdr.undp.org/en/content/human-development-report-2009.

United Nations Development Programme. (2010). *Wenchuan Quake: Undp response*. Retrieved November 30 http p://www.undp.org.cn/modules.php?op=modload&name=News&file=article&catid=52&sid=4307.

United Nations Office for Disaster Risk Reduction (UNISDR). (2011). *Global assessment report on disaster risk reduction*. Geneva: United Nations Office for Disaster Risk Reduction.

United Nations Office for Disaster Risk Reduction (UNISDR). (2015a). *Global assessment report on disaster risk reduction 2015*. Geneva: United Nations International Strategy on Disaster Reduction. Retrieved from http:// www.preventionweb.net/english/hyogo/gar/2015/en/home/GAR_2015/GAR_2015_1.html.

United Nations Office for Disaster Risk Reduction (UNISDR). (2015b). *Sendai framework for disaster risk reduction 2015–2030*. Geneva: United Nations International Strategy on Disaster Reduction. Retrieved from http:// www.unisdr.org/files/43291_sendaiframeworkfordrren.pdf.

Utz, R. L., Swenson, K. L., Caserta, M., Lund, D., & deVries, B. (2014). Feeling lonely versus being alone: loneliness and social support among recently bereaved persons. *The Journals of Gerontology Series B: Psychological Sciences and Social Sciences, 69B*(1), 85–94. http://doi.org/10.1093/geronb/gbt075.

Uzzi, B. D. (1997). Social structure and competition in interfirm networks: the paradox of embeddedness. *Administrative Science Quarterly, 42*(1), 35–67. Retrieved from http://www.jstor.org/stable/2393808?origin=crossref.

Väänänen, A., Buunk, B. P., Kivimäki, M., Pentti, J., & Vahtera, J. (2005). When it is better to give than to receive: long-term health effects of perceived reciprocity in support exchange. *Journal of Personality & Social Psychology, 89*(2), 176–193.

Valente, T. W., Coronges, K., Lakon, C., & Costenbader, E. (2008). How correlated are network centrality measures? *Connections, 28*(1), 16–26.

Varda, D. M. (2008). *Partner (program to Analyze Record and Track networks to Analyze Relationships)*. Published Software. University of Colorado Denver. www.partnertool.net.

Varda, D. M. (2011). A network perspective on state–society synergy to increase community-level social capital. *Nonprofit and Voluntary Sector Quarterly, 40*(5), 896–923.

Varda, D. M., Forgette, R., Banks, D. D., & Contractor, N. (2009). Social network methodology in the study of disasters: issues and insights prompted by post-Katrina research. *Population Research & Policy Review, 28*(1), 11–29.

* Verroen, S., Gutteling, J. M., & De Vries, P. W. (2013). Enhancing self-protective behavior: efficacy beliefs and peer feedback in risk communication. *Risk Analysis, 33*(7), 1252–1264.

Vieweg, S., Hughes, A. L., Starbird, K., & Palen, L. (2010). Microblogging during two natural hazards events: what Twitter may contribute to situational awareness. In *Proceedings of the 28th international conference on Human factors in computing systems* (pp. 1079–1088). ACM. http://portal.acm.org/citation.cfm?id=1753486.

Vigdor, J. (2008). The economic aftermath of Hurricane Katrina. *Journal of Economic Perspectives*, *22*(4), 135–154.

Voors, M. J., Nillesen, E. E. N., Verwimp, P., Bulte, E. H., Lensink, R., & Van Soest, D. P. (2012). Violent conflict and behavior: a field experiment in Burundi. *American Economic Review*, *102*(2), 941–964.

de Vries, D., Leslie, P. W., & McCabe, J. T. (2006). Livestock acquisitions dynamics in nomadic pastoralist herd demography: a case study among Ngisonyoka herders of South Turkana, Kenya. *Human Ecology*, *34*(1), 1–25. http://dx.doi.org/10.1007/s10745-005-9000-2.

Wagner, S. (2008). *To know where he lies: Dna technology and the search for Srebrenica's missing*. Berkeley, CA: University of California Press.

Wallace, A. F. C. (1956). *The tornado in Worcester: An exploratory study of individual and community behavior in an extreme situation. Disaster study number 3, committee on disaster Studies, division of Anthropology and psychology*. Washington, DC: National Academy of Sciences - National Research Council.

Walsh, F. (2007). Traumatic loss and major disasters: strengthening family and community resilience. *Family Process*, *46*(2), 207–227.

Wang, M. (2007). Current development status of China's NGOs and policy analysis. *China Public Administration Review*, *1*.

Wang, Z. (2009). China's path towards democratic development on the base of historical tradition and practical needs. *Journal of Social Sciences*, *5*, 21–187.

Warheit, G. J. (1985). A propositional paradigm for estimating the impact of disasters on mental health. *International Journal of Mass Emergencies & Disasters*, *3*(2), 29–48.

Wasserman, S., & Faust, K. (1994). *Social network Analysis: Methods and applications*. Cambridge: Cambridge University Press.

* Wasserman, S., & Pattison, P. E. (1996). Logit models and logistic regressions for social networks: I. An introduction to Markov graphs and p. *Psychometrika*, *61*(3), 401–425.

Wasserman, S., Scott, J., & Carrington, P. J. (2005). Introduction. In P. J. Carrington, J. Scott, & S. Wasserman (Eds.), *Models and methods in social network analysis* (pp. 1–7). New York, NY: Cambridge University Press.

Waugh, W. L. (2003). Terrorism, homeland security and the national emergency management network. *Public Organization Review*, *3*(4), 373–385.

Weick, K. E. (1993). The collapse of sensemaking in organizations: the Mann Gulch disaster. *Administrative Science Quarterly*, *38*(4), 628. http://dx.doi.org/10.2307/2393339.

Weick, K. E., Sutcliffe, K. M., & Obstfeld, D. (2005). Organizing and the process of sensemaking. *Organization Science*, *16*(4), 409–421.

* Weil, F., Lee, M. R., & Shihadeh, E. S. (2012). The burdens of social capital: how socially-involved people dealt with stress after Hurricane Katrina. *Social Science Research*, *41*(1), 110–119.

Westerling, A. L., Hidalgo, H. G., Cayan, D. R., & Swetnam, T. W. (2006). Warming and earlier spring increase western US forest wildfire activity. *Science*, *313*(5789), 940–943.

Whatley, M. A., Webster, J. M., Smith, R. H., & Rhodes, A. (1999). The effect of a favor on public and private compliance: how internalized is the norm of reciprocity? *Basic & Applied Social Psychology*, *21*(3), 251–259.

White, C. (1990). Changing animal ownership and access to land among the Wodaabe (Fulani) of central Niger. In P. W. T. Baxter, & R. Hogg (Eds.), *Property, poverty and people: Changing rights in property and problems of pastoral development* (pp. 240–251). Manchester, UK: Department of Social Anthropology & International Development Center.

Whitman, Z., Stevenson, J. R., Kachali, H., Seville, E., Vargo, J., & Wilson, T. (2014). Organisational resilience following the Darfield earthquake of 2010. *Disasters*, *38*(1), 148–177. http://dx.doi.org/10.1111/disa.12036.

Wilkinson, A. V., Spitz, M. R., Strom, S. S., Prokhorov, A. V., Barcenas, C. H., Cao, Y., et al. (2005). Effects of nativity, age at migration, and acculturation on smoking among adult Houston residents of Mexican descent. *American Journal of Public Health*, *95*(6), 1043–1049.

Williams, J. (2013). Exploring the onset of high-impact mega-fires through a forest land management prism. *Forest Ecology & Management, 294*, 4–10.

Wilson, T. (1994). What determines where transnational labor migrants go? Modifications in migration theories. *Human Organization, 53*(3), 269–278.

Wisner, B. (2001). Risk and neoliberal state: why post-Mitch lessons didn't reduce El Salvador's earthquake losses. *Disasters, 25*(3), 251–268.

Wood, D. P., & Cowan, M. L. (1991). Crisis intervention following disasters: are we doing enough? (A Second Look). *American Journal of Emergency Medicine, 9*(6), 598–602.

World Health Organization. (1997). *Composite international diagnostic inventory (Version 2.1).* Geneva, Switzerland: World Health Organization.

Wright, K. M., Ursano, R. J., Bartone, P. T., & Ingraham, L. H. (1990). The shared experience of catastrophe: an expanded classification of the disaster community. *American Journal of Orthopsychiatry, 60*(1), 35–42.

Wu, Z., & Gong, J. (2008). A study on the relations between the building of civil society and China's modernization. *Journal of Kunming University of Science & Technology, 8*(12), 30–33.

Yang, Y., & Almquist, Z. W. (2015). *networkDyn: R Package.* University of Minnesota.

Yu, K. (2009). *Democracy is a good thing: Essays on politics, society, and culture in Contemporary China.* Washington, DC: The Brookings Institution.

* Zakour, M. J., & Gillespie, D. F. (1998). Effects of organizational type and localism on volunteerism and resource sharing during disasters. *Nonprofit & Voluntary Sector Quarterly, 27*(1), 49–65. http://dx.doi.org/10.1177/0899764098271004.

Zane, D. F., Bayleyegn, T. M., Hellsten, J., Beal, R., Beasley, C., Haywood, T., et al. (2011). Tracking deaths related to Hurricane Ike, Texas, 2008. *Disaster Medicine & Public Health Preparedness, 5*(1), 23–28.

Zhang, H. (2009). Social consensus of devoted love and the concept of civil society: an ideological analysis of earthquake recovery and disaster relief. *Academy, 2*, 45–49.

Zimet, G. D., Dahlem, N. W., Zimet, S. G., & Farley, G. K. (1988). The multidimensional scale of perceived social support. *Journal of Personality Assessment, 52*(1), 30–41.

Zylberberg, Y. (2010). *Do tropical typhoons smash community ties? Theory and Evidence from Vietnam.* Working Papers 11 Vietnam: Development and Policies Research Center (DEPOCEN). https://ideas.repec.org/p/dpc/wpaper/1110.html.

Index

Note: Page numbers followed by "f" indicate figures, "t" indicate tables and "b" indicate boxes.

Printed in the United States
By Bookmasters